OXFORD STATISTICAL SCIENCE SERIES

OXFORD STATISTICAL SCIENCE SERIES

Estimating Functions

Edited by

V. P. GODAMBE

Department of Statistics and Actuarial Science,
University of Waterloo, Waterloo, Canada

CLARENDON PRESS · OXFORD

1991

Oxford University Press, Walton Street, Oxford OX2 6DP

Oxford New York Toronto
Delhi Bombay Calcutta Madras Karachi
Petaling Jaya Singapore Hong Kong Tokyo
Nairobi Dar es Salaam Cape Town
Melbourne Auckland
and associated companies in
Berlin Ibadan

Oxford is a trade mark of Oxford University Press

Published in the United States
by Oxford University Press, New York

A catalogue record for this book is available from the British Library

Library of Congress Cataloging in Publication Data
Estimating functions / edited by V. P. Godambe.
—(Oxford statistical science series; 7)
1. Estimation theory. I. Godambe, V. P. II. Series.
QA276.8.E76 1991   519.5'44—dc20   91-7676

ISBN 0-19-852228-2

Typeset by Integral Typesetting, Gorleston, Norfolk
Printed in Great Britain by
St. Edmundsbury Press,
Bury St. Edmunds, Suffolk

Preface

A need for a book-length treatment of *estimating functions* covering its important developments, theoretical and applied, has been felt for quite some time. This volume of collected papers on the subject is a result of consultations with many colleagues working in this area in different countries. Of course this is not a text book on the subject. This volume aims at two types of audience: research statisticians who are interested in the subject but may not be specialists, and specialists. An attempt therefore has been made to keep a proper balance between reviews and research papers in the volume. A few articles are primarily reviews. Many others also contain some review material but they primarily present new research. Yet a large number of papers consist exclusively of new and original investigations.

Generally, long before concepts are scientifically abstracted or formalized, they are used intuitively or informally in practical applications. This is certainly true for statistical concepts such as sufficiency, ancillarity, power, and robustness. The concept of estimating function is no exception to this. Instances of informal or intuitive uses of 'estimating functions' are found not only in statistics but also in other sciences like physics and economics. A scientific abstraction or formalization of a concept gives a much clearer understanding and renders it applicable very generally, far beyond its intuitive uses. This happened, with my formulation (1960) of the *optimality criterion* for estimating functions, which, as observed initially, is satisfied by the *score function* in a 'fully parametric' family. Rather unexpectedly, the 'optimality criterion' proved to be all the more versatile in dealing with 'semi-parametric' models defined only up to the first two or more moments of the distribution. These models are routinely employed in biostatistics, stochastic processes, survey sampling, and elsewhere. In these areas *optimal estimating functions* provided a practical handy tool for statistical analysis and brought about a conceptual unification. This story is told in the pages that follow.

The chapters in this volume, apart from an 'overview' chapter, are divided into four sections: biostatistics, stochastic processes, survey sampling, and theory. This division, at least in the case of some chapters, is somewhat arbitrary; they could as well belong to some other section than the one they are put into. I hope this will not cause any confusion or misunderstanding. As is always the case, the editor may not be in agreement with the authors' views.

Finally, I would like to thank all the contributors to this volume and referees for their helpful co-operation, without which this publication would not have been possible. I would especially like to mention in this connection C. Dean, M. Lesperance, H. Mantel, S. Rai, and for editorial assistance, D. Chapman.

Waterloo   V.P.G.
September 1990

Contents

Contributors

P. Basak
Department of Mathematics
Penn State University
Ivy Side
Altoona
Pennsylvania 16603
USA

I. V. Basawa
Department of Statistics
University of Georgia
Athens
Georgia 30602
USA

Vasant P. Bhapkar
Department of Statistics
University of Kentucky
Lexington
Kentucky 40506
USA

S. R. Chamberlin
Department of Statistics and
   Actuarial Science
University of Waterloo
Waterloo
Ontario
Canada
N2L 3G1

I-Shou Chang
Department of Mathematics
National Central University
Chung-Li
Taiwan
Republic of China

D. R. Cox
Nuffield College
Oxford
OX1 1NF
UK

C. B. Dean
Department of Mathematics and
   Statistics
Simon Fraser University
Burnaby
British Columbia
Canada
V5A 1S6

A. F. Desmond
Department of Mathematics and
   Statistics
University of Guelph
Guelph
Ontario
Canada
N1G 2W1

H. Ferguson
Department of Statistics and
   Actuarial Science
University of Waterloo
Waterloo
Ontario
Canada
N2L 3G1

Malay Ghosh
Department of Statistics
University of Florida
512C Nuclear Sciences Center
Gainsville
Florida 32611
USA

V. P. Godambe
Department of Statistics and
   Actuarial Science
University of Waterloo
Waterloo
Ontario
Canada
N2L 3G1

P. E. Greenwood
Department of Mathematics
University of British Columbia
Vancouver
British Columbia
Canada
V6T 1Y4

C. C. Heyde
Statistics Research Section
School of Mathematical Sciences
Australian National University
GPO Box 4
Canberra
ACT 2601
Australia

Chao A. Hsiung
Institute of Statistical Science
Academia Sinica
Taipei
Taiwan
Republic of China

J. E. Hutton
Department of Mathematics
Bucknell University
Lewisberg
Pennsylvania 17837
USA

B. K. Kale
Department of Statistics
University of Poona
Pune-411007
India

Subhash Lele
Department of Biostatistics
The Johns Hopkins University
615 N. Wolfe Street
Baltimore
Maryland 21205
USA

K.-Y. Liang
Department of Biostatistics
The Johns Hopkins University
615 N. Wolfe Street
Baltimore
Maryland 21205
USA

Y.-X. Lin
Statistics Research Section
School of Mathematical Sciences
Australian National University
GPO Box 4
Canberra
ACT 2601
Australia

B. G. Lindsay
Department of Statistics
Penn State University
219 Pond Lab
University Park
Pennsylvania 16802
USA

X.-H. Liu
Department of Biostatistics
The Johns Hopkins University
615 N. Wolfe Street
Baltimore
Maryland 21205
USA

Chris J. Lloyd
Department of Statistics
La Trobe University
Bundoora
Victoria
Australia-3083

D. L. McLeish
Department of Statistics and
   Actuarial Science
University of Waterloo
Waterloo
Ontario
Canada
N2L 3G1

H. Mantel
Social Survey Methods Division
Statistics Canada
4th Floor, Jean Tallon Building
Tunney's Pasture
Ottawa
Canada
K1A 0T6

P. I. Nelson
Department of Statistics
Kansas State University
Manhattan
Kansas 66506
USA

O. T. Ogunyemi
Department of Mathematics
Oakland University
Rochester
Michigan 48063
USA

R. L. Prentice
Division of Public Health Sciences
Fred Hutchinson Cancer Research
   Center
1124 Columbia Street
Seattle
Washington 98104
USA

N. Reid
Department of Statistics
University of Toronto
100 St George St.
Toronto
Ontario
Canada
M5S 1A1

Y. Ritov
Department of Statistics
The Hebrew University of Jerusalem
Jerusalem 91905
Israel

Christopher G. Small
Department of Statistics and
   Actuarial Science
University of Waterloo
Waterloo
Ontario
Canada
N2L 3G1

D. A. Sprott
Department of Statistics and
   Actuarial Science
University of Waterloo
Waterloo
Ontario
Canada
N2L 3G1

A. Thavaneswaran
Department of Statistics
The University of Manitoba
Winnipeg
Manitoba
Canada
R3T 2N2

K. Vijayan
Department of Mathematics
The University of Western Australia
Nedlands
Western Australia 6009

W. Wefelmeyer
Department of Mathematics
University of Cologne
5000 Cologne 41
Germany

Eiji Yamamoto
Department of Applied Mathematics
Okayama University of Science
1-1 Ridai-cho
Okayama
Japan 700

Takemi Yanagimoto
The Institute of Statistical
   Mathematics
4-6-7 Minami-Azabu
Minato-Ku
Tokyo
Japan 106

Paul Yip
Department of Statistics
La Trobe University
Bundoora
Victoria
Australia 3083

L. P. Zhao
Epidemiology Programme
Cancer Research Centre of Hawaii
University of Hawaii at Manoa
1236 Lauhala St.
Honolulu
Hawaii 96813
USA

PART 1
Overview

1
Estimating functions: an overview
V. P. Godambe and B. K. Kale

ABSTRACT

In a wide variety of statistical applications, mainly two methods of estimation
are used—least squares (LS) and maximum likelihood (ML). Each method has
its strengths and weaknesses. The LS method is much more widely applicable
than the ML method. The latter requires full distributional specification while
the former is applicable only under the assumption of the first two moments.
On the other hand, ML estimation is generally optimal while LS estimation
is optimal only when the underlying distribution is normal. With the advent
of estimating functions in recent years, the two methods, ML and LS, have
been merged into a single method of estimation. This estimation has the
strengths of both ML and LS methods and the weaknesses of neither. Its varied
areas of application include biostatistics, stochastic processes, survey sampling,
and so on. This paper provides an overview of the modern developments in
theory and applications of estimating functions, with a historical perspective.

1.1 Historical background

For nearly two centuries now, the two most prevalent methods of estimation
are the method of 'least squares' (LS) and the method of 'maximum
likelihood' (ML), using the present-day terminology. Legendre put forward
the LS method in 1805, on more or less intuitive grounds. He also coined
the term 'LS method'. Gauss (1809, 1823) provided two statistical justifications
to the LS method. (I) He related it to the ML estimation and the normal
model by showing that within the location family of distributions, the two
methods LS and ML coincide *uniquely* for the normal family. This indicated
that the acceptance of both LS and ML methods was in line with the
acceptance of the normal model. (II) He showed that with conditions only
on the first two moments of the distribution but otherwise irrespective of its
distributional form, the LS estimate has minimum variance within the class of
'linear unbiased estimates'. This is the well-known Gauss–Markov (GM)
theorem. For our subsequent discussion, it is important to note that Gauss
did not invoke the present-day concept of 'unbiased estimates'; instead he
used the concept of 'consistent estimates' that is, 'estimates which equal the

true value when the observations are without error' (Bertrand 1888; Sprott 1983). This 'consistency' which is a 'finite sample' property should be distinguished from the 'asymptotic consistency' often mentioned in the literature. It is of course true that for linear models, the linear estimates which are consistent in the Gauss sense are also unbiased estimates and conversely (Heyde and Seneta 1977, Chapter 4).

As seen above, Gauss provided justifications for the LS method by relating it to the ML and unbiased minimum variance (UMV) methods of estimation. These 'relationships' among the three methods LS, ML and UMV, when exploited fully, it is argued below, provides a 'theory' of estimation which combines the strengths of the three methods and at the same time eliminates their weaknesses. This theory, called the *theory of estimating functions*, will be reviewed in the subsequent sections.

1.2 Estimating functions and the Gauss–Markov (GM) theorem

To emphasize the basic concepts, we consider the simplest case of the LS estimation and the GM theorem, the extensions to the general linear models being straightforward. Let $y_1, y_2, \ldots, y_n$ be independent real random variates with common expectation and variance given by

$$E(y_i) = \theta, \quad \text{Var}(y_i) = \sigma^2, \quad i = 1, 2, \ldots, n. \tag{1.1}$$

The unknown parameter θ is known to belong to a specified real interval. To estimate θ, on the basis of observations $y_1, y_2, \ldots, y_n$ by the LS method, we minimize $\sum (y_i - \theta)^2$, for the variations of θ. The minimum is attained at the LS estimate of θ, namely the sample mean $\bar{y} = (\sum y_i)/n$. Again as a very special case we have here, the following theorem.

GM Theorem   *Let $(y_1, y_2, \ldots, y_n)$ be independent real random variates satisfying (1.1) and l be any linear unbiased estimate of θ. That is*

$$l = \sum a_i y_i, \tag{1.2}$$

where a_i, $i = 1, \ldots, n$, are any constants such that the expectation $E(l) = \theta$, implying

$$\sum a_i = 1. \tag{1.3}$$

Then the variance of l is minimized for $a_i = 1/n$, $i = 1, \ldots, n$. That is, the LS estimate sample mean $\bar{y}$ is the linear UMV estimate for θ.

Now instead of restricting our attention to the functions of observations only, that is estimates, we also consider real functions g of observations and the parameter θ, $g = g(y_1, y_2, \ldots, y_n, \theta)$. Such functions g are called *estimating functions* in view of their central role in estimation, which will be evident from what follows. Consider an estimating function g of the form

$$g = \sum (y_i - \theta)b_i \qquad (1.4)$$

where b_i, $i = 1, \ldots, n$, are any constants. It is easy to see that any linear unbiased estimate in (1.2) can be obtained as a solution in θ of the equation

$$g = 0 \qquad (1.5)$$

for suitably chosen constants b_i, $i = 1, \ldots, n$, in (1.4). Noting that because of (1.1) in (1.4) the expectation

$$E(g) = 0, \qquad (1.6)$$

so we have the following alternative version of the GM theorem.

GM Theorem (A) *If $(y_1, y_2, \ldots, y_n)$ are independent real random variates satisfying condition (1.1), the variance of the estimating function g in (1.4) is minimized for the variations of b_i, $i = 1, \ldots, n$, subject to the condition*

$$\sum b_i = C, \quad \text{a constant,} \qquad (1.7)$$

for $b_i = C/n, i = 1, \ldots, n$. With this substitution, $b_i = C/n$ in (1.4), the equation (1.5) provides the sample mean $\bar{y}$ as the estimate of θ.

In the above theorem we first consider shifting the emphasis from the criteria of UMV-ness within linear unbiased estimates to an optimality criteria applicable to the estimating function $g = \sum b_i(y_i - \theta)$. In order that $g = 0$ defines an estimator $\hat{\theta}_b = \sum b_i y_i / \sum b_i$, we must have $\sum b_i \neq 0$. Thus we consider the class G_0 of estimating functions given by

$$g(y, \theta) = \sum b_i(y_i - \theta) \qquad (1.8)$$

for any constants b_i satisfying $\sum b_i \neq 0$. Note that g is Gauss consistent in the sense that if all $y_i = \theta$ then $g = 0$. Further we call an estimating function g *unbiased*, if it satisfies $E(g) = 0$. Thus $g \in G_0$ is Gauss consistent as well as unbiased and $\text{Var}(g) = \sigma^2 \sum b_i^2$. The equations $g = 0$ and $kg = g' = 0$, where $k \neq 0$, are equivalent and define the same estimate $\hat{\theta}_b$ which is unbiased and has variance $\sigma^2 \sum b_i^2 / (\sum b_i)^2$. But $\text{Var}(g') = k^2 \sum \text{Var}(g)$ can however be made arbitrarily small and thus the comparison of two estimating functions on the basis of variance alone is not meaningful unless some standardization is introduced. The standardized version of g is defined by

$$g_s = \sum b_i(y_i - \theta)/\{-\sum b_i\}. \qquad (1.9)$$

Note that $g = 0$ and $g_s = 0$ determine the same estimate $\hat{\theta}_b$ and $\text{Var}(g_s) = \sigma^2 \sum b_i^2 / (\sum b_i)^2 = \text{Var}(g_s')$. An estimating function $g^* \in G_0$ is said to be *optimal* in G_0 if

$$\text{Var}(g_s^*) \leqslant \text{Var}(g_s) \qquad \text{for any } g \in G_0. \tag{1.10}$$

We observe that $\text{Var}(g_s) = \text{Var}(\hat{\theta}_b)$ and the optimal estimating function g^* is unique up to a constant multiple.

Motivation behind the standardization (1.9) and the optimality criteria in (1.10) is provided by the following argument. In order to be used as an estimating equation the estimating function g should be as near to zero as possible when θ is the true value. This requires that $\text{Var}(g)$ be made as small as possible. Further $g(y, \theta + \delta\theta)$ should be as far away from zero as possible when θ is the true value. This requires that $[E(\partial g/\partial \theta)]^2 = (\sum b_i)^2$ be as far away from zero as possible. Both of these requirements can be combined into one requiring that $E[g/E(\partial g/\partial \theta)]^2 = \text{Var}(g_s)$ be made as small as possible. In the above set-up this is equivalent to minimizing $\text{Var}(\hat{\theta}_b)$ as every linear unbiased estimate can be written as a solution of $\sum b_i(y_i - \theta) = 0$ and conversely. The GM theorem can therefore be reformulated as follows.

GM Theorem (A)   *If* $(y_1, y_2, \ldots, y_n)$ *are independent random variates with* $E(y_i) = \theta$ *and* $\text{Var}(y_i) = \sigma^2$, *then* $g^* = \sum (y_i - \theta)$ *is an optimum estimating function in* G_0. *The equation* $g^* = 0$ *provides the sample mean* $\bar{y}$ *as an estimate of* θ.

We emphasize that the above GM theorem (A) is logically equivalent to the GM theorem and as such it lends as much justification to the LS estimate $\bar{y}$ as the GM theorem does. However, unlike the GM theorem its alternative version GM theorem (A) admits the following extension.

1.3   An extension of the Gauss–Markov theorem

As said before, we can get all the linear unbiased estimates of the form (1.2) by solving for θ, the equations (1.5), for different estimating functions g in G_0. Conversely every member of G_0 corresponds to some unbiased linear estimate in (1.2). But it is easy to see that, unlike the GM theorem, the GM theorem (A) remains valid if the b_i in (1.4) are allowed to be functions of the parameter θ, $b_i = b_i(\theta)$, $i = 1, 2, \ldots, n$. However, in this case the solutions in θ, of the equations (1.5), provide a much *wider* class of estimates than that of linear unbiased estimates implied by G_0. In this class of estimating functions every implied estimate need not be unbiased but it will necessarily satisfy the property of Gauss consistency discussed in Section 1.1, i.e. when all y_i

are observed without any error, that is $y_i = \theta$, $i = 1, \ldots, n$, $g = 0$. Thus the extension of the GM theorem (A), namely GM theorem (B), given subsequently, where the coefficients b_i are allowed to depend on θ, provides a greater justification for the LS estimate $\bar{y}$ than the GM theorem does.

Thus we now consider an extended class of estimating functions $G_1 = \{g\}$ where $g(y, \theta) = \sum b_i(\theta)(y_i - \theta)$ and where $b_i(\theta)$ are differentiable functions of θ. Note that $g(y, \theta)$ is Gauss consistent with $\text{Var}(g) = \sigma^2 \sum b_i^2(\theta)$. Further $E(\partial g/\partial \theta) = -\sum b_i(\theta)$ and the standardized version of g is given by

$$g_s = \sum b_i(\theta)(y_i - \theta)/\{-\sum b_i(\theta)\}. \tag{1.11}$$

The only difference between (1.9) and (1.11) is that the b_i are now allowed to depend on θ. We observe that $\text{Var}(g_s) = \sigma^2 \sum b_i^2(\theta)/[\sum b_i(\theta)]^2$ and is minimized for every $\theta \in \Omega$ when $b_1(\theta) = \cdots = b_n(\theta) = k(\theta) \neq 0$. Thus $k(\theta) \sum (y_i - \theta)$ is the optimal estimating function in the extended class G_1 satisfying the optimality criteria (1.10) with G_0 replaced by G_1. Hence $(\bar{y} - \theta)$ can be taken as the optimal estimating function up to a multiplicative constant depending only on θ. We thus have a more general version of the GM Theorem (A):

GM Theorem (B) *If $(y_1, y_2, \ldots, y_n)$ are independent random variates with $E(y_i) = \theta$ and $\text{Var}(y_i) = \sigma^2$, then $g^* = \sum (y_i - \theta)$ is an optimal estimating function in the class G_1 and the equation $g^* = 0$ provides $\bar{y}$ as an estimate of θ.*

As said before the GM theorem (B) provides a greater justification to the LS estimate $\bar{y}$ than the classical GM theorem does. This added versatility is clearly the result of shifting emphasis from estimates to estimating functions. It is pointed out that not all estimates covered by the GM Theorem (B) need be unbiased, yet the corresponding estimating function is unbiased. We can thus in general define any function $g(y_1, y_2, \ldots, y_n, \theta)$ such that $E(g) = 0$ is an unbiased estimating function. The argument for standardization of g in the previous section can be generalized to define the standardized version of g as

$$g_s = g/E\left(\frac{\partial g}{\partial \theta}\right). \tag{1.12}$$

The optimality criteria based on minimization of $\text{Var}(g_s)$ leads us to define g^* to be optimal in the class G if $g^* \in G$ and

$$\text{Var}(g_s^*) \leqslant \text{Var}(g_s) \qquad \text{for any } g \in G. \tag{1.13}$$

In the next section we show how the flexibility provided by the standardization and the above optimality criteria allows us to obtain an optimal estimating function in a situation in which the classical Gauss–Markov approach does not correspond to LS approach.

1.4 A failure of the Gauss–Markov (GM) approach

Now we discuss a situation where the GM theorem fails completely to apply to the LS estimate, yet the extended GM theorem (B) of Section 3 shows a way out. Let $(y_1, y_2, \ldots, y_n)$ be independent real random variates.

$$E(y_i) = \alpha_i(\theta), \quad \mathrm{Var}(y_i) = \sigma^2, \qquad i = 1, 2, \ldots, n, \qquad (1.14)$$

where α_i are differentiable functions of θ with unique inverse functions α_i^{-1}. The LS estimate is obtained by minimizing $\sum (y_i - \alpha_i(\theta))^2$, which is as intuitive in this case as it is when $\alpha_i(\theta) = \theta$. Yet it is justified by the GM theorem only for those functions $\alpha_i(\theta)$ which are linear in θ. The failure of the GM approach is mainly due to the fact that even though y_i is unbiased for $\alpha_i(\theta)$, $\alpha_i^{-1}(y_i)$ (except for a linear function) is not unbiased for θ. However, if y_i is Gauss consistent for $\alpha_i(\theta)$, *then* $\alpha_i^{-1}(y_i)$ is Gauss consistent for θ. The LS approach gives immediately the estimating equation

$$\sum (y_i - \alpha_i(\theta)) \frac{\partial \alpha_i}{\partial \theta} = 0. \qquad (1.15)$$

Then following the same argument as used in deriving the GM Theorem (B) of Section 1.3, we consider the class G_2 of estimating functions $\tilde{g}$ of the type

$$\tilde{g} = \sum b_i(\theta)(y_i - \alpha_i(\theta)). \qquad (1.16)$$

Using (1.12) we have,

$$\tilde{g}_s = \sum b_i(\theta)(y_i - \alpha_i(\theta)) \bigg/ \left(-\sum b_i \frac{\partial \alpha_i}{\partial \theta} \right). \qquad (1.17)$$

Now $\mathrm{Var}(\tilde{g}_s) = \sigma^2 \sum b_i^2 / (\sum b_i \, \partial \alpha_i / \partial \theta)^2$ is minimized for $b_i(\theta) = k(\theta) \, \partial \alpha_i / \partial \theta$. This shows that an optimal estimating function within G_2 is given by $\sum (y_i - \alpha_i(\theta)) \, \partial \alpha_i / \partial \theta$, leading to the LS estimating equation in (1.15).

It is important to note that in the case when $(y_1, y_2, \ldots, y_n)$ are normally distributed, all three methods LS, ML and optimal estimation leads to the same estimating equation. On the other hand the approach based on UMV-ness of the estimate of θ fails in case where $\alpha_i(\theta)$ are non-linear even if we assume the normality of $(y_1, y_2, \ldots, y_n)$. Yet by transferring the UMV-ness of an estimate to that of a standardized estimate function we could establish the optimality of the LS estimating equation. However, the estimate θ obtained from (1.15) would be biased for θ. But this is at most a small price to pay, as in many instances unbiased estimates of θ, even under normality, would not exist. This brings out the flexibility of the approach to estimation by estimating functions. In the next section we show how this approach suggests a modification to the LS approach in some situations where the classical LS approach fails.

1.5 A failure of the least squares (LS) approach

In all the previous discussion we have assumed tacitly that $\mathrm{Var}(y_i) = \sigma^2$ is independent of θ. Now consider independent random variates $y_i, y_2, \ldots, y_n$ such that

$$E(y_i) = \alpha_i(\theta) \quad \text{and} \quad \mathrm{Var}(y_i) = c\sigma_i^2(\theta) \qquad (1.18)$$

where $\sigma_i^2(\theta)$ are specified differentiable functions of θ and c is an *unknown* positive constant not depending on θ. To estimate θ, the LS approach requires minimization of $\sum (y_i - \alpha_i(\theta))^2/\sigma_i^2(\theta)$, leading to the LS estimating equation

$$\tilde{g}_1^* + B = 0, \qquad (1.19)$$

where $\tilde{g}_1^* = \sum (y_i - \alpha_i(\theta))(\partial\alpha_i/\partial\theta)\sigma_i^{-2}$ and $B = \sum (y_i - \alpha_i(\theta))^2(\partial\sigma_i/\partial\theta)\sigma_i^{-3}(\theta)$. We first note that whereas $E(\tilde{g}_1^*) = 0$, $E(B) = \sum (\partial\sigma_i/\partial\theta)\sigma_i^{-1} = \sum (\partial \log \sigma_i/\partial\theta)$ is in general non-zero and thus $\tilde{g}_1^* + B$ is not an unbiased estimating function. Even for large samples, although $\tilde{g}_1^*/n$ converges in probability to zero this may not be the case with B/n, and in fact B/n could diverge to $\pm\infty$ depending on the nature of $(1/n) \sum (\partial \log \sigma_i/\partial\theta)$. Thus for large n, the solution of the equation $\tilde{g}_1^* = 0$, under some regularity conditions, will converge in probability to the true value whereas the solution of the LS estimating equation may not. The latter in fact could converge to a value far away from the true value. It is often suggested that we take $\tilde{g}_1^* = 0$ as an estimating equation, a sort of modified LS estimating equation. We now show that $\tilde{g}_1^*$ is in fact an optimal estimating function in the class G_2 defined by (1.16). Let

$$\tilde{g}_1 = \sum (y_i - \alpha_i(\theta))b_i(\theta).$$

Now using the standardization (1.12) we have

$$\mathrm{Var}(\tilde{g}_{1s}) = c \sum b_i^2(\theta)\sigma_i^2(\theta)/\{\sum b_i(\theta)(\partial\alpha_i/\partial\theta)\}^2.$$

Further, $\mathrm{Var}(\tilde{g}_{1s})$ is minimized for $b_i(\theta) = k(\theta) \cdot (\partial\alpha_i/\partial\theta)\sigma_i^{-2}(\theta)$, $i = 1, 2, \ldots, n$. This leads to the optimum estimating function in G_2 given by

$$\tilde{g}_1^* = \sum (y_i - \alpha_i(\theta)) \frac{\partial\alpha_i}{\partial\theta} \Big/ \sigma_i^2(\theta). \qquad (1.20)$$

Hence the modified LS estimating equation $\tilde{g}_1^* = 0$ corresponds to the optimum estimating function in G_2.

It is interesting to compare the optimum estimating equation $\tilde{g}_1^* = 0$, with the ML equation for a specified value $c = c_0$ say, in (1.18) when $(y_1, y_2, \ldots, y_n)$ are assumed to be normally distributed. The likelihood equation then is the same as LS equation corrected for bias and is given by

$$g = \tilde{g}_1^*/c_0 + \left\{ \frac{B}{c_0} - \sum \frac{\partial \log \sigma_i}{\partial\theta} \right\} = 0. \qquad (1.21)$$

We note that $E(g) = 0$ only when $c = c_0$ in (1.18). Otherwise

$$E(g) = \left\{ \frac{c}{c_0} - 1 \right\} \sum \frac{\partial \log \sigma_i}{\partial \theta}.$$

Thus if the true value of c in (1.18) is such that c/c_0 is far away from one then the likelihood equation is again biased and leads to the same problems as in the case of the LS estimating equation even for large n. On the other hand the optimum estimating equation $\tilde{g}_1^* = 0$ is unaffected by the value of c. Of course if c is not specified, the ML method is undefined in principle even if we assume normality and it is also in general undefined for the model specifying only the first two moments of random variables (see Section 1.8). Yet for the parametric sub-model obtained from (1.18) by assuming normality and a specified value of c, the ML equation is reasonably well approximated by the optimal estimating equation $\tilde{g}_1^* = 0$. This is in line with the connection, established by Gauss, between the LS and ML estimating equations. Further, the above optimality property $\tilde{g}_1^*$ is mathematically and statistically analogous to the optimality property of the score function $\partial \log p/\partial \theta$ first established by Godambe (1960) in case of a (full) parametric model given by $\{p(x, \theta), \theta \in \Omega \subseteq R_1\}$.

In the following section we will consider the estimating functions and equations in the context of parametric models and indicate how this approach provides a logical frame for estimation of parameter(s) of interest in the presence of nuisance parameter(s). This also shows that the estimating functions provide a common connecting link for estimation in parametric and semi-parametric models. Our restriction to the classes G_0, G_1 and G_2 of estimating functions in the context of the LS method (which assumes only the first two moments) is justified by the nature of the corresponding semi-parametric models (Godambe 1985; Godambe and Heyde 1987; Godambe and Thompson 1989). The classes G_0, G_1, G_2 are of the type $\sum a_i(\theta)y_i + b_i(\theta)$, i.e. linear in $(y_1, y_2, \ldots, y_n)$ but possibly non-linear in θ. Durbin (1960) considered linear optimum estimating functions with many applications to time series. In the parametric models we will define a more general class G of estimating functions and search for an optimal estimating function in G.

1.6   Estimating functions—optimality

Let x be an abstract random variable on a sample space (χ, β) with a probability density function $p(x, \theta)$, w.r.t. a σ-finite measure μ on (χ, β). Here θ is a real- or vector-valued parameter which is assumed to be unknown. If we know θ, then the probability distribution of x becomes completely known and hence our object is to estimate θ on the basis of the observed value of x. Conventionally the problem of estimation is tackled by proposing an estimator $T(x)$ and then studying the properties of the estimator $T(x)$ which depend on the sampling distribution of x. These optimal properties include sufficiency, unbiasedness, minimum variance or minimum mean squared error, etc. The estimator $T(x)$ is obtained by some standard methods such as the least squares, maximum likelihood, minimum chi-square, method of moments, among others. Most often these methods, though intuitive, are *ad hoc* and do not directly follow from the optimality properties demanded or expected from the 'best' estimator, a notable exception being the construction of minimum variance unbiased estimator following the Rao–Blackwell, Lehmann–Scheffé approach.

A common feature of methods such as least squares, maximum likelihood, moments, minimum chi-square is that these methods lead to an estimating equation $g(x, \theta) = 0$, in case θ is real, or a set of estimating equations $g_i(x, \theta) = 0$, $i = 1, 2, \ldots, m$ in case of a vector-valued parameter $\theta = (\theta_1, \theta_2, \ldots, \theta_m)$. Indeed the phrase 'equation for estimation' occurs as early as Fisher (1935). Another important early reference is Kendall (1951).

Following the ideas of the earlier sections we now consider an estimating function $g(x, \theta)$ defined $\chi \times \Omega$ rather than a statistic $T(x)$. We will first consider θ real valued and we assume that $g(x, \theta)$, for each $\theta \in \Omega$, is such that $E(g) = 0$ and $\text{Var}(g)$ is finite. The idea of using an estimating function, a function of observations as well as the parameters, has been around for quite some time. Pivotal quantities or 'pivots' used by Fisher (1935) constitute a prime example of this. The distribution of a pivot does not depend on θ and this property is exploited for making inference on θ. On the other hand we require that only the first moment of g be independent of θ, which we can take to be zero without loss of generality. Our assumption that $\text{Var}(g)$ is finite can be rephrased by requiring that $\text{Var}(g') = 1$ by redefining $g' = g/\sqrt{\text{Var}(g)}$, equivalently assuming that the first two moments of g' are independent of θ. However, we will not pursue this line of thought considered by Barnard (1973) but instead follow Godambe's (1960) optimality criteria informally discussed and illustrated in previous sections.

Since the estimating function g is initially used to obtain an estimator by solving the equation $g = 0$, the unbiasedness condition

$$E_\theta(g) = 0 \qquad \forall \theta \in \Omega \tag{1.22}$$

becomes a natural one. It is not very restrictive since as mentioned above we are anyway assuming the existence of the first two moments of g. In order that the estimating equation $g(x, \theta) = 0$ should determine an estimator $\hat{\theta}_g$ as a function of x, it is necessary that the implicit function theorem holds. For this, a sufficient condition is $\partial g/\partial \theta \neq 0$. As $g(x, \theta)$, for each θ fixed, is a random variable, we must have for each θ, $P\{\partial g/\partial \theta \neq 0\} = 1$. A weaker assumption is that

$$E\left(\frac{\partial g}{\partial \theta}\right) \neq 0 \qquad \forall \theta \in \Omega. \tag{1.23}$$

Let G be the class of all estimating functions which satisfy (1.22), (1.23) and such that $\mathrm{Var}(g) = E(g^2)$ is finite. Godambe (1960) defined $g^* \in G$ as an *optimal* function if for any $g \in G$

$$\frac{E\{(g^*)^2\}}{\{E(\partial g^*/\partial \theta)\}^2} \leqslant \frac{E\{(g^2)\}}{\{E(\partial g/\partial \theta)\}^2} \qquad \forall \theta \in \Omega. \tag{1.24}$$

Again the motivation behind this optimality criteria is similar to the one discussed earlier: We require that $\mathrm{Var}(g) = E(g^2)$ is as small as possible and $[E\{g(x, \theta + \delta\theta)\}]^2$ is as large as possible. Both these objectives are achieved if we minimize $E(g^2)/\{E((\partial g/\partial \theta)\}^2$ for all $g \in G$ uniformly in $\theta \in \Omega$. In terms of the standardization given by (1.12) the criterion (1.24) is equivalent to

$$\mathrm{Var}(g_s^*) \leqslant \mathrm{Var}(g_s) \qquad \forall \theta \in \Omega. \tag{1.25}$$

The interpretation of the standardization (1.12) in terms of the variances of the estimating functions g and 'constant' $\times g$, discussed earlier is due to Barnard. Godambe's (1960) main result was that under mild regularity conditions on the class of estimating functions G and the class of density functions $\{p(x, \theta), \theta \in \Omega\}$ the score function $\partial \log p/\partial \theta$ is optimal. In this sense the likelihood equation is an optimal estimating equation. It is easy to demonstrate that an optimal estimating function is essentially unique in the sense that if g_1^* and g_2^* are both optimal, then $g_{1s}^* = g_{2s}^*$.

The optimality criterion of Godambe given above and the standardization (1.12) it introduces can be viewed in many different ways. Kale (1962a) following the general theory of estimating functions as developed by Kimball (1946) and Wilks (1948) derived an extension of the Cramer–Rao inequality for estimating functions. Under mild regularity conditions, Kale (1962a) proved that for any $g \in G$

$$\mathrm{Var}(g) \geqslant \left\{E\left(\frac{\partial g}{\partial \theta}\right)\right\}^2 \Big/ I(\theta), \tag{1.26}$$

where $I(\theta)$ is the Fisher information. He also pointed out that $\partial \log p/\partial \theta$, attains the extended Cramer–Rao lower bound to the $\mathrm{Var}(g)$ inequality for $g \in G$ and is therefore optimal. The extended Cramer–Rao inequality given by (1.26) can be written as

$$\mathrm{Var}(g_s) \geqslant \frac{1}{I(\theta)}, \tag{1.27}$$

where $g_s = g/E(\partial g/\partial \theta)$, the standardized version of g. The advantage of (1.27) over (1.26), i.e. of standardized version of g over g itself, is that the lower bound $1/I(\theta)$ is independent of $g \in G$.

Another justification of Godambe's optimality criteria (1.24) is provided by the fact that under mild regularity conditions the estimator $\hat{\theta}_g^*$, provided by solving the optimal estimating equation $g^*(x, \theta) = 0$, minimizes, at least asymptotically, the mean square error $E(\hat{\theta}_g - \theta)^2$ where $\hat{\theta}_g$ is the estimator provided by $g(x, \theta) = 0$ for $g \in G$. This follows from the argument that g_s and $\hat{\theta}_g - \theta$ are stochastically equivalent as $n \to \infty$ and minimizing $\text{Var}(g_s)$ is the same as the minimizing mean squared error of $\hat{\theta}_g$. For details we refer to Kale (1985) as well as Small and McLeish (1988). A further justification in terms of asymptotically shortest confidence intervals—in a very general setting of stochastic processes—is due to Godambe and Heyde (1987). An early work in this direction is due to Wilks (1939).

A different kind of justification of Godambe's optimality criteria and the standardization of the estimating function is provided by its connection with the Newton–Raphson process for solving the corresponding estimating equation $g(x, \theta) = 0$. In many cases $\hat{\theta}_g$, the estimator defined by the solution of the estimating equation cannot be obtained explicitly. It is to be determined by an iterative procedure. A commonly employed procedure is the Newton–Raphson procedure, given by

$$\theta_{r+1} = \theta_r - \left\{ g(x, \theta) \Big/ \frac{\partial g}{\partial \theta} \right\}_{\theta = \theta_r} \tag{1.28}$$

with a trial value θ_1 as an consistent estimator of θ. A modification of the Newton–Raphson process suggested by Fisher (1925) in the context of the likelihod equation, leads to the well-known method of scoring for parameters. Thus the Fisher–Newton–Raphson iterative procedure is given by

$$\theta_{r+1} = \theta_r - \left\{ g(x, \theta) \Big/ E\left(\frac{\partial g}{\partial \theta}\right) \right\}_{\theta = \theta_r}. \tag{1.29}$$

Thus the correction term to the successive iterates is $g_s(x, \theta)$, the standardized version of g. Minimizing $\text{Var}(g_s)$ thus corresponds to choosing the estimating function g with smallest correction term on the average. The optimality of the likelihood equation now translates into the fact that the method of scoring for parameters converges very fast, even though it is only a first-order process, a phenomenon observed by Kale (1962b).

1.7 Multiparametric case

Durbin (1960) considered linear estimating functions for the vector-valued parameter. When the indexing parameter θ is vector valued, say $(\theta_1, \theta_2, \ldots, \theta_m) \in \Omega_m \subset R_m$, then we consider a vector-valued estimating function $(g_1(x, \theta), \ldots, g_m(x, \theta))' = g(x, \theta)$ with $E(g) = 0$, and such that the variance–covariance matrix $M_g = E(g \cdot g')$ exists and is positive definite. In the single-parameter case we imposed the condition that $E(\partial g/\partial \theta) \neq 0$.

Analogously in the vector-valued situation, if D_g denotes the $m \times m$ matrix with elements $E(\partial g_i / \partial \theta_j)$, we assume that D_g is non-singular.

Assuming that the matrix M_g is positive definite and D_g is non-singular, Kale (1962a) proved that for a vector-valued $g \in G^{(m)}$

$$M_g - D_g J^{-1} D_g' \qquad \text{is non-negative definite,} \qquad (1.30)$$

where J is the Fisher information matrix and $G^{(m)}$ is the class of m-dimensional vector-valued estimating functions. The standardization (1.12) in the present case immediately leads to the standardized version of g, given by $D_g^{-1} g = g_s$, and the extended Cramer–Rao inequality for a regular standardized estimating function is now given by

$$M_{g_s} - J^{-1} \qquad \text{is non-negative definite.} \qquad (1.31)$$

The above standardization was first proposed by Ferreira (1982) and also independently by Chandrasekar (1983).

For the vector-valued g, several different optimality criteria can be proposed. The most common among these are

(i) matrix optimality: $M_{g_s} - M_{g_s^*}$ is non-negative definite;
(ii) trace optimality: $\text{Tr}(M_{g_s}) \geqslant \text{Tr}(M_{g_s^*})$;
(iii) determinant optimality: $|M_{g_s}| \geqslant |M_{g_s^*}|$.

Chandrasekar and Kale (1984) proved that the three criteria are equivalent in the sense that if g^* is optimal w.r.t. any one of these criteria then it is also optimal w.r.t. the remaining two.

From the extended Cramer–Rao inequality (1.30) or (1.31) it immediately follows that, in the class of all regular unbiased estimating functions, the vector score function

$$\frac{\partial \log p}{\partial \theta} = \left(\frac{\partial \log p}{\partial \theta_1}, \ldots, \frac{\partial \log p}{\partial \theta_m} \right)$$

is optimal. The essential uniqueness of the optimal estimating function follows from the fact that if g_1 and g_2 are both optimal then their standardized versions, g_{1s} and g_{2s}, are identical. In particular this implies that the likelihood equation is an optimal estimating equation as the estimating equations $g = 0$ and $g_s = 0$ are identical. Bhapkar (1972) defined the concept of the efficiency of an estimating equation corresponding to the trace and determinant optimality criteria and he also introduced the Rao–Blackwellization of an estimating function with respect to minimal sufficient statistics. Another interesting reference in this connection is Morton (1981).

1.8 Invariance and nuisance parameters

Another important property of the estimating functions is the invariance under one-to-one transformation of the parameter θ. Thus if $g(x, \theta)$ is an estimating function, then under any one-to-one differentiable transformation $\Phi = \alpha(\theta)$, where $\partial\alpha/\partial\theta$ is non-singular, then $g(x, \alpha^{-1}(\Phi)) = g_1(x, \Phi)$ is an estimating function for Φ. If $\hat{\theta}_g$ and $\hat{\Phi}_{g_1}$ are the estimates obtained from the equations $g = 0$ and $g_1 = 0$ respectively, then $\hat{\Phi}_{g_1} = \alpha(\hat{\theta}_g)$. It is well known that this invariance property is enjoyed by the maximum likelihood estimation but does not hold for the unbiased minimum variance estimation (Godambe 1976; Godambe and Thompson 1978). On the other hand unlike unbiased estimation, the maximum likelihood estimation faces difficulties when nuisance parameters are involved, i.e. when we are interested in $\theta^{(1)} = (\theta_1, \theta_2, \dots, \theta_r)$ 'only', while $\theta^{(2)} = (\theta_{r+1}, \dots, \theta_m)$ acts as a nuisance parameter. In fact the maximum likelihood estimator of only $\theta^{(1)}$, the parameter of interest, is technically undefined. A naïve approach, based on obtaining the maximum likelihood estimator of the entire parameter $\hat{\theta} = (\hat{\theta}^{(1)}; \hat{\theta}^{(2)})$ and then using $\hat{\theta}^{(1)}$, in general leads to anomalies, as is well known from the Neyman–Scott problem. We will now show how the estimating function approach alleviates the difficulties arising in the nuisance-parameter case. This will also show that the approach based on estimating functions unifies maximum likelihood estimation as well as unbiased minimum variance estimation by eliminating their respective weaknesses: the nuisance parameters in the case of the former and non-invariance under a one-to-one transformation of the parameter in the case of the latter.

A discussion of invariance of estimating equations from a different perspective is given by Okuma (1976). In the next section we present the case of nuisance parameters.

1.9 Optimality: nuisance-parameter case

Consider now an abstract random variable x with p.d.f. belonging to $\{p(x, \theta), \theta \in \Omega_m\}$, where $\theta = (\theta^{(1)}, \theta^{(2)})$ with $\theta^{(1)} = (\theta_1, \dots, \theta_r)$ as the parameter of interest and $\theta^{(2)} = (\theta_{r+1}, \dots, \theta_m)$ as nuisance parameter. A regular estimating function for $\theta^{(1)}$ is an r-dimensional function $g(x, \theta^{(1)})$ such that $E(g) = 0$, $\forall \theta \in \Omega_m$ and $M_g = E(gg')$ is positive definite and the $r \times r$ matrix $D_g = E(\partial g/\partial\theta^{(1)})$ is non-singular. Let $g_s = D_g^{-1}g$ denote the standardized version of g. Note that g_s may depend on $\theta^{(2)}$, the nuisance parameter, although the corresponding estimating equations $g_s = 0$ and $g = 0$ are same. An estimating function $g^*(x, \theta^{(1)})$ is optimal in $G(\theta^{(1)})$, the class of all regular estimating functions for estimating $\theta^{(1)}$ in the presence of the nuisance parameter $\theta^{(2)}$ if

$M_{g_s} - M_{g_s^*}$ is non-negative definite, $\forall \theta \in \Omega_m$, and $\forall g \in G(\theta^{(1)})$. As seen earlier this is equivalent to trace and determinant optimality.

Godambe and Thompson (1974) considered the case $r = 1$ and $m = 2$, a single parameter of interest and a real-valued nuisance parameter and showed that in the case of $N(\theta_1, \theta_2)$ the optimal estimating function for θ_1 ignoring θ_2 is $(\bar{x} - \theta_1)$, while $s^2 - (n - 1)\theta_2$ is the optimal estimating function for θ_2 ignoring θ_1. An interesting reference in this connection is Barnard (1973). Godambe (1976), considering the case $r = 1$ and $m > 1$ with $\theta = (\theta_1, \theta^{(2)}) \in \Omega_1 \times \Omega_2$, introduced an interesting structure for estimating θ_1 ignoring $\theta^{(2)}$. Godambe (1976) assumed that there exists a statistic $T(x)$ such that

$$p(x, \theta) = f_t(x; \theta_1)h(t, \theta_1, \theta^{(2)}),$$

where h is the p.d.f. of T and $f_t(x, \theta_1)$ is the conditional p.d.f. of x given t, which depends only on θ_1, the parameter of interest. Further, he assumed that the class $\{h(t, \theta_1, \theta^{(2)}), \theta_1 \text{ fixed}, \theta^{(2)} \in \Omega_2\}$ is complete. Under mild regularity conditions on p, f_t, h and $G(\theta^{(1)})$ Godambe (1976) showed that the conditional score function $\partial \log f_t / \partial \theta_1$ is the optimal estimating function for θ_1 ignoring $\theta^{(2)}$. Using this theory Godambe (1976) showed how the Neyman–Scott problem can be resolved and how the optimum estimating function for estimating the error variance ignoring block means leads to the minimum variance unbiased estimator, which is consistent when the number of blocks goes to infinity. Ferreira (1982) and Chandrasekar (1983) generalized these results for $r > 1$ and showed that the conditional vector score function $\partial \log f_t / \partial \theta^{(1)}$ is an optimal estimating function. Kale (1987a) pointed out the analogy of Godambe's structure with the Neyman structure used in the construction of UMV tests in the presence of nuisance parameters.

Chandrasekar and Kale (1984) proved a Cramer–Rao type inequality, namely $M_{g_s} - J^{11}$ is non-negative definite $\forall \theta \in \Omega$, $\forall g \in G(\theta^{(1)})$, where J^{11} is the $r \times r$ matrix when J^{-1} is partitioned corresponding to $(\theta^{(1)}, \theta^{(2)})$. Kale (1987b) proved the essential uniqueness of the optimal estimating function. Following the extension of the Cramer–Rao inequality approach, Subramanyam and Naik-Nimbalkar (1990) obtained a generalization of (1.31) for a Hilbert space valued parameter and proved that Aalen's (1978) estimator of the cumulative intensity function emerges as a solution of the optimal estimating equation.

Godambe (1976, 1980) uses the optimality of the conditional score function to define partial sufficiency and ancillarity of a statistic $T(x)$ for $\theta^{(1)}$, the parameter of interest when $\theta^{(2)}$ acts as a nuisance parameter. Kale (1987a) has shown that for a multiparameter exponential family, Godambe's approach for defining partial sufficiency and ancillarity succeeds whereas many other approaches fail. Recently Bhapkar (1988) has extensively studied this problem along with the problem of defining Fisher information about $\theta^{(1)}$ ignoring $\theta^{(2)}$.

In the above set-up we assumed that the statistic $T(x)$ exists *uniquely* for all values of $\theta^{(1)} \in \Omega_1$. Lindsay (1982) deals with the case when T depends on $\theta^{(1)}$.

1.10 Extensions

It is thus clear that the approach based on estimating functions unifies the method of maximum likelihood and the method of minimum variance unbiased estimation in the case of parametric models. It is no wonder that this theory has been applied successfully for estimation problems in such diverse fields from survey-sampling to time-series and stochastic processes as exemplified by the papers of Godambe (1985) and Godambe and Thompson (1986), Thavaneswaran and Abraham (1988).

We have also seen how the estimating function theory successfully tackles the situation where the usal GM or LS theory fails to give a reasonable solution. Now, as mentioned previously, the GM or LS approach is primarily meant for semi-parametric models where we do not assume the exact form of the density $p(x, \theta)$ but assume only the knowledge of the first two moments. Since the form of p is not known, estimating functions based on $\partial \log p/\partial\theta$ are not available here. However, just as Halmos (1946) showed that $\bar{x}$ is minimum variance unbiased estimator of $E(x) = \theta$ in the class $\mathscr{F}_0$ of all continuous distribution functions with mean θ, Godambe and Thompson (1978) proved that $(\bar{x} - \theta)$ is an optimal estimating function within the subclass $\mathscr{F}_1 \subset \mathscr{F}_0$ with location parameter θ.

In general a parameter of a distribution in a semi-parametric model is a well-defined functional of the underlying population distribution function and therefore the definition of such a parameter is closely connected with the method of estimation of this parameter. Suppose this parameter $\theta(F)$ for $F \in \mathscr{F}$ is a parameter of interest and $\psi(F)$ is a nuisance parameter such that $(\theta(F), \psi(F))$ is a labelling parameter for $\mathscr{F}$. Godambe and Thompson (1984) obtained an optimal estimating function for estimating $\theta(F)$ which in turn could also be used to define the parameter $\theta(F)$. This line of work has not been followed very vigorously and deserves more attention.

The theory of optimum estimating functions has provided a new and fruitful perspective on 'quasi-likelihood' (Wedderburn 1974) by identifying the 'quasi-score function' with the 'optimal estimating function' (Godambe 1985; Godambe and Heyde 1987; Godambe and Thompson 1989; McCullagh and Nelder 1989). This is also true in connection with 'partial likelihood' (Cox 1975; Godambe 1985).

To indicate the varied applications of estimating function theory to the area of biostatistics, we just refer to Kalbfliesch and Lawless (1988), Liang and Zeger (1986), and Prentice (1988).

It is now clear that, among researchers in different areas of statistics, there is an increasing trend to utilize estimating function theory for statistical model building, inference and the like. A purpose of this 'Overview' is to encourage further investigations concerning the theory, both of foundational and applied nature. Another important review of the subject that has recently appeared, is that by Heyde (1989).

1.11   Acknowledgements

Both authors are grateful to the National Science and Engineering Council of Canada and the University Grants Commission, India, for providing financial support during their leave periods. They also want to thank Drs B. Abraham and C. Dean for comments on the earlier draft of the paper.

References

Aalen, O. (1978). Nonparametric inference for a family of counting processes. *Ann. Stat.*, **6**, 701–6.

Barnard, G. A. (1973). Maximum likelihood and nuisance parameters. *Sankhya A*, **35**, 133–8.

Bertrand, J. (1888). *Calcul de probabilités* (2nd edn 1972), Chelsea, New York.

Bhapkar, V. P. (1972). On a measure of efficiency of an estimating equation. *Sankhya A*, **34**, 467–72.

Bhapkar, V. P. (1988). On generalized principles for inference in the presence of nuisance parameter, Tech. Report No. 266, Department of Statistics, University of Kentucky.

Chandrasekar, B. (1983). Contributions to the theory of unbiased statistical estimation functions. Ph.D. thesis submitted to University of Poona, Pune-7, India.

Chandrasekar, B. and Kale, B. K. (1984). Unbiased statistical estimation functions in presence of nuisance parameter. *J. Stat. Plan. Inf.*, **9**, 45–54.

Cox, D. R. (1975). Partial likelihood. *Biometrika*, **62**, 269–76.

Durbin, J. (1960). Estimation of parameters in time series regression models. *J. Roy. Stat. Soc. Ser. B*, **22**, 139–53.

Ferreira, P. E. (1982). Multiparametric estimating equations. *Ann. Stat. Math.*, **34A**, 423–31.

Fisher, R. A. (1925). Theory of statistical estimation. *Proc. Cambridge Phil. Soc.*, **22**, 700–6.

Fisher, R. A. (1935). The fiducial argument in statistical Inference. *Ann. Eugenics*, **6**, 391–6.

Gauss, C. F. (1809). Theoria motus corporum coelestum, *Werke*, 7. Translated into English by C. H. Davis (1963). Dover, New York.

Gauss, C. F. (1823). Combinationes erroribus minimis obnoxiae. Parts 1, 2 and Suppl., *Werke*, **4**, 1–108.

Godambe, V. P. (1960). An optimum property of regular maximum likelihood estimation. *Ann. Math. Stat.*, **31**, 1208–12.

Godambe, V. P. (1976). Conditional likelihood and unconditional optimum estimating equations. *Biometrika*, **63**, 277–84.

Godambe, V. P. (1980). On the sufficiency and ancillarity in the presence of nuisance parameter, *Biometrika*, **67**, 269–76.

Godambe, V. P. (1985). The foundations of finite sample estimation in stochastic processes. *Biometrika*, **72**, 419–28.

Godambe, V. P. and Heyde, C. C. (1987). Quasi-likelihood and optimal estimation. *Int. Stat. Rev.*, **55**, 231–44.

Godambe, V. P. and Thompson, M. E. (1974). Estimating equations in the presence of nuisance parameters. *Ann. Stat.*, **2**, 568–71.

Godambe, V. P. and Thompson, M. E. (1978). Some aspects of the theory of estimating equations. *J. Stat. Plan. Inf.*, **2**, 95–104.

Godambe, V. P. and Thompson, M. E. (1984). Robust estimation through estimating equations. *Biometrika*, **71**, 115–25.

Godambe, V. P. and Thompson, M. E. (1985). Logic of least squares revisited. Pre-print.

Godambe, V. P. and Thompson, M. E. (1986). Parameters of superpopulation and survey population, their relationship and estimation. *Int. Stat. Rev.*, **54**, 127–38.

Godambe, V. P. and Thompson, M. E. (1989). An extension of quasilikelihood estimation (with discussion). *J. Stat. Plan. and Inf.*, **22**, 137–72.

Halmos, P. (1946). The theory of unbiased estimation. *Ann. Math. Stat.*, **17**, 43–54.

Heyde, C. C. (1989). Quasi-likelihood and optimality of estimating functions: some current unifying themes. *Bull. Int. Stat. Inst.*, Book 1, 19–29.

Heyde, C. C. and Seneta, E. (1977). *I.J. Bienayme: statistical theory anticipated.* Springer, New York.

Kalbfliesch, J. D. and Lawless, J. F. (1988). Likelihood analysis of multi-state models for disease incidence and mortality. *Statistics and Medicine*, **7**, 149–60.

Kale, B. K. (1962a). An extension of Cramer–Rao inequality for statistical estimation functions. *Skand. Aktur.*, **45**, 60–89.

Kale, B. K. (1962b). On the solution of likelihood equations by iteration processes. *Biometrika*, **49**, 479–86.

Kale, B. K. (1985). Theory of unbiased statistical estimation functions. Lecture Notes, Dept. of Statistics, Iowa State University, Ames, Iowa, USA 50011, and Tech. Report 81, Dept. of Statistics University of Poona, Pune 411007, India.

Kale, B. K. (1987a). Optimal estimating function in multi-arameter exponential family. Tech. Report 87-02 Dept. of Statistics and Actuarial Science, University of Waterloo, Waterloo, Canada.

Kale, B. K. (1987b). Essential uniqueness of optimal estimating functions. *J. Stat. Plan. Inf.*, **17**, 405–7.

Kendall, M. G. (1951). Regression, structure and functional relationship—I. *Biometrika*, **38**, 11–25.

Kimball, B. F. (1946). Sufficient statistical estimation functions for the parameters of the distribution of maximum values. *Ann. Math. Stat.*, **17**, 299–309.

Liang, K. Y. and Zeger, S. L. (1986). Longitudinal data analysis using generalized linear models. *Biometrika*, **73**, 13–22.

Lindsay, B. (1982). Conditional score functions: some optimality results. *Biometrika*, **69**, 503–12.

McCullagh, P. and Nelder, J. A. (1989). *Generalized linear models* (2nd edn). Chapman and Hall, London.

Morton, R. (1981). Efficiency of estimating equations and the use of pivots. *Biometrika*, **68**, 227–33.

Okuma, A. (1976). On invariance of estimating equations. *Bull. Kyushu Inst. Tech.*, **23**, 11–16.

Prentice, R. L. (1988). Correlated Binary regression with covariates specific to each binary observation. *Biometrics*, **44**, 1033–48.

Small, C. and McLeish, D. L. (1988). *The theory and applications of statistical inference functions*. Lecture Notes in Statistics No. 44, Springer Verlag, Heidelberg, New York, London.

Sprott, D. A. (1983). Gauss Carl Friedrich, *Encyclopaedia of Statistical Sciences*, **3**, Eds. Kotz Johnson, Wiley, pp. 305–8.

Subramanyam, A. and Naik-Nimbalkar, U. V. (1990). Optimal unbiased statistical estimating functions for Hilbert space valued parameters. *J. Stat. Plan. Inf.*, **24**, 95–105.

Thavaneswaran, A. and Abraham, B. (1988). Estimation for non-linear time series models using estimating equations. *J. Time Series Analysis*, **9**, 99–108.

Wedderburn, R. W. H. (1974). Quasi-likelihood functions, generalized linear models, and Gauss–Newton method. *Biometrika*, **61**, 439–47.

Wilks, S. S. (1938). Shortest average confidence intervals from large samples. *Ann. Math. Stat.* **9**, 166–75.

Wilks, S. S. (1948). *Mathematical statistics*. Wiley, New York.

PART 2
Biostatistics

2
Applications of estimating function theory to replicates of generalized proportional hazards models

I-Shou Chang and Chao A. Hsiung

ABSTRACT

Four types of proportional hazards models for counting processes are introduced according to the assumptions on the baseline functions. Estimation of the relative risk coefficients is considered in the framework of estimating function theory for each of these models. It is argued, when one multivariate counting process is observed, the optimal estimating functions are the same for each of these four models. However, when replicates of the multivariate counting processes are observed, the optimal estimating functions are different for these four models.

2.1 Introduction and summary

The proportional hazards model of survival analysis and its analysis by the method of partial likelihood originate in the work of Cox (1972, 1975). Since its introduction, it has been at the centre of many important statistical developments. In particular, Andersen and Gill (1982) formulated Cox's regression model for counting processes, which studied multivariate failure time data using martingale methods. On the other hand, Prentice *et al.* (1981) proposed regression models for multivariate failure time data in which the baseline function may depend on the past history of the process.

To be precise, let $N_1(t) = (N_{11}(t), \ldots, N_{1K}(t))$ be a K-variate counting process for which the intensity of $N_{1k}(t)$, relative to the self-excited filtration of $N_1(t)$, has the form

$$\lambda_{1k}(t) = \lambda_{10}(t) Y_{1k}(t) \, e^{\theta' Z_{1k}(t)}, \qquad (2.1)$$

where $Y_{1k}(\cdot) \geqslant 0$ is a bounded predictable process, $Z_{1k}(\cdot)$ is an R^d-valued bounded predictable process and $\theta \in \Theta \in R^d$ is the relative risk regression coefficient. When the baseline hazard,

$$\lambda_{10}(t) = h_0(t), \qquad (2.2)$$

is a deterministic function, (2.1) is the Cox's regression model discussed in Andersen and Gill (1982).

In the counting process language, what Prentice *et al.* (1981) suggests is that there are practical situation in which $\lambda_{10}(\cdot)$ of (2.1) is a non-deterministic predictable process. A simple example is

$$\lambda_{10}(t) = \sum_{i=0}^{\infty} h_i(t - T_{1i})I_{(T_{1i}, T_{1i+1}]}(t), \tag{2.3}$$

where $T_{1i} = \inf\{t > 0| \sum_{k=1}^{K} N_{1k}(t) = i\}$, and h_i is a deterministic function. In the industrial context, (2.3) specifies that components of a machine share a common baseline hazard rate, which depends on the total number of events experienced by the machine.

An important special case of (2.3) is

$$\lambda_{10}(t) = \sum_{i=0}^{\infty} h_i I_{(T_{1i}, T_{1i+1}]}(t), \tag{2.4}$$

where the h_i are constants.

More generally, $\lambda_{10}(t)$ can be an arbitrary predictable process,

$$\lambda_{10}(t) = \sum_{i=0}^{\infty} h_i(T_{11}, \ldots, T_{1i}, X_{11}, \ldots, X_{1i}, t)I_{(T_{1i}, T_{1i+1}]}(t), \tag{2.5}$$

where $X_{1i} = k$ if $N_{1k}(T_{1i}) - N_{1k}(T_{1i} -) = 1$ and h_i is a non-negative deterministic function.

It follows from the representation theorem for predictable processes (cf. Brémaud (1981), p. 58, p. 59, p. 307, p. 309) that any non-negative predictable process not depending on the covariates admits the representation (2.5). On the other hand, the Radon–Nikodym theorem (cf. Brémaud (1981), p. 165, p. 166, p. 168, p. 187) implies that the intensities of any family of mutually absolutely continuous probability measures for a multivariate counting process have a predictable process as their common factor. Thus, (2.5) is a model for multivariate counting process for which members of the family of probability measures share a most general predictable process as their common hazard rate apart from a factor depending on covariates. In short, (2.5) is about the most general baseline hazard rate one would use for a multivariate counting process based on likelihood considerations.

Like model (2.2), model (2.3), model (2.4) and model (2.5) will also be called proportional hazards models for the following reasons. A superficial reason is that the univariate counting processes involved have proportional hazard rate. A deeper reason emerges when we study the statistical problem of estimating θ, treating the h_i as nuisance parameters.

In fact, a generalization of Chang and Hsiung (1990) shows that from the viewpoint of estimating function theory, the optimal estimates for θ are the same in each of the models (2.2)–(2.5). (A brief discussion of this optimality concept is given in Section 2.3.) The classes of estimating functions considered are also the same. In this sense, we are encouraged to use the more flexible model (2.5), rather than (2.2), (2.3) or (2.4). In other words, we need not worry about the validity of the model assumptions (2.2), (2.3), or (2.4).

In this paper, we are concerned with the same estimation problem when replicates from proportional hazards models are observed. We show that models (2.2)–(2.5) are all different in this respect, unlike the situation of one replicate.

Let $N_1, \ldots, N_J$ be independent and identically distributed K-variate counting processes, which implies that they have common parameters θ and h_i. The statistical problem is to estimate θ based on $N_1, \ldots, N_J$ and related observables, treating the h_i as nuisance parameters. We note that these related observables, including the Y and Z, are different from each $j = 1, \ldots, J$.

When (2.2) is assumed, $(N_1, \ldots, N_J)$ itself remains a proportional hazards model as a JK-variate counting process. Hence Chang and Hsiung (1990) implies that the optimal estimating function is the maximum partial likelihood estimation (MPLE) for the JK-variate process, which is not the sum of the individual MPLEs for each N_j. In fact, using the notation introduced in Section 2.2, the optimal estimating function is

$$\tilde{G}_l(t) = \sum_{j=1}^{J} \sum_{k=1}^{K} \int_0^t \left(Z_{jkl}(s) - \frac{\sum_{j=1}^{J} \sum_{k=1}^{K} Y_{jk}(s)\, e^{\theta' Z_{jk}(s)} Z_{jkl}(s)}{\sum_{j=1}^{J} \sum_{k=1}^{K} Y_{jk}(s)\, e^{\theta' Z_{jk}(s)}} \right) dN_{jk}(s). \quad (2.6)$$

When (2.3), (2.4), or (2.5) is assumed, the situation is different. Since $(N_1, \ldots, N_J)$ putting together is no longer a JK-variate counting process with proportional hazard rate, we need a guide in searching for good estimators. In fact, the main theme of this paper is to demonstrate that estimating function theory initiated by Godambe (1960) serves such a purpose in these semi-parametric models.

Following Godambe (1985), Godambe and Thompson (1974) and Godambe (1984), Chang and Hsiung (1990) adapted the concept of Fisher information in the presence of nuisance parameters to the proportional hazards model and formulated the corresponding optimality criterion. We shall see in Section 2.3 that this framework is also valid for replicates from proportional hazards models, and forms the basis of this work.

The findings are as follows. For models (2.3) and (2.5), the optimal estimating function is the sum of the optimal estimating functions for each proportional hazards model $N_j, j = 1, 2, \ldots, J$. Based on this, one might be led to not use the more restricted model (2.3) at all. But our results indicate that, although the optimal estimating functions are the same in these two models, the classes of estimating functions over which they are optimal are different. The one for model (2.5) is smaller (cf. (2.12)). Thus the estimation problem is not quite the same. We note that the optimal estimating function for models (2.3) and (2.5) is given in (2.14), which is different from (2.6). Hence the estimation problem for model (2.2) is completely different from those for models (2.3) and (2.5).

As for model (2.4), we may also like to combine the MPLE based on each separate N_j, $1 \leqslant j \leqslant J$, to form an estimate for θ, as was the situation for models (2.3) and (2.5). But reasoning with estimating function theory indicates that none of these combinations can be optimal. This remarkable contrast suggests that we put together $N_1, \ldots, N_J$ and then derive from scratch an estimate for θ. The resulting optimal estimating function is given in (2.25).

Our results seem to suggest that the optimal estimating function for θ is the sum of the individual optimal estimating functions for each $N_j, j = 1, \ldots, J$, when the nuisance parameter space is big, like models (2.3) and (2.5). When the nuisance parameter space is not big, like models (2.2) and (2.4), there may be a possibility of getting an estimating function better than the sum of the individual optimal estimating functions.

This paper is organized as follows. The main results are contained in Section 2.2. Section 2.3 gives a brief account of estimating function theory used in this paper. In Section 2.2.1, we present the case for model (2.3) in detail and indicate only the necessary changes when model (2.5) is considered. In Section 2.2.2, we treat the model (2.4). Martingales, stochastic integrals and related concepts used freely in this paper can be found, for example, in Brémaud (1981) and Elliott (1982). Some of them were explained in Chang and Hsiung (1990).

2.2 Optimal estimating functions

2.2.1 REPLICATES OF MODEL (2.3) AND (2.5)

Let $N_1, \ldots, N_J$ be independent and identically distributed K-variate counting processes. Assume that, relative to its self-excited filtration $\mathscr{F}_{j,t}$, $N_j(t) = (N_{j1}(t), \ldots, N_{jK}(t))$ has intensity $\lambda_j(t) = (\lambda_{j1}(t), \ldots, \lambda_{jK}(t))$ of the form

$$\lambda_{jk}(t) = \lambda_{j0}(t) Y_{jk}(t) e^{\theta' Z_{jk}(t)}. \tag{2.7}$$

Here $\theta \in \Theta \subset R^d$, $\lambda_{j0}(\cdot) \geqslant 0$, $Y_{jk}(\cdot) \geqslant 0$ and $Z_{jk}(\cdot)$ are bounded predictable stochastic processes of class SP_j defined as follows.

Definition 2.1 A stochastic process h is said to be of class SP_j if

$$h(t) = \sum_{i=0}^{\infty} h_i(t - T_{ji}) I_{(T_{ji}, T_{ji+1}]}(t), \tag{2.8}$$

where $T_{ji} = \inf\{t > 0| \sum_{k=1}^{K} N_{jk}(t) = i\}$, $h_i: [0, \infty) \to [0, \infty)$ is a deterministic function. An R^d-valued process is of class SP_j if each of its components is of class SP_j.

Here, we assume that

$$\lambda_{j0}(t) = \sum_{i=0}^{\infty} \lambda_i(t - T_{ji}) I_{(T_{ji}, T_{ji+1}]}(t)$$

for some deterministic function $\lambda_i(\cdot)$. Let $\lambda = (\lambda_1, \lambda_2, \ldots)$. The statistical problem is to estimate θ based on the data

$$\{N_j(t), Y_j(t), Z_j(t) | 0 \leqslant t \leqslant T, 1 \leqslant j \leqslant J\},$$

treating the λ_i as nuisance parameters. Here $Y_j = (Y_{j1}, \ldots, Y_{jK})$, $Z_j = (Z_{j1}, \ldots, Z_{jK})$, and T is an $\mathscr{F}_t = \sigma\{\mathscr{F}_{j,t} | 1 \leqslant j \leqslant J\}$ stopping time.

The class of estimating functions we will consider is $\mathscr{G} = \{G(t, \theta)\}$, where $G(\cdot, \theta)$ is a mean-zero square-integrable right continuous $\mathscr{F}_t$-martingale for the parameter (θ, λ). We note that $G(\cdot, \theta)$ is independent of λ.

Before applying Corollary 2.1 in the Appendix to find the optimal estimating function, we will discuss the structure of elements in $\mathscr{G}$.

Let

$$M_{jk}(t) = N_{jk}(t) - \int_0^t \lambda_{jk}(s) \, ds. \tag{2.9}$$

We note that, because λ_{jk} depends on both λ and θ, so does M_{jk}. Since $N_1, \ldots, N_J$ are independent, M_{jk} is also an $\mathscr{F}_t$-martingale for every j,k. Let $G(\cdot, \theta) \in \mathscr{G}$, then $G(\cdot, \theta)$ admits the integral representation

$$G(t, \theta) = \sum_{j=1}^{J} \sum_{k=1}^{K} \int_0^t H_{jk}(s, \theta) \, dM_{jk}(s) \qquad \text{a.s.,} \tag{2.10}$$

for the parameter (θ, λ). Here $H_{jk}(\cdot, \theta)$ is a predictable process satisfying

$$E \int_0^t |H_{jk}(s, \theta)|^2 \lambda_{jk}(s) \, ds < \infty.$$

for every $t > 0$.

Observe that

$$G(t, \theta) = \sum_{j=1}^{J} \sum_{k=1}^{K} \int_0^t H_{jk}(s, \theta) \, dN_{jk}(s)$$

$$- \int_0^t \sum_{j=1}^{J} \lambda_{j0}(s) \sum_{k=1}^{K} H_{jk}(s, \theta) Y_{jk}(s) \, e^{\theta' Z_{jk}(s)} \, ds, \tag{2.11}$$

where the first term is an integral with counting measure and the second term is a Lebesgue integral. This together with the assumption that $G(t, \theta)$

is a function of the observation up to time t and θ implies that H_{jk} can be chosen to be independent of λ and, on $[0, \infty)$,

$$\sum_{j=1}^{J} h_j(t) \sum_{k=1}^{K} H_{jk}(t, \theta) Y_{jk}(t) e^{\theta' Z_{jk}(t)} = 0 \qquad \text{a.s.,} \qquad (2.12)$$

for every bounded process h_j in SP_j, $j = 1, \ldots, J$.

Now, let

$$A_{jl}(t) = \frac{\sum_{k=1}^{K} Y_{jk}(t) e^{\theta' Z_{jk}(t)} Z_{jkl}(t)}{\sum_{k=1}^{K} Y_{jk}(t) e^{\theta' Z_{jk}(t)}}, \qquad (2.13)$$

$$G_l(t) = \sum_{j=1}^{J} \sum_{k=1}^{K} \int_0^t (Z_{jkl}(s) - A_{jl}(s)) \, dN_{jk}(s), \qquad (2.14)$$

$$U_l(t) = \sum_{j=1}^{J} \sum_{k=1}^{K} \int_0^t A_{jl}(s) \, dM_{jk}(s), \qquad (2.15)$$

where Z_{jkl} is the lth component of Z_{jk}. We note that both A_{jl} and G_l are independent of the nuisance parameter λ, while U_l does depend on both θ and λ.

One can easily verify that

$$G_l(t) = \sum_{j=1}^{J} \sum_{k=1}^{K} \int_0^t (Z_{jkl}(s) - A_{jl}(s)) \, dM_{jk}(s), \qquad (2.16)$$

which, combining with the fact that $G_l(t)$ is independent of λ, shows $G_l \in \mathcal{G}$.

Using (2.12) and the fact that $A_{jl} \cdot \lambda_{j0}$ is in SP_j, we know the quadratic variation

$$\langle U_l(\cdot), G(\cdot, \theta) \rangle_t = \sum_{j=1}^{J} \sum_{k=1}^{K} \int_0^t A_{jl}(s) H_{jk}(s, \theta) \lambda_{j0}(s) Y_{jk}(s) e^{\theta' Z_{jk}(s)} \, ds$$

$$= \int_0^t \sum_{j=1}^{J} A_{jl}(s) \lambda_{j0}(s) \sum_{k=1}^{K} H_{jk}(s, \theta) Y_{jk}(s) e^{\theta' Z_{jk}(s)} \, ds$$

$$= 0, \qquad (2.17)$$

which implies that $U_l(\cdot)$ is orthogonal to every G in $\mathcal{G}$.

It follows from (2.17) and the fact

$$\frac{\partial}{\partial \theta_l} \log L(t, \theta, \lambda) = G_l(t) + U_l(t)$$

that the following Theorem 2.1 holds. For readers' convenience an explicit likelihood ratio formula is given as follows (cf. Brémaud, 1981, p. 187; Gill, 1980, p.14).

$$\log L(t, \theta, \lambda) = \sum_{j=1}^{J} \sum_{k=1}^{K} \int_0^t \theta' Z_{jk}(s) \, dN_{jk}(s)$$

$$+ \sum_{j=1}^{J} \sum_{k=1}^{K} \int_0^t (1 - e^{\theta' Z_{jk}(s)}) Y_{jk}(s) \lambda_{j0}(s) \, ds, \quad (2.18)$$

which implies

$$\frac{\partial}{\partial \theta_l} \log L(t, \theta, \lambda) = \sum_{j=1}^{J} \sum_{k=1}^{K} \int_0^t Z_{jkl}(s) \, dM_{jk}(s)$$

$$= G_l(t) + U_l(t). \quad (2.19)$$

Theorem 2.1 $\{G_1, \ldots, G_d\}$ *defined in (2.14) satisfy the condition in Corollary 2.1. Therefore, it is optimal in estimating θ, eliminating λ, as an element of $\mathscr{G}$.*

Finally we shall consider the general proportional hazards model (2.5). Suppose now that λ_{j0}, Y_{jk}, Z_{jk} in (2.7) satisfy all the conditions discussed in this subsection except being of class SP_j. In other words, they need only to be ordinary predictable processes like (2.5). Then Theorem 2.1 still holds, with only a different $\mathscr{G}$. In this case, the h_i in (2.12) can be an arbitrary bounded predictable process.

2.2.2 REPLICATES OF MODEL (2.4)

Notation in this subsection bears the same meaning as those in Section 2.2.1, unless otherwise stated. We assume $Y_{jk}(\cdot) \geq 0$ and Z_{jk} are bounded predictable processes, not necessarily of class SP_j. We assume

$$\lambda_{j0}(t) = \sum_{i=0}^{\infty} \lambda_i I_{(T_{ji}, T_{ji+1}]}(t), \quad (2.20)$$

where λ_i are constants.

Let $G(\cdot, \theta) \in \mathscr{G}$. Observe that

$$G(t, \theta) = \sum_{i=0}^{\infty} \sum_{j=1}^{J} \sum_{k=1}^{K} \int_0^t H_{jk}(s, \theta) I_{(T_{ji}, T_{ji+1}]}(s) \, dM_{jk}(s)$$

$$= \sum_{i=0}^{\infty} \sum_{j=1}^{J} \sum_{k=1}^{K} \int_0^t H_{jk}(s, \theta) I_{(T_{ji}, T_{ji+1}]}(s)(dN_{jk}(s) - \lambda_i Y_{jk}(s) e^{\theta' Z_{jk}(s)} \, ds)$$

$$= \sum_{i=0}^{\infty} \sum_{j=1}^{J} \sum_{k=1}^{K} \int_0^t H_{jk}(s, \theta) I_{(T_{ji}, T_{ji+1}]}(s) \, dN_{jk}(s)$$

$$- \sum_{i=0}^{\infty} \lambda_i \int_0^t \sum_{j=1}^{J} \sum_{k=1}^{K} H_{jk}(s, \theta) I_{(T_{ji}, T_{ji+1}]}(s) Y_{jk}(s) e^{\theta' Z_{jk}(s)} \, ds. \quad (2.21)$$

It follows from (2.21) and the arguments below (2.11) that H_{jk} can be chosen to be independent of λ and, on $[0, \infty)$,

$$\sum_{j=1}^{J} \sum_{k=1}^{K} H_{jk}(t, \theta) I_{(T_{ji}, T_{ji+1}]}(t) Y_{jk}(t) e^{\theta' Z_{jk}(t)} = 0 \qquad \text{a.s.,} \qquad (2.22)$$

for every $i = 0, 1, 2, \ldots$.

Let

$$A_{il}(t) = \frac{\sum_{j=1}^{J} \sum_{k=1}^{K} I_{(T_{ji}, T_{ji+1}]}(t) Y_{jk}(t) e^{\theta' Z_{jk}(t)} Z_{jkl}(t)}{\sum_{j=1}^{J} \sum_{k=1}^{K} I_{(T_{ji}, T_{ji+1}]}(t) Y_{jk}(t) e^{\theta' Z_{jk}(t)}}, \qquad (2.23)$$

$$G_{il}(t) = \sum_{j=1}^{J} \sum_{k=1}^{K} \int_{0}^{t} (Z_{jkl}(s) - A_{il}(s)) I_{(T_{ji}, T_{ji+1}]}(s) \, dN_{jk}(s), \qquad (2.24)$$

$$G_l(t) = \sum_{i=0}^{\infty} G_{il}(t). \qquad (2.25)$$

It is obvious that G_{il} is a martingale and independent of λ. Consequently $G_l(\cdot) \in \mathcal{G}$. Similarly to (2.19), we have also

$$\frac{\partial}{\partial \theta_l} \log L(t, \theta, \lambda) = G_l(t) + U_l(t), \qquad (2.26)$$

where $U_l(t) = \sum_{i=0}^{\infty} \sum_{j=1}^{J} \sum_{k=1}^{K} \int_{0}^{t} A_{il}(s) I_{(T_{ji}, T_{ji+1}]}(s) \, dM_{jk}(s)$.

Using (2.22), we know the quadratic variation

$$\langle U_l(\cdot), G(\cdot, \theta) \rangle_t = \sum_{i=0}^{\infty} \sum_{j=1}^{J} \sum_{k=1}^{K} \int_{0}^{t} A_{il}(s) H_{jk}(s, \theta) I_{(T_{ji}, T_{ji+1}]}(s) \lambda_i Y_{jk}(s) e^{\theta' Z_{jk}(s)} \, ds$$

$$= \sum_{i=0}^{\infty} \lambda_i \int_{0}^{t} A_{il}(s) \sum_{j=1}^{J} \sum_{k=1}^{K} H_{jk}(s, \theta) I_{(T_{ji}, T_{ji+1}]}(s) Y_{jk}(s) e^{\theta' Z_{jk}(s)} \, ds$$

$$= 0,$$

which implies that $U_l(\cdot)$ is orthogonal to every G in $\mathcal{G}$. Therefore, we have shown the following theorem.

Theorem 2.2 $\{G_1, \ldots, G_d\}$ *defined in (2.25) satisfy the condition in Corollary 2.1. Therefore, it is optimal in estimating θ, eliminating λ, as an element of $\mathcal{G}$.*

Remark 1 Theorem 2.2 also holds when the λ_i in (2.20) are deterministic functions. In this case, model (2.4) is not a special case of model (2.3).

Remark 2 We would like to point out that the optimal estimating function (2.25) for model (2.4) is obtained from stratifying the observations on the number of preceding failures, which is fundamentally different from the optimal estimating function (2.14) for model (2.3) or (2.5).

Remark 3 This remark concerns an interesting example suggested for consideration by a referee. Assume that

$$Y_{jk}(t) = 0 \tag{2.27}$$

for $t \in (T_{j,k-1}, T_{j,k}]$. If one is willing to accept model (2.4) and observes $J > 1$ replicates of it, (2.25) is the desired estimating function suggested by this work. If one can only accept model (2.3) or (2.5), one sees from (2.29) and (2.14) that our theory is not applicable because of the regularity condition. This and many other related interesting questions are under investigation from various viewpoints. We will report them in a separate paper.

2.3 Appendix

In this appendix, the Fisher information and Godambe's optimality criterion developed in Chang and Hsiung (1990) for the Cox model are extended to a semi-parametric model for counting processes, which contains as special cases all the semi-parametric models for replicates discussed in this paper.

Let $N(t) = (N_1(t), \ldots, N_K(t))$ be a multivariate counting process adapted to a filtration $\mathscr{F}_t$. Assume that the intensity of $N_k(t)$ relative to $\mathscr{F}_t$ is $\mu(t, \theta, \lambda)$, where $\theta \in \Theta \subset R^d$ and $\lambda \in \Lambda$. Here Λ is an abstract set. We are interested in estimating θ, treating λ as a nuisance parameter.

The class of estimating functions we consider is $\mathscr{G} = \{G(t, \theta)\}$, where $G(\cdot, \theta)$ is a mean-zero square-integrable right continuous $\mathscr{F}_t$-martingale for the parameter (θ, λ). We emphasize that $G(\cdot, \theta)$ is independent of λ.

Let $\mathscr{U} = \{U(t, \theta, \lambda)\}$, where $U(t, \theta, \lambda)$ is a square-integrable $\mathscr{F}_t$ martingale orthogonal to every martingale $G(t, \theta)$ in $\mathscr{G}$ for the parameter (θ, λ); an equivalent condition for the orthogonality is

$$E_{(\theta, \lambda)} G(T, \theta) U(T, \theta, \lambda) = 0, \tag{2.28}$$

for every stopping time T.

Let $\mathscr{P}_T^{(\theta, \lambda)}$ denote the probability measure on $\mathscr{F}_T$ for the parameter (θ, λ). Assume that $0 \in \Theta$ and $\mathscr{P}_T^{(\theta, \lambda)}$ is absolute continuous with respect to $\mathscr{P}_T^{(0, \lambda)}$ for every $\theta \in \Theta$, $\lambda \in \Lambda$, and stopping time T. Let

$$L(T) = L(T, \theta, \lambda) = \frac{d\mathscr{P}_T^{(\theta, \lambda)}}{d\mathscr{P}_T^{(0, \lambda)}}$$

denote the Radon–Nikodym derivative.

With some regularity condition, we can show, using (2.28) and the arguments in Godambe (1984), that

$$E_{(\theta,\lambda)}\left\{\frac{G(T,\theta)}{E_{(\theta,\lambda)}(DG(T,\theta))}\right\}^2 \geqslant \left[\inf_{U \in \mathcal{U}} E_{(\theta,\lambda)}\{D \log L(T,\theta,\lambda) - U(T,\theta,\lambda)\}^2\right]^{-1},$$

(2.29)

where D is any directional differentiation in $\Theta \subset R^d$. This leads to the following definitions and propositions. Since they parallel those in Chang and Hsiung (1990) closely, we shall give only the statements.

Definition 2.2   Let D be any directional differentiation in R^d. The Fisher information $I(\theta, L(T), D)$ for $L(T)$ about θ in the direction D, eliminating λ, is given by

$$I(\theta, L(T), D) = \inf_{U \in \mathcal{U}} E_{(\theta,\lambda)}(D \log L(T,\theta,\lambda) - U(T,\theta,\lambda))^2. \qquad (2.30)$$

Theorem 2.3   *A sufficient condition for $U_D \in \mathcal{U}$ to satisfy*

$$I(\theta, L(T), D) = E_{(\theta,\lambda)}(D \log L(T,\theta,\lambda) - U_D(T,\theta,\lambda))^2 \qquad (2.31)$$

for every stopping time T is that $D \log L(t,\theta,\lambda) - U_D(t,\theta,\lambda) \in \mathcal{G}$. When $U_D \in \mathcal{U}$ satisfying (2.31) exists, it is unique. Furthermore, the lower bound in (2.29) is attained for

$$G_D(t,\theta) = D \log L(t,\theta,\lambda) - U_D(t,\theta,\lambda). \qquad (2.32)$$

Definition 2.3   Let $G_1, \ldots, G_d \in \mathcal{G}$. We say $\{G_1, \ldots, G_d\}$ is optimal in estimating θ, eliminating λ, if, for every directional differentiation D, G_D is a linear combination of $G_1, \ldots, G_d$, whenever G_D exists as in (2.32).

Theorem 2.4   *Let $D_1, \ldots, D_d$ be any d linearly independent directional differentiations on R^d. Suppose $G_i \equiv G_{D_i}$ exists as in (2.32) for every $i = 1, \ldots, d$. Then $\{G_1, \ldots, G_d\}$ is optimal in estimating θ, eliminating λ.*

Corollary 2.1   *Let $G_1, \ldots, G_d \in \mathcal{G}$ be given. Assume that, for every $l = 1, \ldots, d$,*

$$\frac{\partial}{\partial \theta_l} \log L(t,\theta,\lambda) - G_l(t,\theta)$$

is orthogonal to every martingale in $\mathcal{G}$ for every parameter (θ, λ). Then $\{G_1, \ldots, G_d\}$ is optimal in estimating θ, eliminating λ, in the sense of Definition 2.3.

2.4 Acknowledgements

We are grateful to the referees for their critical reading and valuable suggestions.

The research was partly supported by National Science Council of the Republic of China.

References

Andersen, P. K. and Gill, R. (1982). Cox's regression model for counting processes: A large sample study. *Ann. Stat.*, **10**, 1100–20.

Brémaud, P. (1981). *Point processes and queues, martingale dynamics.* Springer-Verlag, New York.

Chang, I. S. and Hsiung, C. A. (1990). Finite sample optimality of maximum partial likelihood estimation in Cox's model for counting processes. *J. Stat. Plan. Inf.*, **25**, 35–42.

Cox, D. R. (1972). Regression models and life tables (with discussion). *J. Roy. Stat. Soc. Ser.* B **34**, 187–220.

Cox, D. R. (1975). Partial likelihood. *Biometrika*, **62**, 269–76.

Elliott, R. J. (1982). *Stochastic calculus and applications.* Springer-Verlag, New York.

Gill, R. D. (1980). *Censoring and stochastic integrals.* Mathematical Centre Tracts 124, Mathematisch Centrum, Amsterdam.

Godambe, V. P. (1960). An optimum property of regular maximum likelihood estimation. *Ann. Math. Stat.*, **31**, 1208–12.

Godambe, V. P. (1984). On ancillarity and Fisher information in the presence of a nuisance parameter. *Biometrika*, **71**, 626–9.

Godambe, V. P. (1985). The foundations of finite sample estimation in stochastic processes. *Biometrika*, **72**, 419–28.

Godambe, V. P. and Thompson, M. E. (1974). Estimating equations in the presence of a nuisance parameter. *Ann. Stat.*, **2**, 568–71.

Prentice, R. L., Williams, B. J., and Peterson, A. V. (1981). On the regression analysis of multivariate failure time data. *Biometrika*, **68**(2), 373–9.

3
Estimating equations for mixed Poisson models

C. B. Dean

ABSTRACT

Count data analysed under a Poisson assumption often exhibit overdispersion. To accommodate the extra-Poisson variation, mixed Poisson models are frequently used. Inference using maximum likelihood techniques is possible assuming the Poisson mixture to be, for example, the negative binomial or the Poisson–log normal distribution. However, interest here focuses on the use of estimating equations and in particular, quadratic and quasi-likelihood estimating equations. A general discussion of optimal quadratic estimation is provided by Crowder (1987) and Godambe and Thompson (1989). The estimators obtained from the use of optimal quadratic estimating equations are shown to be very efficient under a variety of distributions. The structure of the optimal quadratic estimating equations is used to identify simpler estimating equations that are close to optimal, in terms of efficiency, in many practical situations, but are robust under misspecification of the third and fourth moments of the underlying distribution. These use quasi-likelihood estimation for the regression parameters. Even if the variance form is misspecified these methods yield consistent parameter estimates and the variances of the estimates are very close to correct. The use of the variance correction implemented by Liang and Zeger (1986) is discussed. More complicated Poisson random effects models are also considered. One special model has a block random-effects structure. Data from this panel structure may consist of repeat observations on an individual where the random effects are 'individual specific' block effects. Quadratic estimating functions for this type of model are discussed.

3.1 Introduction

Extra-Poisson variation is a common occurrence in the analysis of count data. To accommodate the overdispersion, mixed Poisson models can be used. A Poisson mixture is obtained when the Poisson parameter, μ, the mean of the distribution, is itself a random variable with density $p(\mu)$ on the range $(0, \infty)$. Statistical methods for Poisson mixtures include (i) maximum likelihood under a parametric model; here the distribution $p(\mu)$ is specified, or (ii) the use of quasi-likelihood/weighted least-squares estimating techniques where only certain moments of $p(\mu)$ are specified.

The first approach may be appropriate in certain situations where the estimation of tail probabilities is of importance. The negative binomial model is a particularly flexible mixed Poisson distribution and the estimators of covariate effects enjoy certain desirable properties as will be discussed later. Insurance applications often demand a heavier tailed alternative to the negative binomial, such as the Poisson-inverse Gaussian mixture. Estimation of the parameters of this model does not require numerical quadrature, unlike that for many Poisson mixtures. For a discussion of these models see, for example, Lawless (1987) and Willmot (1987).

For inference in a regression situation where we are primarily interested in the effects of covariates, the use of estimating equations is a viable, robust approach. The quasi-likelihood equations proposed by Wedderburn (1974) are unbiased linear estimating equations that yield the maximum likelihood estimates when the underlying distribution is a member of the exponential family. Unbiased quadratic estimating equations have been proposed as extensions of Wedderburn's quasi-likelihood equations. Here the assumptions regarding the counts Y_i, $i = 1, \ldots, n$, assign forms to the first two moments of the distribution. See Crowder (1987), Firth (1987) and Godambe and Thompson (1989) for a discussion of quadratic estimating equations.

This paper studies the use of certain estimating equations for inference in mixed Poisson models. Section 3.2 discusses testing for extra-Poisson variation. Estimating equations for simple Poisson mixtures are discussed in Section 3.3 and in Section 3.4 estimators for certain models for longitudinal count data are considered.

3.2 Tests for extra-Poisson variation

Testing for the Poisson assumption by fitting a parametric model, such as the negative binomial, and then testing for a reduction to the Poisson using asymptotic maximum likelihood methods, has been noted as an unreliable procedure since it tends to underestimate the evidence against the Poisson model; see Lawless (1987). Score tests for overdispersion have been proposed by Fisher (1950), Collings and Margolin (1985), Cameron and Trivedi (1986), and Dean and Lawless (1989a). Dean (1991) develops a unifying theory for all these score tests. Tests corresponding to two commonly assumed types of extra-Poisson variation will be discussed.

The tests are score tests against an alternative class of mixed Poisson models derived by considering a model with random effects $v_i > 0$, where conditional on v_i the distribution of Y_i is Poisson with a mean of $v_i \mu_i$; the v_i are continuous, independent variates with $E(v_i) = 1$ and finite variance depending on a paramter τ. Covariates $\mathbf{x}_i$ play a role in determining μ_i so $\mu_i = \mu_i(\mathbf{x}_i; \boldsymbol{\beta})$, where $\boldsymbol{\beta}$ is a vector of regression parameters.

The distribution of Y_i given $\mathbf{x}_i$ in this random effects model is $\Pr\{Y_i = y_i|\mathbf{x}_i\} = E_{v_i}\{f_P(y_i; v_i\mu_i)\}$, where $f_P(y; \mu) = \mu^y e^{-\mu}/y!$ is the Poisson probability density function. Expanding about the mean of μ_i yields

$$\Pr[Y_i = y_i|\mathbf{x}_i] = f_P(y_i; \mu_i)\left\{1 + \sum_{r=2}^{\infty} \frac{\alpha_r}{r!} D_r(y_i; \mu_i)\right\},$$

where $\alpha_r = E(v_i - 1)^r$, assumed to be $o(\tau)$ for $r \geqslant 3$, and

$$D_r(y_i; \mu_i) = \left\{\frac{\partial^{(r)}}{\partial v_i^{(r)}} f_P(y_i; v_i\mu_i)|_{v_i = 1}\right\} / f_P(y_i; \mu_i)$$

$$= \sum_{j=0}^{r} \binom{r}{j} y_i^{(r-j)}(-\mu_i)^j.$$

The contribution from the random ith observation Y_i to the log-likelihood function under the mixed model is

$$l_i(\boldsymbol{\beta}, \tau) = \log f_P(Y_i; \mu_i) + \log\left\{1 + \sum_{r=2}^{\infty} \frac{\alpha_r}{r!} D_r(Y_i; \mu_i)\right\}. \qquad (3.1)$$

The first type of overdispersed model sets $\alpha_2 = \mathrm{var}(v_i) = \tau < \infty$. In this case $E(Y_i) = \mu_i$ and $\mathrm{var}(Y_i) = \mu_i(1 + \tau\mu_i)$. The test of $H: \tau = 0$ is based on

$$\sum_{i=1}^{n} \left\{\frac{\partial l_i(\boldsymbol{\beta}, \tau)}{\partial \tau}\right\}\bigg|_{\tau=0} = \frac{1}{2} \sum_{i=1}^{n} \{(Y_i - \hat{\mu}_i)^2 - Y_i\},$$

where $\hat{\mu}_i$ is the maximum likelihood estimate of μ_i under the Poisson assumption. The standardized form of this statistic is

$$T_a = \frac{\sum_{i=1}^{n} \{(Y_i - \hat{\mu}_i)^2 - Y_i\}}{(2 \sum_{i=1}^{n} \hat{\mu}_i^2)^{1/2}},$$

which converges in distribution to standard normal as $n \to \infty$ and under usual regularity conditions. Dean and Lawless (1989a) provide a 'small sample' corrected version of T_a where the numerator is replaced by $\sum_{i=1}^{n} \{(Y_i - \hat{\mu}_i)^2 - Y_i + \hat{h}_{ii}\hat{\mu}_i\}$, and $\hat{h}_{ii} = h_{ii}(\hat{\mu}_i)$. Here h_{ii} is the ith diagonal element of $H = W^{1/2}(X^T W X)^{-1} X^T W^{1/2}$, where $W = \mathrm{diag}(\mu_1, \ldots, \mu_n)$ and X has ijth entry $\mu_i^{-1}(\partial\mu_i/\partial\beta_j)$, $i = 1, \ldots, n; j = 1, \ldots, p$. The matrix H is the leverage matrix for Poisson regression and is usually computed for diagnostics. This corrected version of T_a converges very quickly to normality. Dean and Lawless also derive the asymptotic distribution of T_a as $\mu_i \to \infty$, $i = 1, \ldots, n$, and this is a linear combination of chi-squared variates.

A second way of incorporating extra-Poisson variation specifies that $\mathrm{var}(Y_i) = \mu_i(1 + \tau)$. See, for example, Jorgensen (1987) and McCullagh and Nelder (1989). The random effects derivation corresponding to this model may not be appealing since it specifies that $\mathrm{var}(v_i) = \tau/\mu_i$, $\tau < \infty$, i.e. that the random effects depend on the fixed effects. The 'small sample' adjusted score test statistic for testing that $\tau = 0$ is

$$T_b = \frac{1}{\sqrt{(2n)}} \sum_{i=1}^{n} \left\{ \frac{(Y_i - \hat{\mu}_i)^2 - Y_i + \hat{h}_{ii}\hat{\mu}_i}{\hat{\mu}_i} \right\},$$

which converges to $N(0, 1)$ in distribution under standard maximum likelihood asymptotic theory. Note that when $\mu_i \to \infty$, $i = 1, \ldots, n$, a test based on T_b is equivalent to one based on the Pearson statistic, $\sum \{(Y_i - \hat{\mu}_i)^2/\hat{\mu}_i\}$, since $(Y_i/\hat{\mu}_i) \xrightarrow{P} 1$ in this case.

These two mixed Poisson models are simply different parametrizations of the same model in the single sample problem. When the μ_i are not too different, a test based on either of the statistics would be appropriate for detecting overdispersion as discussed here. However, if the μ_i vary widely, then both tests should be performed. They correspond to different variance structures in the mixed model, and the choice of alternative model to be used should be based on diagnostics for checking the goodness-of-fit of the model such as residual and probability plots. Collings and Margolin (1985) and Dean and Lawless (1989a) provide evidence for the gains in power that can be achieved by using these tests to detect overdispersion instead of certain usual goodness-of-fit tests.

3.3   Estimating equations for Poisson mixtures

Estimation for mixed Poisson models discussed here uses quadratic and quasi-likelihood estimating equations. The quasi-likelihood equations proposed by Wedderburn (1974) are unbiased linear estimating equations for β that yield the maximum likelihood estimates when the Poisson assumption is correct, or when the Poisson mixture is negative binomial with τ known. Unbiased quadratic estimating equations have been proposed by Crowder (1987), Firth (1987) and Godambe and Thompson (1989) as extensions of Wedderburn's quasi-likelihood equations. They are of the general form

$$g_r^*(\beta, \tau) = \sum_{i=1}^{n} [a_{ir}(\beta, \tau)\{Y_i - \mu_i\} + d_{ir}(\beta, \tau)\{(Y_i - \mu_i)^2 - \sigma_i^2\}] = 0, \quad (3.2)$$

where a_{ir} and d_{ir} are functions of β and τ. The use of unbiased quadratic estimating equations is a competing alternative to maximum likelihood estimation. Some such estimators have high efficiency under a general class of models which includes the negative-binomial and Poisson-inverse Gaussian mixture. This will be discussed further.

Consider independent counts Y_i arising from a mixed Poisson distribution with

$$E(Y_i) = \mu_i, \text{ var}(Y_i) = \sigma_i^2 = \mu_i(1 + \tau\mu_i),$$

and define

$$\gamma_{1i} = E\{(Y_i - \mu_i)\sigma_i^{-1}\}^3, \qquad \gamma_{2i} = E\{(Y_i - \mu_i)\sigma_i^{-1}\}^{-4} - 3,$$

and

$$\gamma_i = \gamma_{2i} + 2 - \gamma_{1i}^2, \qquad i = 1, \ldots, n,$$

so γ_{1i} and γ_{2i} represent the skewness and kurtosis coefficients respectively. The optimal quadratic estimating equations for $\boldsymbol{\beta}$ and τ provide estimators with asymptotic minimal variance among those produced by (3.2). They are

$$g_{1,r} = \sum_{i=1}^{n} \left[\frac{Y_i - \mu_i}{\sigma_i^2} + \left\{ \frac{(Y_i - \mu_i)^2}{\sigma_i^2} - 1 - \frac{\gamma_{1i}(Y_i - \mu_i)}{\sigma_i} \right\} \right.$$

$$\left. \times \left\{ \frac{(1 + 2\tau\mu_i)}{\sigma_i} - \gamma_{1i} \right\} \frac{1}{\gamma_i \sigma_i} \right] \left(\frac{\partial \mu_i}{\partial \beta_r} \right) = 0, \qquad r = 1, \ldots, p,$$

and

$$g_{1,p+1} = \sum_{i=1}^{n} \left\{ \frac{(Y_i - \mu_i)^2}{\sigma_i^2} - 1 - \frac{\gamma_{1i}(Y_i - \mu_i)}{\sigma_i} \right\} \frac{\mu_i^2}{\gamma_i \sigma_i^2}.$$

The term $((1 + 2\tau\mu_i)\sigma_i^{-1} - \gamma_{1i})$ appearing in g_{ir}, $r = 1, \ldots, p$, is also a factor in the contribution of the ith observation to the asymptotic covariance of the resulting estimators of $\boldsymbol{\beta}$ and τ. Hence when this asymptotic covariance is close to zero, $g_{1,r}$, $r = 1, \ldots, p$ can be well approximated by the quasi-likelihood equations,

$$g_{2,r} = \sum_{i=1}^{n} \left\{ \frac{(Y_i - \mu_i)}{\sigma_i^2} \right\} \left(\frac{\partial \mu_i}{\partial \beta_r} \right) = 0, \qquad r = 1, \ldots, p.$$

There are some important benefits in using the quasi-likelihood equations for estimation of $\boldsymbol{\beta}$. The resulting estimator, $\tilde{\boldsymbol{\beta}}$, has asymptotic variance F^{-1}, where $F = \lim_{n \to \infty} n^{-1} F_n$, and

$$F_n = \left\{ U^{\mathrm{T}} \text{ diag} \left(\frac{\mu_1^2}{\sigma_1^2}, \ldots, \frac{\mu_n^2}{\sigma_n^2} \right) U \right\},$$

with the matrix U having ijth element $\mu_i^{-1}(\partial\mu_i/\partial\beta_j)$, $i = 1, \ldots, n, j = 1, \ldots, p$. Hence F has rs-element

$$F_{rs} = \lim_{n \to \infty} \frac{1}{n} \sum_{i=1}^{n} \frac{1}{\sigma_i^2} \left(\frac{\partial\mu_i}{\partial\beta_r} \right) \left(\frac{\partial\mu_i}{\partial\beta_s} \right), \qquad r, s = 1, \ldots, p.$$

This is independent of the choice of the $(p + 1)$st estimating equation for τ, and this asymptotic variance is correct regardless of the true values of γ_{1i} and γ_{2i}. In addition, for an arbitrary mixed model with log-likelihood function $l(\boldsymbol{\beta}, \tau) = \sum_{i=1}^{n} l_i(\boldsymbol{\beta}, \tau)$, and $l_i(\boldsymbol{\beta}, \tau)$ given by (3.1) we have

$$
E\left\{-\frac{\partial^2 l_i}{\partial \beta_r \, \partial \beta_s}\right\} = -\left(\frac{\partial \mu_i}{\partial \beta_r}\right)\left(\frac{\partial \mu_i}{\partial \beta_s}\right)E\left[\frac{\partial^2 \log f_P(Y_i; \mu_i)}{\partial \mu_i^2}\right.
$$

$$
\left. + \frac{\sum_{r=2}^{\infty} \alpha_r (r!)^{-1}(\partial^2 D_r/\partial \mu_i^2)}{1 + \sum_{r=2}^{\infty} \alpha_r (r!)^{-1} D_r} - \left\{\frac{\sum_{r=2}^{\infty} \alpha_r (r!)^{-1}(\partial D_r/\partial \mu_i)}{1 + \sum_{r=2}^{\infty} \alpha_r (r!)^{-1} D_r}\right\}^2\right]
$$

$$
= -\left(\frac{\partial \mu_i}{\partial \beta_r}\right)\left(\frac{\partial \mu_i}{\partial \beta_s}\right)\left[E\left\{-\frac{\partial^2 \log f_P(y_i; \mu_i)}{\partial \mu_i^2}\right\} + E_P\left\{\sum_{r=2}^{\infty} \frac{\alpha_r}{r!}\left(\frac{\partial^2}{\partial \mu_i^2} D_r\right)\right\}\right.
$$

$$
\left. - E_P\left\{\sum_{r=2}^{\infty} \frac{\alpha_r}{r!}\left(\frac{\partial}{\partial \mu_i} D_r\right)\right\}^2\left\{1 + \sum_{r=2}^{\infty} \frac{\alpha_r}{r!} D_r\right\}^{-1}\right]
$$

$$
= \left(\frac{\partial \mu_i}{\partial \beta_r}\right)\left(\frac{\partial \mu_i}{\partial \beta_s}\right)\left[\frac{1}{\mu_i} - \tau + o(\tau)\right],
$$

where E_p refers to expectation taken under the Poisson model. For $\tau\mu_i$ small, this is approximately $\sigma_i^{-2}(\partial \mu_i/\partial \beta_r)(\partial \mu_i/\partial \beta_s)$. If, in addition, $E\{-\partial^2 l/\partial \beta_r \, \partial \tau\} \simeq 0$, $r = 1, \ldots, p$, then the asymptotic variance of the maximum likelihood estimator of $\boldsymbol{\beta}$ will be close to F^{-1}, the asymptotic variance of the quasi-likelihood estimator of $\boldsymbol{\beta}$. This holds for many practical situations. For the negative binomial model, the quasi-likelihood equations for $\boldsymbol{\beta}$ are the maximum likelihood equations, and the quasi-likelihood estimators are fully efficient. Dean and Lawless (1989b) have discussed efficiency values in certain illustrative regression scenarios.

The structure of the optimal estimating equations can also be used to identify simpler, efficient choices for $g_{1,p+1}$, the estimating equation for τ. Often Gaussian estimation will work well as an estimation procedure; applying Gaussian estimation for τ and setting $\gamma_i = \gamma$ yields the equation

$$
g_{2,p+1} = \sum_{i=1}^{n}\left\{\frac{(Y_i - \mu_i)^2 - \sigma_i^2}{(1 + \tau\mu_i)^2}\right\} = 0 \tag{3.3}
$$

and the combination of quasi-likelihood equations with $g_{2,p+1}$ has been shown to be efficient and robust for Poisson mixtures. In practice, γ_i often varies little, and under the negative binomial model γ_i is a constant, $i = 1, \ldots, n$. See Breslow (1990), Davidian and Carroll (1988) and Dean et al. (1989) for a discussion of the use of $g_{2,p+1}$ with the quasi-likelihood equations.

The estimating equations discussed above assume $\mathrm{var}(Y_i) = \mu_i + \tau\mu_i^2 = \sigma_{1i}^2$. Estimating equations for τ assuming $\mathrm{var}(Y_i) = \mu_i(1 + \tau) = \sigma_{2i}^2$ can be similarly obtained. Choice of variance form is an important question in inference for

Poisson mixtures. If the range of the μ_i is small, then σ_{2i}^2 will provide a good approximation for σ_{1i}^2. However, if the μ_i vary greatly, then assumptions regarding variance form should be supported by diagnostic checks such as residual and probability plots. Liang and Zeger (1986) have adopted an approach which allows for misspecification of σ_i^2 where an 'empirical' covariance matrix is estimated from the data. The asymptotic variance of $\hat{\beta}$ would be estimated by

$$F_n^{-1} G_n F_n^{-1},$$

where the rs-element of G_n is

$$G_{n,rs} = \sum_{i=1}^{n} \frac{(Y_i - \mu_i)^2}{\sigma_i^4} \left(\frac{\partial \mu_i}{\partial \beta_r}\right) \left(\frac{\partial \mu_i}{\partial \beta_s}\right),$$

and μ, τ are replaced by their corresponding maximum likelihood estimates. Under misspecification of the variance form, the joint solution $(\hat{\beta}, \tilde{\tau})$ to $g_{2,r} = 0, r = 1, \ldots, p + 1$, converges in probability to (β^*, τ^*), where β^* is the true value of β, and τ^* satisfies

$$\lim_{n \to \infty} \frac{1}{n} \sum_{i=1}^{n} \left\{ \frac{\text{var}(Y_i) - \sigma_i^2}{(1 + \tau \mu_i)^2} \right\} = 0.$$

Substitution of $\tilde{\tau}$ does not affect the asymptotic distribution of $\hat{\beta}$, or of the test statistics discussed below; see Lawless (1987), Moore (1986).

A final consideration is that of testing procedures for β or a subset of β. Breslow (1990) derives model based and 'empirical' score tests. Partition β as $\beta^T = (\beta_1^T, \beta_2^T)$ where β_1 is $k \times 1$ and β_2 is $(p - k) \times 1$, and similarly partition $h^T = (g_{2r}, r = 1, \ldots, p)$ as $h^T = (h_1^T, h_2^T)$ and F_n and G_n into submatrices of dimension $k \times k$, $k \times (p - k)$, etc. For a hypothesized $\beta_2 = \beta_{20}$, let $\tilde{\beta}_1(\beta_{20})$ be the value of β_1 that solves the equation $h_1(\beta_1^T, \beta_{20}^T) = 0$. Then the score test for testing $H: \beta_2 = \beta_{20}$ is based on $u = h_2[\tilde{\beta}_1(\beta_{20}), \beta_{20}^T]$. Here τ is evaluated at $\tilde{\tau}_0$ which satisfies $g_{2,p+1} = 0$ when $\beta^T = (\tilde{\beta}_1^T(\beta_{20}), \beta_{20}^T)$. The asymptotic variance of the test statistic is

$$\text{var}\left[\frac{1}{\sqrt{n}} u\right] = F_{22} - F_{21} F_{11}^{-1} F_{12} = \Sigma,$$

and the score test statistic is $u^T \Sigma u$ which is asymptotically distributed as $\chi_{(k)}^2$. Uncertainty in the variance form may be accommodated using a similar approach to that described above. The 'empirical' covariance matrix is given in Breslow (1990).

3.4   Models for longitudinal count data

It is sometimes more natural in a Poisson analysis to include random effects in a more complicated form than that presented in the previous section. For example, data in a panel structure may consist of repeat observations on an individual where the random effects are 'individual specific' block effects. Hausman et al. (1984) discuss a problem where the response variate is the total number of patents applied for by a firm in a given year, and data for several firms and years are available. One of the models that they propose contains firm specific random effects and was supported by the fact that residuals from a simple Poisson analysis showed a 'within-firm' correlation. Thall (1988) also considers maximum likelihood estimation in this type of model. Morton (1987), Zeger (1988), and Firth, in a recent technical report, discuss random-effect Poisson regression models for longitudinal data. Zeger's paper discusses the use of estimating equations for time series of counts.

Consider then a mixed Poisson model where given the random effect v_i, the distribution of Y_{ij} is Poisson with mean $v_i \mu_{ij}(\mathbf{x}_{ij}; \boldsymbol{\beta})$, where $\mathbf{x}_{ij}$ is a $p \times 1$ vector of fixed covariates including a constant term. Here i might index individuals and j represent the different times that the observations were recorded, $j = 1, \ldots, m_i$, $i = 1, \ldots, n$, so the covariates are time dependent. In some cases i might index 'groups' and j, individuals within the groups. Given v_i, the Y_{ij}, $j = 1, \ldots, m_i$, are independent variates. Assume that the v_i are positive, i.i.d. random variables, with probability density function $p(v; \tau)$, depending on a parameter τ, and with $E(v_i) = 1$ and $\text{var}(v_i) = \tau < \infty$. The variable v_i represents a random effect that is specific to the ith individual. Unconditionally, we have $E(Y_{ij}) = \mu_{ij}$, $\text{var}(Y_{ij}) = \mu_{ij} + \tau \mu_{ij}^2$, $\text{cov}(Y_{ij}, Y_{ik}) = \tau \mu_{ij} \mu_{ik}$, $j \neq k$, $\text{cov}(Y_{ij}, Y_{lk}) = 0$, $i \neq l$, for $j, k = 1, \ldots, m_i$, and $i, l = 1, \ldots, n$. Thus $E(Y_{i.}) = \mu_{i.}$, $\text{var}(Y_{i.}) = \mu_{i.} + \tau \mu_{i.}^2 = \sigma_{i.}^2$, $\text{cov}(Y_{i.}, Y_{l.}) = 0$, where $Y_{i.} = \sum_{j=1}^{m_i} Y_{ij}$ and $\mu_{i.} = \sum_{j=1}^{m_i} \mu_{ij}$, $i, l = 1, \ldots, n$.

The distribution of $(Y_{i1}, \ldots, Y_{im_i})$ in the mixed model is

$$\Pr\{y_{i1}, \ldots, y_{im_i}\} = E_v\{\Pr(y_{i1}, \ldots, y_{im_i} | v_i)\} = \frac{C_i}{D_i} E_v\{f_P(y_{i.}; v_i \mu_{i.})\}, \quad (3.4)$$

where $C_i = \prod_{j=1}^{m_i} (\mu_{ij}^{y_{ij}} / y_{ij}!)$, $D_i = (\mu_{i.}^{y_{i.}} / y_{i.}!)$, and $f_P(y; \mu)$ is the Poisson probability function. The log likelihood function for $(\boldsymbol{\beta}, \tau)$ from (3.4) is $l(\boldsymbol{\beta}, \tau) = \sum_{i=1}^{n} l_i(\boldsymbol{\beta}, \tau)$, where

$$l_i(\boldsymbol{\beta}, \tau) = \log \Pr\{y_{i1}, \ldots, y_{im_i}\}$$

$$= \log\left(\frac{C_i}{D_i}\right) + \log\left[\int_0^\infty (v_i \mu_{i.})^{y_{i.}} \, e^{v_i \mu_{i.}} \frac{1}{y_{i.}!} p(v_i) \, dv_i\right].$$

Hence the likelihood factors into two parts; the first is a function of y_{ij}, μ_{ij} and $\boldsymbol{\beta}$, and the second is a function of $y_{i.}$, $\boldsymbol{\beta}$, and τ. When $\mu_{ij} = \mu_i$, $j = 1, \ldots, m_i$, the first part is independent of $\boldsymbol{\beta}$, so the second contains all the information concerning the parameters.

To test for a reduction to the Poisson model we test that $\tau = 0$. The partial score test statistic is easily derived because of the factorization of the likelihood. We have

$$\frac{\partial l}{\partial \tau}\bigg|_{\tau=0} = \frac{1}{2}\sum_{i=1}^{n}\{(Y_{i.} - \hat{\mu}_{i.}) - Y_{i.}\},$$

$$E\left\{-\frac{\partial^2 l}{\partial \beta_r \, \partial \tau}\right\}\bigg|_{\tau=0} = 0, \qquad r = 1, \ldots, p,$$

$$E\left\{-\frac{\partial^2 l}{\partial \tau^2}\right\} = \frac{1}{2}\sum_{i=1}^{n}\mu_{i.}^2,$$

where $\hat{\mu}_{i.}$ is the maximum likelihood estimate of $\mu_{i.}$ under the Poisson assumption. The score test statistic is

$$\frac{\sum_{i=1}^{n}\{(Y_{i.} - \hat{\mu}_{i.})^2 - Y_{i.}\}}{(2\sum_{i=1}^{n}\hat{\mu}_{i.}^2)^{1/2}}.$$

Under the Poisson and usual regularity assumptions this statistic has an asymptotic ($n \to \infty$) standard normal distribution, and the limiting null distribution when $\mu_{i.} \to \infty$, $i = 1, \ldots, n$, n fixed, can also be derived using similar arguments as in Dean and Lawless (1989a). Notice that only the aggregate data $(Y_{1.}, \ldots, Y_{n.})$ are required in order to test for a reduction to the Poisson model.

Robust estimation of (β, τ) using only moment assumptions concerning the distribution of v_i is also possible. First assume τ known. We have a vector of observations $\mathbf{Y} = (Y_{11}, \ldots, Y_{1m_1}, \ldots, Y_{n1}, \ldots, Y_{nm_n})$, with mean $\boldsymbol{\mu} = (\mu_{11}, \ldots, \mu_{1m_1}, \ldots, \mu_{n1}, \ldots, \mu_{nm_n})$, and variance V in block diagonal form,

$$V = \begin{bmatrix} V_{(1)} & 0 & \cdots & 0 \\ & \ddots & & \vdots \\ 0 & 0 & \cdots & V_{(n)} \end{bmatrix}, \tag{3.5}$$

where $V_{(k)}$ has ith diagonal element $V_{(k)ii} = \mu_{ki} + \tau\mu_{ki}^2$, and ijth off-diagonal element $V_{(k)ij} = \tau\mu_{ki}\mu_{kj}$, $j = 1, \ldots, m_i$, $i \neq j$, and $k = 1, \ldots, n$. The mean μ_{ij} is a function of a vector of fixed covariates $\mathbf{x}_{ij}$, and a $p \times 1$ parameter vector $\boldsymbol{\beta}$, $\mu_{ij} = \mu_{ij}(\mathbf{x}_{ij}; \boldsymbol{\beta})$, and we require to estimate $\boldsymbol{\beta}$. Consider using optimal linear unbiased estimating equations, that is, unbiased estimation equations that are linear in $(y_{ij} - \mu_{ij})$ and that provide estimators with minimum variance among all other estimators produced by linear unbiased estimating equations. The theory of estimating equations developed for uncorrelated observations can be applied to the transformed functions h_i, $i = 1, \ldots, N$, $N = \sum_{i=1}^{n} m_i$, where h_i is the ith element of the vector $\mathbf{h}$,

$$\mathbf{h} = V^{-1/2}(\mathbf{Y} - \boldsymbol{\mu}). \tag{3.6}$$

Let the vector estimating function $\mathbf{f}$ be defined by $\mathbf{f} = Q\mathbf{h}$, where $Q = Q(\boldsymbol{\beta})$ is a $p \times n$ matrix, and let $J_Q = E(\mathbf{ff}')$, and $H_Q = E(-\partial \mathbf{f}/\partial \boldsymbol{\beta})$. As Q varies, $\mathbf{f}$ defines the class of linear unbiased estimating functions. The optimal estimating function in this class, $\mathbf{f}^*$, satisfies $\mathbf{f}^* = Q^*\mathbf{h}$, $Q^* = Q^*(\boldsymbol{\beta})$, with the matrix

$$J_Q - H_Q(H_{Q^*})^{-1}J_Q(H'_{Q^*})^{-1}H'_Q \tag{3.7}$$

being positive semi-definite for all other functions, $\mathbf{f} = Q\mathbf{h}$. Godambe and Thompson (1989) show that (3.7) is positive semi-definite if $H_Q = E(\mathbf{ff}^{*\prime})$. Since

$$E\left(-\frac{\partial \mathbf{f}}{\partial \boldsymbol{\beta}}\right) = -QE\left(\frac{\partial \mathbf{h}}{\partial \boldsymbol{\beta}}\right) \quad \text{and} \quad E(\mathbf{ff}^{*\prime}) = QQ^{*\prime},$$

Q^* satisfies $Q^{*\prime} = E(-\partial \mathbf{h}/\partial \boldsymbol{\beta}) = V^{-1/2}(\partial \boldsymbol{\mu}/\partial \boldsymbol{\beta})$. Define the matrix D as $D = (\partial \boldsymbol{\mu}/\partial \boldsymbol{\beta})$. Then the optimal linear unbiased estimating function is

$$\mathbf{f}^* = D'V^{-1}(\mathbf{Y} - \boldsymbol{\mu}), \tag{3.8}$$

the quasi-likelihood function for $\boldsymbol{\beta}$ (see McCullagh and Nelder 1989, Section 9.2). Since V, (3.5), is a block diagonal matrix then V^{-1} is also block diagonal,

$$V^{-1} = \begin{bmatrix} V_{(1)}^{-1} & 0 & \cdots & 0 \\ & \ddots & & \vdots \\ 0 & 0 & \cdots & V_{(n)}^{-1} \end{bmatrix} \tag{3.9}$$

where $V_{(i)}^{-1} = \mathrm{diag}(\mu_{i1}^{-1}, \ldots, \mu_{im_i}^{-1}) - \tau(1 + \tau\mu_{i.})^{-1}\mathbf{1}_{m_i}\mathbf{1}'_{m_i}$, and $\mathbf{1}_k$ is a $k \times 1$ unit vector. Let $\mathbf{Y}_i$ and $\boldsymbol{\mu}_i$ be the vectors $\mathbf{Y}_i = (Y_{i1}, \ldots, Y_{im_i})$, $\boldsymbol{\mu}_i = (\mu_{i1}, \ldots, \mu_{im_i})$, and let $D_i = (\partial \boldsymbol{\mu}_i/\partial \boldsymbol{\beta})$, $i = 1, \ldots, n$. Then the function $\mathbf{f}^*$ may be written

$$\mathbf{f}^* = \sum_{i=1}^{n} D'_i V_{(i)}^{-1}(\mathbf{Y}_i - \boldsymbol{\mu}_i). \tag{3.10}$$

The ith element in the sum in (3.10) is a vector with rth entry

$$\begin{aligned} f_i^{(r)*} &= \sum_{j=1}^{m_i} \left\{ \frac{Y_{ij} - \mu_{ij}}{\mu_{ij}}\left(\frac{\partial \mu_{ij}}{\partial \beta_r}\right) - \frac{\tau}{1 + \tau\mu_{i.}}(Y_{i.} - \mu_{i.})\frac{\partial \mu_{ij}}{\partial \beta_r} \right\} \\ &= \frac{Y_{i.} - \mu_{i.}}{\sigma_{i.}}\left(\frac{\partial \mu_{i.}}{\partial \beta_r}\right) + \sum_{j=1}^{m_i} Y_{ij}\left\{ \frac{1}{\mu_{ij}}\left(\frac{\partial \mu_{ij}}{\partial \beta_r}\right) - \frac{1}{\mu_{i.}}\left(\frac{\partial \mu_{i.}}{\partial \beta_r}\right) \right\}. \end{aligned} \tag{3.11}$$

The equation $\mathbf{f}^* = \mathbf{0}$ can be shown to be the maximum likelihood equation for estimating $\boldsymbol{\beta}$ when v is assumed to be distributed as a gamma variate and in addition, when $\mu_{ij} = \mu_i$, $j = 1, \ldots, m_i$, $i = 1, \ldots, n$, then the function f^* depends on the data only through the aggregate totals and $f^{(r)*}$ becomes $f^{(r)*} = \sum_{i=1}^{n} \{(Y_{i.} - \mu_{i.})/\sigma_{i.}\}(\partial \mu_{i.}/\partial \beta_r)$.

When τ is unknown, we may use the estimating equations $\mathbf{f}^* = \mathbf{0}$ to estimate $\boldsymbol{\beta}$, with an additional estimating equation, $f^{(p+1)} = 0$, for τ, where $E\{f^{(p+1)}\} = 0$. Let the new vector of estimating functions be denoted by $\mathbf{f}_1$,

$$\mathbf{f}_1 = \begin{pmatrix} \mathbf{f}^* \\ f^{(p+1)} \end{pmatrix}. \tag{3.12}$$

The asymptotic variance of $\tilde{\boldsymbol{\beta}}$, the estimator of $\boldsymbol{\beta}$, does not depend on $f^{(p+1)}$, and $\tilde{\boldsymbol{\beta}}$ will be fully efficient if the underlying distribution of v is gamma.

Since $(Y_1, \ldots, Y_n)$ is sufficient for τ, one choice of estimating functions $f^{(p+1)}$ is derived from using these totals to base inference on τ, and the estimating function (3.3) can be adapted by replacing Y_i by $Y_{i.}$, μ_i by $\mu_{i.}$ and σ_i^2 by $\sigma_{i.}^2$ to obtain

$$f_1^{(p+1)} = \sum_{i=1}^{n} \left\{ \frac{(Y_{i.} - \mu_{i.})^2 - \sigma_{i.}^2}{(1 + \tau\mu_{i.})^2} \right\}.$$

Recall that the estimator of τ obtained using this function has high efficiency when v is gamma or inverse-Gaussian. When v has a gamma distribution the estimators of $\boldsymbol{\beta}$ and τ, $\tilde{\boldsymbol{\beta}}_L$ and $\tilde{\tau}_L$, are asymptotically independent, which mimics the property of the maximum likelihood estimates in this situation. Thus the estimation function $\mathbf{f}_1$ with $f^{(p+1)} = f_1^{(p+1)}$ provides an estimator $(\tilde{\boldsymbol{\beta}}_L, \tilde{\tau}_L)$ with similar properties to the estimator obtained from using $g_{2,r}$, $r = 1, \ldots, p+1$ as estimating functions in the simple mixed Poisson model. To allow for misspecification of the variance form, 'empirical' covariance matrices discussed earlier can also be employed for inference regarding the parameters. Finally note that the estimating function $\mathbf{f}^*$ was derived as the optimal linear unbiased estimating function for a vector of observations $\mathbf{Y}$ with arbitrary variance matrix V, and hence is applicable for estimation of $\boldsymbol{\beta}$ in a variety of more complicated random effects models.

References

Breslow, N. (1990). Tests of hypotheses in overdispersed Poisson regression and other quasi-likelihood models. *J. Amer. Stat. Assoc.*, **85**, 565–71.

Cameron, A. C. and Trivedi, P. K. (1986). Econometric models based on count data: comparisons and applications of some estimators and tests. *J. Appl. Economet.*, **1**, 29–53.

Collings, B. J. and Margolin, B. H. (1985). Testing goodness of fit for the Poisson assumption when observations are not identically distributed. *J. Amer. Stat. Assoc.*, **80**, 411–18.

Crowder, M. J. (1987). On linear and quadratic estimating functions. *Biometrika*, **74**, 591–7.

Davidian, M. and Carroll, R. J. (1988). A note on extended quasilikelihood. *J. Roy. Stat.*, Soc. B, **50**, 74–82.

Dean, C. B. (1991). Testing for extra-Poisson and extra-bionomial variation. *J. Amer. Stat. Assoc.* (In press.)

Dean, C. and Lawless, J. F. (1989a). Testing for overdispersion in Poisson regression models. *J. Amer. Stat. Assoc.*, **84**, 467–72.

Dean, C. and Lawless, J. F. (1989b). Comments on 'An extension of quasilikelihood estimation', by Godambe and Thompson. *J. Stat. Plan. Inf.*, **22**, 155–8.

Dean, C., Lawless, J. F., and Willmot, G. E. (1989). A mixed Poisson-inverse Gaussian regression model. *Canad. J. Stat.*, **17**, 171–81.

Firth, D. (1987). On the efficiency of quasi-likelihood estimation. *Biometrika*, **74**, 223–45.

Fisher, R. A. (1950). The significance of deviations from expectation in a Poisson series. *Biometrics*, **6**, 17–24.

Godambe, V. P. and Thompson, M. E. (1989). An extension of quasi-likelihood estimation. *J. Stat. Plan. Inf.*, **22**, 137–52.

Hausman, J., Hall, B. H., and Griliches, L. (1984). Econometric models for count data with an application to the patents—R and D relationship. *Econometrica*, **52**, 909–38.

Jorgensen, B. (1987). Exponential dispersion models. *J. Roy. Stat. Soc.*, B, **49**, 127–62.

Lawless, J. F. (1987). Negative binomial regression models. *Canad. J. Stat.*, **15**, 209–26.

Liang, K. Y. and Zeger, S. L. (1986). Longitudinal data analysis using generalized linear models. *Biometrika*, **73**, 13–22.

McCullagh, P. and Nelder, J. A. (1989). *Generalized linear models*. Chapman and Hall, London.

Moore, D. F. (1986). Asymptotic properties of moment estimators for overdispersed counts and proportions. *Biometrika*, **73**, 583–8.

Morton, R. (1987). A generalized linear model with nested strata of extra Poisson variation. *Biometrika*. **74**, 247–57.

Thall, P. F. (1988). Mixed Poisson likelihood regression models for longitudinal interval count data. *Biometrics*, **44**, 179–209.

Wedderburn, R. W. M. (1974). Quasi-likelihood functions, generalized linear models and the Gauss–Newton method. *Biometrika*, **61**, 439–47.

Willmot, G. E. (1987). The Poisson-inverse Gaussian distribution as an alternative to the negative binomial. *Scand. Actuar. J.*, **4**, 113–27.

Zeger, S. L. (1988). A regression model for time series of counts. *Biometrika*, **75**, 621–9.

4
Estimating equations in generalized linear models with measurement error
Kung-Yee Liang and Xin-Hua Liu

ABSTRACT

This paper concerns the estimation of regression coefficients, β, when some of the study variables x are measured with error. The approach of replacing x with the James–Stein estimate suggested by Whittemore (1989) is examined in the context of generalized linear models. This approach is shown to produce consistent estimates of β for the identity and log links with Gaussian measurement error and x. For binary nonlinear regression, the same approach produces inconsistent estimate of β which, however, can be corrected by a quasi-likelihood method. A simulation study on a simple logistic regression model was conducted to compare numerically the above two approaches, the naïve approach with x replaced by the observed covariates and a bias-corrected estimator proposed by Stefanski (1985). The results for the quasi-likelihood method are promising.

4.1 Introduction

It is common, especially in epidemiologic studies, that some of the study variables are measured with error. Examples include the measurement of the dose level of non-ionizing radiation in an occupational cohort study and the measurement of daily food intake in studying its association with major cancers. To fix the idea, denote by x a $p \times 1$ vector of true exposure variables which are thought to be related to the outcome variable y. However, z instead of x is observed subject to error, i.e. $z = x + e$, where e is a random vector with mean zero and covariance matrix Ω. In the literature, the following three assumptions have been made either explicitly or implicitly:

(i) Ω is a known matrix,
(ii) x, when treated as a random vector, and e are statistically independent, and
(iii) $f(y|z, x) = f(y|x)$, where f denotes the appropriate probability (density) function.

Assumption (i) is needed to avoid the problem of identifiability for parameters. This assumption, however, can be relaxed if Ω is available either through previous studies or through repeated measurements of x in the same study. Assumption (ii) means that the magnitude of the error, e, is uncorrelated with the true value of x. Assumption (iii), known as the surrogacy assumption, deserves special attention. It means that once the true value of x is observed, no additional information on y is obtained by observing z. There are at least two other equivalent expressions for (iii),

$$f(y, z|x) = f(y|x)f(z|x), \tag{4.1}$$

or

$$f(z|x, y) = f(z|x) \tag{4.2}$$

which will prove useful for later development.

It is fairly well known now that the naïve approach of estimating β, the regression coefficient relating x to y, by simply ignoring the measurement error and replacing x with z in the estimation functions for β, is generally inconsistent (e.g. Stefanski 1985). The major problem with this approach, as pointed out by Whittemore (1989), is that as a surrogate for x, z is too variable since by assumption (ii), $\text{var}(z) \geqslant \text{var}(x)$. By replacing x with the James–Stein (1961) shrinkage estimator, she found that the resulting estimator of β is consistent in the classical linear errors-in-variables model (e.g. Fuller 1987). She also found through simulation, without formal justification, that the same approach performs well in some non-linear cases.

The purposes of this paper are to (i) examine Whittemore's approach and (ii) propose an alternative approach in the context of generalized linear models (Nelder and Wedderburn 1972). The needed assumptions are given in Section 4.2 along with some efficiency calculations for the proposed method. Relationship with some existing work is described as we progress through this paper. It is also worth noting that the main driving force of the proposed method is on the adoption of the concept of estimating functions advocated by Godambe (1960, 1976). Section 4.3 focuses on the logistic regression model. Connection with known results in the literature regarding misclassification, as well as case-control study applications, are discussed. Some simulation results are presented in Section 4.4 followed by discussion.

4.2   Two sets of estimating equations

4.2.1   THE MODEL

For each $i = 1, \ldots, n$, let $f(y_i|x_i; \theta)$ be the probability (density) function of y_i given x_i indexed by $\theta = (\alpha, \beta)$, where α is a scalar and β is a $p \times 1$ vector of coefficients associated with x. To incorporate the more general situation where only part of the exposure variables are measured with error, we assume $x = (x_1, x_2)$ with x_1 a $p_1 \times 1$ vector observed without error, and x_2 a $p_2 \times 1$ vector observed only indirectly through z_2. Thus $z = (z_1, z_2) = (x_1, z_2)$ is observed for each individual. Further, we assume $x \sim N(m, \Sigma)$ where Σ is a full-rank matrix, and conditional on x, $z \sim N(x, \Omega)$. Here

$$\Omega = \begin{pmatrix} \Omega_{11} & \Omega_{12} \\ \Omega_{12} & \Omega_{22} \end{pmatrix} = \begin{pmatrix} 0 & 0 \\ 0 & \Omega_{22} \end{pmatrix}$$

is, according to assumption (i), a known matrix of rank p_2. This measurement error model has been considered, for example, by Liu (1987). Finally, we denote by $\delta = (\delta_1, \delta_2)$, a vector which characterizes m and Σ, respectively.

4.2.2 WHITTEMORE'S APPROACH

To focus on the main issue, we assume for the moment that all of the variables in the vector x are subject to measurement errors, i.e. $p_1 = 0$. The approach proposed by Whittemore (1989), for which $p_2 = 1$ was considered, can be described as follows. Define

$$U(y, x, \theta) = \sum_{i=1}^{n} U_i(y_i, x_i, \theta) = \sum_{i=1}^{n} \partial \log f(y_i|x_i; \theta)/\partial\theta \qquad (4.3)$$

as the score function for θ based on $f(y|x)$, and note that the naïve approach corresponds to the replacement of x_i in (4.3) by z_i. Define $e(z_i, \delta)$ as the James–Stein estimator of x_i, which has the known property of shrinking z_i towards $E(x_i) = m$. Whittemore (1989) proposes the replacement of x_i in (4.3) by $e(z_i; \delta)$ instead of z_i, i.e. to estimate θ by solving

$$U^*(y, z, \theta, \hat{\delta}) = U(y, e(z; \hat{\delta}), \theta) = 0, \qquad (4.4)$$

where $\hat{\delta}$ is a $\sqrt{n}$-consistent estimate of δ. We note that in accordance with the normality assumptions in Section 4.2.1, $e(z; \delta)$ is identical to $E(x|z; \delta)$. Thus, solving the equation in (4.4) is equivalent to solving

$$U(y, E(x|z; \hat{\delta}), \theta) = 0. \qquad (4.5)$$

One way to critically examine the validity of this approach is through the argument of unbiasedness for the estimating functions U^* (Godambe 1960; Liang 1987). While it is difficult to discuss this approach in general, we restrict $U(y, x, \theta)$ of (4.3) to the form

$$U_i(y_i, x_i, \theta) = A_i(x_i, \theta)(y_i - \mu_i(x_i)), \qquad (4.6)$$

where $\mu_i(x_i) = E(y_i|x_i)$. Note that the generalized linear models for $f(y|x)$ clearly satisfy formula (4.6) (McCullagh and Nelder 1989).

It is easy to see that the unbiasedness of U^* is ensured if the operations between μ_i and expectation are exchangeable, i.e.

$$E(y_i|z_i) = E(E(y_i|x_i)|z_i) = E(\mu_i(x_i)|z_i)$$
$$= \mu_i(E(x_i|z_i)). \qquad (4.7)$$

Note that the first equality in (4.7) is true under assumption (iii), irrespective of the distributional assumptions on (z, x). It is clear that the last equality in (4.7) is not held in general except when μ_i is linear. For example, when y is binary and μ_i^{-1} is the probit link function, we have

$$E(\mu(x)|z) = \mu\{E(x|z)/(1 + \text{Var}(x|z)\beta^2)\}$$
$$\neq \mu(E(x|z)). \qquad (4.8)$$

Thus, Whittemore's approach generally gives rise to inconsistent estimate of θ. On the other hand, the equality in (4.8) suggests that $\mu(E(x|z))$ may be a valid approximation for $E(\mu(x)|z) = E(y|z)$ if β is small. Another situation where the approximation may be valid is when μ is approximately linear in x, as in this case the operations between μ and expectation are indeed exchangeable. For example, Cox (1970) has noted that both logistic and probit curves are approximately linear when probabilities for $y = 1$ are not too extreme. This observation, in fact, forms the motivation of estimating procedures Clark (1982) took, which is equivalent to Whittemore's by replacing x with $E(x|z)$.

In the next subsection an alternative approach quasi-likelihood method is introduced.

4.2.3   A QUASI-LIKELIHOOD METHOD

Under (4.1) of Section 4.1 and the assumptions made in Section 4.2.1, the joint likelihood for (y_i, z_i), $i = 1, \ldots, n$ is simply

$$e^{l_1(\theta, \delta)} = \prod_{i=1}^{n} \int f(y_i, z_i, x_i) \, dx_i = \prod_{i=1}^{n} \int f(y_i|x_i; \theta) f(x_i; \delta) \, dx_i. \qquad (4.9)$$

A major problem with maximizing the full likelihood, e^{l_1}, is computation, as calculation of a p_2-dimensional integral is repeatedly involved. To ease the computational burden, Schafer (1987) proposes the uses of EM algorithm to maximize the likelihood e^{l_1}, which requires the approximation to carry it out. An alternative approach which was suggested by Whittemore and Keller (1988) is to consider

$$e^{l_2(\theta, \delta)} = \prod_{i=1}^{n} f(y_i|z_i; \theta, \delta) = \prod_{i=1}^{n} \int f(y_i|x_i; \theta) f(x_i|z_i; \delta) \, dx_i. \qquad (4.10)$$

Intuitively z contains most of the information about δ, and hence, by conditioning on z, $l_2(\theta, \delta)$ carries very little information about δ even though it does depend on δ. It then makes sense to replace δ by a $\sqrt{n}$-consistent estimator $\hat{\delta}$ in l_2 before maximizing it with respect to θ. In fact this is equivalent to the pseudo-likelihood method proposed by Gong and Samaniego (1981) in dealing with nuisance parameters when the joint likelihood is ill-behaved, since

$$l_1(\theta, \hat{\delta}) = l_2(\theta, \hat{\delta}) + \sum_{i=1}^{n} \log f(z_i; \hat{\delta}),$$

which is proportional to $l_2(\theta, \delta)$ as far as θ is concerned. There are at least two reasons for introducing (4.10) or $l_2(\theta, \hat{\delta})$. First, this corresponds to the conventional approach for making inference about θ when no measurement error occurred, in which case the conditional distribution of y given $z = x$ contains all the information about θ. Second and more importantly, this leads to the next approach, termed the quasi-likelihood method.

The pseudo-likelihood method shares, in general, the same computational problems as the full likelihood approach. However, as will be seen at the end of this section, there is a rich class of distributions $\{f(y|x)\}$ such that the first two conditional moments of y given z can be calculated easily. Consider then

$$S(\theta, \hat{\delta}) = \sum_{i=1}^{n} \left[\frac{\partial}{\partial \theta} \{E(y_i|z_i); \theta, \hat{\delta}\}\{\mathrm{Var}(y_i|z_i; \theta, \hat{\delta})\}^{-1}\{y_i - E(y_i|z_i; \theta, \hat{\delta})\} \right],$$

$$= \sum_{i=1}^{n} \frac{\partial \mu_i(\theta, \hat{\delta})}{\partial \theta} V_i^{-1}(\theta, \hat{\delta})\{y_i - \mu_i(\theta, \hat{\delta})\}, \tag{4.11}$$

an estimating function for θ. Note that (4.11) reduces to the quasi-likelihood score functions (Wedderburn 1974) in the absence of measurement errors. This approach was suggested by Whittemore and Keller (1988) who consider the case where δ is known. Let $\hat{\theta}_Q$ be the solution to $S(\theta, \hat{\delta}) = 0$ and we have

Proposition 4.1 *Let (θ_0, δ_0) be the true values of (θ, δ). Under some regularity conditions and the assumption that*

$$\sqrt{n}\left(\frac{S(\theta_0, \delta_0)/n}{\hat{\delta} - \delta_0} \right) \rightarrow N\left(0, \Sigma = \begin{pmatrix} V_{11} & V_{12} \\ V_{12}^t & V_{22} \end{pmatrix}\right),$$

then

$$\sqrt{n}(\hat{\theta}_Q - \theta_0) \rightarrow N(0, V_Q),$$

where

$$V_Q = B^{-1}(V_{11} + 2AV_{12} + AV_{22}A^t)B^{-1},$$

$$A = \lim_{n \to \infty} E\left(\sum_i \frac{\partial \mu_i}{\partial \theta} V_i^{-1} \left(\frac{\partial \mu_i}{\partial \delta}\right)^t\right)\Big/ n,$$

$$B = \lim_{n \to \infty} E\left(-\frac{\partial S}{\partial \theta}\right)\Big/ n,$$

all evaluated at (θ_0, δ_0).

The proof is similar to that in Gong and Samaniego (1981) and in Liang and Zeger (1986) and is therefore omitted. Note that $V_{12} = 0$ if $\hat{\delta}$ depends on (y, z) only through z since $E(S|z) = 0$.

Note that $E(y_i|z_i) = E(\mu_i(x_i)|z_i)$ in (4.11) also involves p_2-dimensional integral in general and could be computationally intensive as well. Some exceptions are when μ_i^{-1}, the link function, is either an identity, log or probit function (Zeger *et al.* 1988). Specifically, for the identity link, $E(y_i|z_i) = E(\alpha + \beta^t x_i|z_i) = \alpha + \beta^t E(x_i|z_i)$. For the log link,

$$E(y_i|z_i) = E\left(e^{\alpha + \beta^t x_i}|z_i\right) = e^{\alpha + \beta^t \operatorname{Var}(x_i|z_i)\beta/2 + \beta^t E(x_i|z_i)} \tag{4.12}$$

and for the probit link

$$E(y_i|z_i) = E(\Phi(\alpha + \beta^t x_i|z_i)) = \Phi\left(\frac{\alpha + \beta^t E(x_i|z_i)}{(1 + \beta^t \operatorname{Var}(x_i|z_i)\beta)^{1/2}}\right), \tag{4.13}$$

where $\Phi(\cdot)$ is the cumulative distribution function for $N(0, 1)$. Thus, for the above three cases the link function specified by $E(y|x)$ is preserved for $E(y|z)$ except the linear predictor is $E(x|z)$ instead of z, along with some other modifications as seen in (4.12) and (4.13). The variance function, $\operatorname{Var}(y_i|z_i)$, can also be derived in a closed form in a similar manner and is omitted.

The logistic regression model, which is popular in many applications, will be discussed in full details in Section 4.3.

Finally, there is a close connection between the results from the identity and log links and that in Gail *et al.* (1984). In dealing with analysis of the treatment effect where some important covariates are omitted, they found that the identity and the log links are the only links that preserve the consistency of treatment effect estimation. In addition, the identity link ensures the consistency of intercept parameters as well, but the log link does not. In the context of measurement errors, our result suggests that the same two links lead to the consistent estimation of both α and β if (i) the true, unobserved covariates x are replaced by $E(x|z)$ and in addition, (ii) the term $x_1^t \operatorname{var}(x_i|z_i)x_i/2$ is offset when the log link is applied.

4.2.4 EFFICIENCY CALCULATION

Thus far, two easily computed estimating procedures are identified for the log and identity links in the context of generalized linear models: the quasi-likelihood approach whose solutions are termed $\hat{\theta}_Q$ and the approach suggested by Whittemore (1989) with solutions $\hat{\theta}_{JS}$. They are easy to compute since a standard statistical package such as GLIM can be used to estimate θ with $E(x|z; \hat{\delta})$ entered as the independent variables. Only minor modification is needed for the quasi-likelihood approach to modify the weight functions for the ys, which can be done by attaching a simple macro to GLIM.

A natural question to address is how these two estimators compare in terms of efficiency. For the identity link and normal distribution of y given x, it is easily seen that $\hat{\theta}_Q = \hat{\theta}_{JS}$ because $\text{var}(y_i|z_i)$ is independent of i. In fact, these two estimators are identical to that derived from the full likelihood e^{l_1} and the pseudo-likelihood $\exp\{l_2(\theta, \hat{\delta})\}$ if the components of δ are estimated by the conventional moment estimators

$$\hat{\delta}_1 = \sum_{i=1}^{n} z_i/n, \qquad \hat{\delta}_2 = \sum_{i=1}^{n} (z_i - \hat{\delta}_1)(z_i - \hat{\delta}_1)^t/n - \Omega. \qquad (4.14)$$

For the log link, we focus on the exponential distribution $f(y|x)$ mainly because no integration is needed to compute the variances of $\hat{\beta}_Q$ and $\hat{\beta}_{JS}$. Both $\hat{\theta}_Q$ and $\hat{\theta}_{JS}$ can be seen as solutions of

$$\sum_{i=1}^{n} \left\{\frac{\partial}{\partial\theta} E(y_i|z_i; \hat{\delta})\right\} E^{-2}(y_i|z_i)\{y_i - E(y_i|z_i; \hat{\delta})\} = 0,$$

$$\sum_{i=1}^{n} \frac{\partial}{\partial\theta} \{e^{\alpha + \beta^t e(z_i; \hat{\delta})}\} e^{-2\alpha - 2\beta^t e(z_i; \hat{\delta})}\{y_i - e^{\alpha + \beta^t e(z_i; \hat{\delta})}\} = 0,$$

respectively, where $E(y_i|z_i)$ is given in (4.12) and $\hat{\delta}$ in (4.14). Note that the main difference between $E(y_i|z_i)$ and $e^{\alpha + \beta^t(z_i; \delta)}$, and hence that between the above two estimating functions, is that the latter one absorbs the intercept with $w = \beta^t \text{var}(x_i|z_i)\beta/2$, which is a function of β. Thus Whittemore's approach is presumably less efficient due to the failure to capture the information regarding β provided by w. The magnitude of efficiency loss depends on that of β and Ω_{22}. For $p_2 = 1$, simple calculation gives asymptotically

$$\frac{\text{var}(\hat{\beta}_{JS})}{\text{var}(\hat{\beta}_Q)} = 1 + \frac{(1 - e^{\beta^2 F/2})^2 + (\Omega_{22}/\Sigma_{22})^2}{2e^{\beta^2 F} + 2\beta^2\Omega_{22}\left(1 - \dfrac{\Sigma_{22}}{\Sigma_{22} + \Omega_{22}}\right) - 1},$$

where $F = \text{var}(x|z) = \Sigma_{22}\Omega_{22}/(\Sigma_{22} + \Omega_{22})$. Table 4.1 presents the efficiency

Table 4.1 Efficiency comparison between $\hat{\beta}_{JS}$ and $\hat{\beta}_Q$ for exponential distribution with mean $e^{\alpha + \beta x}$. The measurement error distribution is $N(0, \Omega)$ and the distribution of x is $N(0, \Sigma)$. Upper entry: $\Sigma = 1$, lower entry: $\Sigma = 2$.

β	Ω 0.1	0.25	0.5	0.75
0.0	0.99	0.94	0.80	0.64
	0.99	0.98	0.94	0.88
0.5	0.99	0.95	0.83	0.71
	0.99	0.99	0.95	0.91
1.0	0.99	0.95	0.88	0.81
	0.99	0.98	0.95	0.93
1.5	0.98	0.95	0.90	0.86
	0.99	0.96	0.92	0.89
2.0	0.97	0.93	0.87	0.84
	0.98	0.93	0.86	0.82
2.5	0.86	0.89	0.83	0.79
	0.96	0.88	0.80	0.75
3.0	0.93	0.85	0.77	0.74
	0.93	0.83	0.74	0.71

comparison for some selected β, Ω and Σ values. Except when β and Ω_{22} are large, high efficiency for $\hat{\beta}_{JS}$ is maintained in comparison with $\hat{\beta}_Q$; see the comments made before the relative efficiency formula above.

4.2.5   EXTENSION TO $p_1 > 0$

When $p_1 > 0$, i.e. some of the covariates are measured without error, the results stated in Section 4.2.3 can be extended in a straightforward manner. Although the James–Stein estimator is not well developed in situations where Ω is not of full rank, the posterior expectation of x given z is still well defined. In fact, by following the proof and the argument in Proposition 4.1 closely, it can be seen that all the results are retained regardless of whether $p_1 > 0$ or not.

4.3 Logistic regression model

4.3.1 LOGIT LINK

The key reason why the probit link is preserved for $E(y|z)$ in Section 4.2.3 is the conjugated property of normal distributions. While it is not true for logistic distribution, we argue that the logit link is approximately preserved when the approximation $\Phi(x/c)$ for $e^x/(1 + e^x)$ is used, where $c = 15\pi/(16\sqrt{3})$ (Johnson and Kotz 1970). To see this, note first that $\mu_i(x_i) = e^{\alpha + x_i^t\beta}/(1 + e^{\alpha + x_i^t\beta})$. Thus

$$E(y_i|z_i; \theta, \delta) = E(e^{\alpha + x_i^t\beta}/(1 + e^{\alpha + x_i^t\beta})|z_i) \approx E(\Phi((\alpha + x_i^t)/c)|z_i)$$

$$= \Phi\left(\frac{\alpha + \beta^t E(x_i|z_i)}{b(\beta, \delta)}\right) \approx \mu_i(E(x_i|z_i)/b(\beta, \delta)), \qquad (4.15)$$

where $b(\beta, \delta) = (1 + \beta^t \text{var}(x_i|z_i)\beta c^{-2})^{1/2}$. The approximation $\Phi(x/c)$ for $e^x/(1 + e^x)$ was examined by Zeger et al. (1988) and they found that there is excellent agreement between the two terms. To focus on the main issues, the discrepancy resulting from the above approximation is ignored for the rest of the paper.

Just as in the case of the probit link, $\hat{\theta}_{JS}$, the solution of $U(y, E(x|z; \delta), \theta) = 0$, 'converges' to $\theta_{JS} = \theta/b(\beta, \delta)$, which is smaller in magnitude than θ component-wise. Denote by $\hat{\theta}_N$ the naïve estimators of θ, i.e. the solution of $U(y, z; \theta) = 0$. We note the following observations in connection with known results in the literature on misclassifications.

1. While they are inconsistent, Wald-type test statistics based on $\hat{\beta}_{JS}$ for testing $\beta = 0$ remain valid, since β_{JS} is proportional to β. The same is true for $\hat{\beta}_N$ because $E(x|z)$ is a linear function of z. The analogy here is that the standard chi-squared tests for independence in contingency tables remain valid, when some of the categorical variables are subject to classification (Mote and Anderson 1965; Korn 1981).

2. The attenuation phenomenon for $\hat{\theta}_{JS}$ is shared by $\hat{\theta}_N$ except when $p_1 = 0$ and $p_2 = 1$. To see this, we assume $p_1 = p_2 = 1$. Then with

$$\rho = \text{Corr}(x_1, x_2) = \Sigma_{12}/(\Sigma_{11}\Sigma_{22})^{1/2} \qquad \text{and} \qquad \Delta = \Omega_{22} + \Sigma_{22}(1 - \rho^2),$$

$$\text{logit Pr}(y = 1|z; \theta, \delta) \approx \{\alpha + \beta^t E(x_i|z_i)\}/b(\beta, \delta)$$

$$= \{\alpha + \beta_1 z_1 + \beta_2 E(x_2|z)\}/b(\beta, \delta)$$

$$= \left\{\tilde{\alpha}(\beta, \delta) + \left(\beta_1 + \beta_2 \frac{\rho\Omega_{22}}{\Delta}\frac{\Sigma_{22}}{\Sigma_{11}}\right)^{1/2} z_1 + \beta_2\left(1 - \frac{\Omega_{22}}{\Delta}\right)z_2\right\}\bigg/ b(\beta, \delta)$$

$$= \{\tilde{\alpha}(\beta, \delta) + \beta_1^* z_1 + \beta_2^* z_2\}/b(\beta, \delta).$$

where $b(\beta, \delta) = (1 + \beta_2^2 \Omega_{22}(1 - \rho^2)c^{-2})^{1/2} \geq 1$. Thus for x_2, which is measured with error, the naïve estimator $\hat{\beta}_{2N}$ converges to $\beta_2^*/b(\beta, \delta)$ which is smaller in magnitude than β_2. Even if x_1 is measured precisely, the estimation of β_1 by $\hat{\beta}_{1N}$ is subject to bias, whose direction could, however, go either way, depending on the sign of ρ along with the relative magnitudes of β_1 and β_2. This result is consistent with the finding by Liu (1987).

4.3.2   CASE-CONTROL STUDY APPLICATIONS

Recall that the surrogacy assumption (iii) in Section 4.1 is equivalent to (4.2), which means, in the context of case-control studies, that the distribution of measurement errors, $f(z|x)$, is independent of y, the disease status. This is known as non-differential misclassification when dealing with discrete risk factors x (e.g., Bross, 1954). To focus on the issues of bias resulting from measurement errors, we assume each case was matched with a control with risk factors x_1 and x_0 respectively. The extension to more complicated matched designs and non-matched designs is straightforward. Adopting the approach of Breslow *et al.* (1978), the probability that z_1 corresponds to the risk factors of the case, given that z_1 and z_0 are observed without knowing the disease status, is simply

$$\frac{f(z_1|y = 1)f(z_2|y = 0)}{f(z_1|y = 1)f(z_0|y = 0) + f(z_1|y = 0)f(z_0|y = 1)}$$

$$= \frac{f(y = 1|z_1)f(y = 0|z_0)}{f(y = 1|z_1)f(y = 0|z_0) + f(y = 0|z_1)f(y = 1|z_0)}$$

$$\approx \frac{e^{(\alpha + \beta^t E(x_1|z_1))/b(\beta, \delta)}}{e^{(\alpha + \beta^t E(x_1|z_1))/b(\beta, \delta)} + e^{(\alpha + \beta^t E(x_2|z_0))/b(\beta, \delta)}}$$

$$= \frac{e^{\beta_{JS}^t E(x_1|z_1)}}{e^{(\beta_{JS})^t E(x_1|z_1)} + e^{\beta_{JS}^t E(x_0|z_0)}}, \tag{4.16}$$

where b and β_{JS} are defined in (4.15). The expression in (4.16) is equivalent to that in Breslow *et al.* (1978) except that β and x are placed by β_{JS} and $E(x|z)$, respectively. Thus the same statistical package for matched case-control analyses can be used whether or not the risk factors are measured with error. Furthermore, consistent estimators of β are achieved by converting β_{JS} to β.

4.4   A simulation study

4.4.1   GENERAL REMARKS

To examine the performance of the proposed procedures, a simulation study for a simple binary logistic regression with $p_1 = 0$ and $p_2 = 1$ was conducted under the following conditions: $\alpha = 0$, $\beta = 1, 2$; sample size $n = 100$ and 200; distribution of x, $N(0, 1)$; distribution of measurement errors, $N(0, \Omega)$, with $\Omega = 0.1, 0.25, 0.5$, and 0.75. For each combination, 500 runs were repeated to calculate the mean, median, mean squared error and 95 per cent coverage probabilities for each procedure.

The estimators of β considered were the naïve estimator $\hat{\beta}_N$, the estimator suggested by Whittemore (1989), $\hat{\beta}_{JS}$, and the quasi-likelihood estimator, $\hat{\beta}_Q$, which is equivalent to the maximum pseudo-likelihood estimator for binary outcomes. They can be seen as the solutions of

$$
\left.
\begin{aligned}
\sum_{i=1}^{n} \binom{1}{z_i} \left(y_i - \frac{e^{\alpha + \beta z_i}}{1 + e^{\alpha + \beta z_i}} \right) &= 0, \\[1ex]
\sum_{i=1}^{n} \binom{1}{e(z_i; \hat{\delta})} \left(y_i - \frac{e^{\alpha + \beta e(z_i; \hat{\delta})}}{1 + e^{\alpha + \beta e(z_i; \hat{\delta})}} \right) &= 0, \\[1ex]
\sum_{i=1}^{n} \binom{1}{e(z_i; \hat{\delta})} \left(y_i - \frac{e^{\{\alpha + \beta e(z_i; \hat{\delta})\}/b(\beta, \hat{\delta})}}{1 + e^{\{\alpha + \beta e(z_i; \hat{\delta})\}/b(\beta, \hat{\delta})}} \right) &= 0,
\end{aligned}
\right\}
\tag{4.17}
$$

respectively, where with $\hat{\delta}$ defined in (14),

$$
e(z_i; \hat{\delta}) = \frac{\Omega}{\hat{\delta}_2 + \Omega} \hat{\delta}_1 + \left(1 - \frac{\Omega}{\hat{\delta}_2 + \Omega} \right) z_i
$$

$$
b(\beta, \hat{\delta}) = \left(1 + \frac{\beta^2 \Omega \hat{\delta}_2}{c^2(\hat{\delta}_2 + \Omega)} \right)^{1/2}
$$

Note that from (4.15), we observed relationships among three estimators

$$
\hat{\beta}_{JS} = \left(1 - \frac{\Omega}{\hat{\delta}_2 + \Omega} \right)^{-1} \hat{\beta}_N = \hat{\beta}_Q / b(\hat{\beta}_Q, \hat{\delta})
\tag{4.18}
$$

Thus only the naïve estimator needs to be estimated and (4.18) can then be applied to get $\hat{\beta}_{JS}$ and $\hat{\beta}_Q$. Note also from (4.18) that β_{JS}, the convergent value of $\hat{\beta}_{JS}$, satisfies

$$
\beta_{JS}^2 = \frac{\beta^2}{1 + \beta^2 c^{-2} \, \text{Var}(x|z)} \leqslant \frac{c^2}{\text{Var}(x|z)}.
\tag{4.19}
$$

This has two important implications. First, the value of β_{JS} to which $\hat{\beta}_{JS}$ converges is necessarily bounded. Secondly, although the corresponding parameter space is unrestricted, $\hat{\beta}_Q$ is very sensitive to how close β_{JS} is to the upper bound c^2/F. Figure 4.1 shows, with $\Sigma = \text{var}(x) = 1$, that while there is a close agreement between β_{JS} and β when $\beta \leqslant 1$, β_{JS} reaches its plateau quickly as β departs from one. This suggests that the empirical distribution

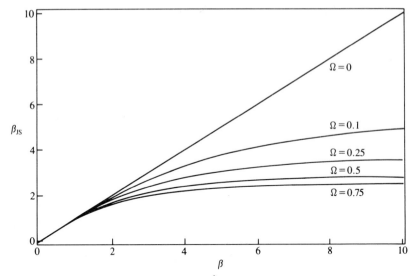

Fig. 4.1   Plot of β_{JS}, the limiting value of $\hat{\beta}_{JS}$, in a simple logistic regression against the true β for various values of $\Omega = \text{var}(z|x)$ with $x \sim N(0, 1)$.

of $\hat{\beta}_Q$ is skewed to the right and the empirical median may serve as a better index than the mean to examine the performance of $\hat{\beta}_Q$ (Burr 1988).

Another estimator of β we considered in the simulation study is the bias-corrected estimator proposed by Stefanski (1985) which is expected to perform well when the error is small.

4.4.2   SIMULATION RESULTS

Simulation results are presented in Table 4.2 for $n = 100$ and in Table 4.3 for $n = 200$. As expected, the naïve estimator is biased toward zero by a factor which increases with $\Omega = \text{var}(z|x)$. Consequently, the lower bounds of the 95 per cent coverage probability which is based on $\hat{\theta} \pm 1.96$ s.e.$(\hat{\theta})$ are much greater than the nominal level, whereas there is a deficit on the upper bound. The estimator suggested by Whittemore, $\hat{\theta}_{JS}$, performs well for $\beta = 1$ and the values of Ω considered. This is consistent with the observations made in the previous paragraph regarding Fig. 4.1. However, when $\beta = 2$, the downward bias of $\hat{\beta}_{JS}$ becomes apparent and the corresponding coverage probability is severely affected. The close agreement between the empirical mean and median for both $\hat{\beta}_N$ and $\hat{\beta}_{JS}$ suggests that the degree of skewness for the corresponding empirical distributions might not be too serious. The unreported histograms for both estimators confirm this observation. We also note that although the empirical variances of $\hat{\beta}_{JS}$ are greater than that of $\hat{\beta}_N$

Table 4.2 Simulation results for logistic regression model with $\alpha = 0$, $x \sim N(0, 1)$ and $z|x \sim N(x, \Omega)$. Sample size is 100. Uppermost entry: $\hat{\beta}_N$, upper entry: $\hat{\beta}_S$, middle entry $\hat{\beta}_{JS}$, lower entry: $\hat{\beta}_Q$.

			$\beta = 1$		
				Coverage probability $\times$ 100	
Ω	Mean	Median	MSE	Lower	Upper
	0.93	0.92	0.072	6.8	0.6
0.10	1.04	1.02	0.092	3.6	1.4
	1.03	1.01	0.083	3.2	1.0
	1.05	1.03	0.095	3.2	1.2
	0.80	0.78	0.103	20.4	0.0
0.25	0.99	0.97	0.109	5.0	1.4
	1.00	0.98	0.097	4.0	2.2
	1.05	1.02	0.132	3.0	0.2
	0.63	0.61	0.179	50.2	0.0
0.50	0.88	0.85	0.110	10.8	0.0
	0.96	0.93	0.105	5.0	1.4
	1.04	0.98	0.171	3.6	0.0
	0.56	0.56	0.224	63.2	0.0
0.75	0.85	0.83	0.112	11.4	0.0
	1.02	0.98	0.130	4.6	0.8
	1.15	1.06	0.290	3.4	0.0
			$\beta = 2$		
	1.78	1.76	0.180	14.0	0.2
0.10	2.07	2.02	0.226	3.6	1.2
	1.96	1.94	0.162	5.0	0.4
	2.11	2.06	0.272	2.8	0.0
	1.44	1.40	0.420	49.4	0.0
0.25	1.90	1.84	0.285	10.2	0.8
	1.81	1.76	0.210	11.6	0.6
	2.17	1.98	0.987	4.0	0.0
	1.13	1.11	0.826	87.4	0.0
0.50	1.68	1.60	0.314	18.4	0.0
	1.72	1.67	0.255	18.4	0.0
	2.35	2.02	2.015	6.2	0.0
	0.90	0.90	1.255	99.6	0.0
0.75	1.42	1.40	0.485	41.2	0.0
	1.60	1.58	0.327	29.2	0.0
	2.30	2.00	1.965	7.8	0.0

Table 4.3 Simulation results for logistic regression model with $\alpha = 0$, $x \sim N(0, 1)$ and $z|x \sim N(x, \Omega)$. Sample size is 200. Uppermost entry: $\hat{\beta}_N$, upper entry: $\hat{\beta}_S$, middle entry $\hat{\beta}_{JS}$, lower entry: $\hat{\beta}_Q$.

| | | | | Coverage probability × 100 | |
Ω	Mean	Median	MSE	Lower	Upper
			$\beta = 1$		
	0.99	0.91	0.041	10.2	0.6
0.10	1.02	1.01	0.046	3.4	2.2
	1.01	1.00	0.042	3.6	2.2
	1.03	1.01	0.048	3.6	2.2
	0.78	0.77	0.077	33.6	0.0
0.25	0.96	0.95	0.047	5.8	1.2
	0.97	0.96	0.043	5.0	1.4
	1.01	0.99	0.054	4.4	1.4
	0.64	0.63	0.149	71.4	0.0
0.50	0.89	0.86	0.054	10.8	0.0
	0.97	0.96	0.047	4.2	0.4
	1.03	1.01	0.069	3.2	0.0
	0.53	0.53	0.237	93.2	0.0
0.75	0.79	0.78	0.087	24.4	0.0
	0.94	0.93	0.062	8.8	0.1
	1.02	1.00	0.098	4.6	0.0
			$\beta = 2$		
	1.74	1.71	0.135	21.4	0.2
0.10	2.01	1.97	0.112	5.0	1.6
	1.92	1.89	0.083	7.4	1.0
	2.05	2.01	0.129	3.2	1.2
	1.44	1.42	0.370	69.0	0.0
0.25	1.89	1.84	0.144	10.8	0.6
	1.80	1.78	0.127	17.0	0.4
	2.08	2.01	0.253	5.2	0.0
	1.11	1.10	0.830	98.4	0.0
0.50	1.63	1.60	0.238	30.6	0.0
	1.68	1.66	0.187	28.2	0.2
	2.11	2.00	0.417	5.2	0.0
	0.91	0.89	1.223	100.0	0.0
0.75	1.42	1.39	0.417	57.2	0.0
	1.60	1.58	0.252	39.8	0.0
	2.15	1.98	0.648	6.0	0.0

in all 16 experiments, the reverse is true for the mean squared error due to the more severe biasednessness of $\hat{\beta}_N$. This is consistent with the dominating property that the James–Stein estimator enjoys in comparison with the conventional estimators, even though the parameter of interest here is β, instead of $E(x)$, for which the James–Stein estimator is directly applied.

The bias-corrected estimator by Stefanski (1985), $\hat{\beta}_S$, performs reasonably well when the measurement error Ω is small, irrespective of the true value of β. However, as expected, it does not perform nearly as well as $\hat{\beta}_{JS}$ both in terms of bias and convergence probability when Ω is away from zero.

As for $\hat{\beta}_Q$, the quasi-likelihood estimator, the sharp contrasts between the empirical means and medians when $\beta = 2$ and $n = 100$ confirm the early observations that the distributions of $\hat{\beta}_Q$ tend to be skewed to the right. This is supported by the relatively greater mean squared errors for $\hat{\beta}_Q$. However, although $\hat{\beta}_Q$ appears to over-correct, the bias introduced by $\hat{\beta}_{JS}$ when measured by the average, the corresponding medians are in an excellent agreement with the true β-values. This is further supported by the good overall coverage probability performance of $\hat{\beta}_Q$. When the sample size increases to 200, there is a much better agreement between the empirical mean and median, suggesting that the problem of skewness is much lessened, although the imbalance between the lower and upper bounds is maintained when $\beta = 2$.

4.5   Discussion

The problem of measurement errors is common in a variety of observational studies. In this paper, we examine systematically the bias of the approach suggested by Whittemore (1989) which replaces the unobserved covariates by the James–Stein estimators in the score functions where the error is ignored. As a comparison, the quasi-likelihood method first introduced by Whittemore and Keller (1988) is discussed and extended to allow $p > 1$ and δ unknown. Under some conditions stated in Section 4.2.1, we identify, in the context of generalized linear models, the identity and log links, for which the consistency of $\hat{\beta}_{JS}$ is ensured. Efficiency calculations indicate that high efficiency is generally maintained by using $\hat{\beta}_{JS}$ instead of $\hat{\beta}_Q$, the quasi-likelihood estimator. For the probit and logit links, the bias of $\hat{\beta}_{JS}$, although smaller in magnitude, is in the same order as the naïve estimator $\hat{\beta}_N$; see Section 4.4.1. However, $\hat{\beta}_Q$, which is functionally closely related to $\hat{\beta}_{JS}$ and $\hat{\beta}_N$ as stated in (4.18), is consistent, at least approximately, for the logit link.

The simulation study presented in Section 4.4 is interesting in several regards. First, most of the simulation studies reported in the literature takes $\beta = 1$ only. Figure 4.1 suggests that at least for probit and logit links, β-values greater than one should also be considered. Second, the results on the coverage reveal that the problem of skewness for some estimators, although recognized in the literature (e.g. Schafer 1987; Burr 1988), may be more serious than first thought. More work is needed to correct this problem. A potentially promising approach is to work directly with the estimating function such as $S(\theta, \hat{\delta})$ to invert the distribution of S instead of its solution $\hat{\theta}$ (Godambe and Thompson 1989).

Finally, we note the following two remarks:

(i) For the two approaches considered in this paper, only the first two moments of y given x are indeed required to be specified, a situation similar to the quasi-likelihood proposed by Wedderburn (1974) where no measurement error is assumed.

(ii) The focus of this paper is on the identification of some conditions such as the normality of x and z given x, which justify the approach of replacing x by the James–Stein estimator in terms of unbiasedness of the corresponding estimating functions. An equally challenging question is to identify $e(z_i, y_i, \delta)$, as a replacement for x_i, such that (a) the resulting estimating functions are unbiased and yet (b) weaker assumptions on the distributions of x and z are made. This approach has been taken up by Stefanski (1989).

4.6   Acknowledgement

This work was supported in part by grant GM392621 awarded by the National Institute of Health.

References

Berkson, J. (1970). Are there two regressions? *J. Amer. Stat. Assoc.*, **45**, 164–80.

Breslow, N. E. (1974). Covariance analysis of censored survival data. *Biometrics*, **30**, 89–99.

Breslow, N. E., Day, N. E., Halvorsen, K. T., Prentice, R. L., and Sabai, C. (1978). Estimation of multiple relative risk function in matched case-control studies. *Amer. J. Epidemiol.*, **108**, 299–307.

Bross, I. (1954). Misclassfication in 2×2 tables. *Biometrics*, **10**, 478–86.

Burr, D. (1988). On errors-in-variables in binary regression—Berkson case. *J. Amer. Stat. Assoc.*, **83**, 739–43.

Clark, R. R. (1982). The error-in-variables problem in the logistic regression model. Unpublished Ph.D. thesis, Department of Statistics, University of North Carolina at Chapel Hill.

Cox, D. R. (1970). *The analysis of binary data*. Chapman and Hall, London.

Feigel, P. and Zelen, M. (1965). Estimation of exponential survival probabilities with concomitant informatin. *Biometrics*, **21**, 826–38.

Fuller, W. A. (1987). *Measurement error models*. Wiley, New York.

Gail, M. H., Wieand, S., and Piantadosi, S. (1984). Biased estimates of treatment effect in randomized experiments with nonlinear regressions and omitted covariates. *Biometrika*, **71**, 431–44.

Godambe, V. P. (1960). An optimum property of regular maximum likelihood estimation. *Ann. Stat.*, **31**, 1208–11.

Godambe, V. P. (1976). Conditional likelihood and unconditional optimum estimating equations. *Biometrika*, **67**, 155–62.

Godambe, V. P. and Thompson, M. E. (1989). An extension of quasi-likelihood estimation. *J. Stat. Plan. Inf.*, **22**, 171–6.

Gong, G. and Samaniego, F. J. (1981). Pseudo maximum likelihood estimation: theory and applications. *Ann. Stat.*, **9**, 861–9.

James, W. and Stein, C. (1961). Estimation and quadratic loss. *Proc. Fourth Berkeley Symposium on Mathematical Statistics and Probability*, Volume 1. California: University of California Press, pp. 361–79.

Johnson, N. L. and Kotz, S. (1970). *Distributions in statistics, continuous univariate distributions*—2. Houghton-Mifflin, Boston.

Korn, E. L. (1981). Hierarchical log-linear models not preserved by classification error. *J. Amer. Stat. Assoc.*, **76**, 110–3.

Liang, K.-Y. (1987). Estimating functions and approximate conditional likelihood. *Biometrika*, **74**, 695–702.

Liang, K.-Y. and Zeger, S. L. (1986). Longitudinal data analysis using generalized linezr models. *Biometrika*, **73**, 13–22.

Liu, K. (1987). Measurement error and its impact on partial correlation and multiple linear regression analyses. *Amer. J. Epidemiol.*, **127**, 864–74.

McCullagh, P. and Nelder, J. A. (1989). *Generalized linear models* (2nd edn). Chapman and Hall, London.

Mote, V. L. and Anderson, R. L. (1965). An investigation of the effect of misclassification on the properties of χ^2-tests in the analysis of categorical data. *Biometrika*, **52**, 95–109.

Nelder, J. A. and Wedderburn, R. W. M. (1972). Generalized linear models. *J. Roy. Stat. Soc.* A, **135**, 370–84.

Schafer, D. W. (1987). Covariate measurement error in generalized linear models. *Biometrika*, **74**, 385–91.

Stefanski, L. A. (1985). The effects of measurement error on parameter estimation. *Biometrika*, **72**, 583–92.

Stefanski, L. A. (1989). Unbiased estimation of a nonlinear function of a normal mean with application to measurement-error models. Technical Report No. 1929, Institute of Statistics, North Carolina State University.

Wedderburn, R. W. M. (1974). Quasi-likelihood functions, generalized linear models, and the Gauss–Newton method. *Biometrika*, **61**, 439–48.

Whittemore, A. S. and Keller, J. B. (1988). Approximations for regression with covariate measurement error. *J. Amer. Stat. Assoc.*, **83**, 1057–66.

Whittemore, A. S. (1989). Errors-in-variables regression using Stein estimates. *Amer. Statistician*, **43**, 226–8.

Zeger, S. L., Liang, K.-Y., and Albert, P. S. (1988). Models for longitudinal data: a generalized estimating equation. *Biometrics*, **44**, 1049–60.

5
A unification of inference from capture–recapture studies through martingale estimating functions

Chris J. Lloyd and Paul Yip

ABSTRACT

This paper attempts to present a unified method of inference for the size of a closed population based on data from a capture–recapture study under varying assumptions about homogeneity across time and individuals. The unifying idea is that of a martingale estimating function. This should ultimately lead to a regression like approach to capture–recapture experiments, incorporating co-variates, goodness-of-fit and diagnostic residuals. The main contention is that what martingale methods may lack in efficiency relative to likelihood methods they more than make up for in computational simplicity, conceptual simplicity and flexibility.

5.1 Introduction

5.1.1 BACKGROUND AND MOTIVATION

The estimation of the size of a population is an important problem in the biological sciences. It is also quite an old problem, over 200 papers having appeared on the subject over the past 25 years. However, as stated by Otis *et al.* (1978)

> one is generally left without a unified approach to the estimation of . . . population.

It is the purpose of this paper to provide such a unified approach to so-called 'capture–recapture' experiments under a wide range of assumptions. The unification is provided by the notion of a weighted martingale estimating function.

The advantages of this approach are threefold. Firstly, for even moderately complex models the calculation of estimates and standard errors is significantly simpler than the likelihood estimates. Indeed, some of our estimates have an explicit form while for some models the ML estimates are extremely difficult to obtain and behave quite erratically (Burnham 1972). A second advantage is that the estimation procedure is easy to understand and qualitative properties may be deduced heuristically. Thirdly, by their nature, martingales are robust to departures from the model apart from mean and variance relations. Departure from variance assumptions will typically ruin standard error estimation but not affect the consistency of the estimator. Changes in the mean assumption can be directly modelled into the martingale differences often with little or no increase in the conceptual complexity of the procedure. This is in marked contrast to that other unifying method of statistical estimation, the likelihood approach.

5.1.2　NOTATION AND DESCRIPTION

Next we will describe capture–recapture experiments in general and establish a consistent notation for the quantities which arise. Consider a population of N individuals, which for definiteness we image as fish in a lake. This unknown parameter N is the object of estimation and is assumed constant over the study period. On each sampling occasion each fish may be sampled exactly once independently of every other fish. Rather than broiling the caught fish in a Hollandaise sauce, they are marked and replaced in the lake. At each sampling occasion then the data comprise a set of caught fish together with their capture history. Roughly speaking as more marked fish appear in the sample we suspect that more of the population of N fish have been captured though making this precise, we will see, depends critically on what is assumed about capture probabilities from fish to fish and from occasion to occasion. It is also possible to sample fish continuously in time. It is equivalent to sampling one at a time, see Becker (1984) and Yip (1989). We establish the following notation for discrete time sampling:

$$N = \text{population size}$$
$$t = \text{number of capture occasions}$$
$$P_j = \text{the capture probability of fish } j \text{ under 'standard' conditions}$$
$$\pi_i = \text{the average capture probability on occasion } i$$
$$n_i = \text{number of fish caught on occasion } i$$
$$m_i(k) = \text{number of fish caught for the } k\text{th time on occasion } i$$
$$C_i(k) = \text{number of fish caught exactly } k \text{ times up to occasion } i$$
$$M_i = \sum_{k=1}^{i} C_i(k) = \text{number of distinct fish caught up to occasion } i$$
$$m_i = n_i - m_i(1) = \text{number of marked fish caught on occasion } i$$
$$U_i = C_i(0) = N - M_i = \text{number of uncaught fish up to occasion } i$$
$$u_i = m_i(1) = \text{number of fish caught for first time on occasion } i.$$

The data comprise the matrix $m_i(k)$, which is triangular since $m_i(k) = 0$ if $k > i$, together with the identifying labels of the fish caught at each occasion. The quantity $C_i(k)$ is calculable from $m_1(k), \ldots, m_i(k)$. We say that

$C_i(k)$ is $\mathscr{F}_i$ measureable where $\mathscr{F}_i$ denotes the σ-field generated by the entire data up to and including occasion i. One may think of $\mathscr{F}_i$ as the 'experimental history up to occasion i'. The object of this paper is to derive efficient estimators of N from the full experimental history matrix $(m_i(k), i = 1, \ldots, t, k = 1, \ldots, i)$ as well as an estimate of the variability of the estimator. We will systematically develop simple and effective estimators starting with the most simplifying assumptions and progressively relaxing them. The notion of a martingale will unify the discussion.

The usual definition of a martingale is in terms of a stochastic process $\{\mathscr{M}_i : i \in I\}$ where I is an index set and an increasing sequence of σ-fields $\{\mathscr{F}_i : i \in I\}$. It is required that $\mathscr{M}_i$ be $\mathscr{F}_i$-measurable and that

$$E(\mathscr{M}_{i+h}|\mathscr{F}_i) = \mathscr{M}_i.$$

If $E(\mathscr{M}_0) = 0$ then $\{\mathscr{M}_i : i \in I\}$ is called a zero mean martingale (ZMM). It is simple to show that the martingale differences $D_i = \mathscr{M}_i - \mathscr{M}_{i-1}$ have zero mean and are $\mathscr{F}_i$-measurable and so write

$$\mathscr{M}_t = \sum_{i=1}^{t} D_i,$$

with the convention $\mathscr{M}_0 = 0$. In statistical inference we look for martingale differences, D_i, which are functions of the unknown parameter θ and then equate the martingale to zero. The reader may consult Godambe (1985) and Lloyd (1987) for the theory of such estimators. The main result relevant to our purposes is the following. For a given sequence of martingale differences $D_i(\theta)$ and $\mathscr{F}_{i-1}$ measurable random variables w_{i-1} the equation

$$\mathscr{M}_t(\theta) = \sum_{i=1}^{t} w_{i-1} D_i(\theta) = 0 \tag{5.1}$$

has solution $\hat{\theta}$ satisfying

$$\frac{\sum_{i=1}^{t} w_{i-1} E(D_i'(\theta)|\mathscr{F}_{i-1})}{\sqrt{(\sum_{i=1}^{t} w_{i-1}^2 \, \mathrm{Var}(D_i|\mathscr{F}_{i-1}))}} (\hat{\theta} - \theta) \to N(0, 1) \qquad \text{as } t \to \infty \tag{5.2}$$

under mild regularity conditions on existence and behaviour of the first and second moments involved. Here the prime denotes differentiation with respect to the parameter θ and we shall continually refer to the $\mathscr{F}_{i-1}$ measurable quantities w_{i-1} as weights. The normalising factor is maximised when we choose

$$w_{i-1}^* = \frac{E(D_i'|\mathscr{F}_{i-1})}{\mathrm{Var}(D_i|\mathscr{F}_{i-1})} \tag{5.3}$$

leading to the largest possible asymptotic normaliser

$$\sqrt{\left(\sum_{i=1}^{t} E^2(D_i'(\theta)|\mathcal{F}_{i-1})/\mathrm{Var}(D_i|\mathcal{F}_{i-1})\right)} \tag{5.4}$$

and hence the narrowest asymptotic confidence intervals. Another interesting small sample interpretation of the optimally weighted ZMM, $\mathcal{M}_t^* = \sum_{i=1}^{t} w_{i-1}^* D_i$ is that it is most highly correlated with the true score function (see numerical evidence of this in Section 5.2.3). We will repeatedly identify sequences of ZMM differences and use the optimality result (5.3) to produce an optimally weighted martingale for inference on the population parameter N.

To solve $\mathcal{M}_t^*(\theta) = 0$ it is typically necessary to iterate since the optimal weights $w_{i-1}^*(\theta)$ are themselves functions of the parameter. If the martingale differences $D_i(\theta)$ are complicated then Newton–Raphson iteration is necessary. This is typical of likelihood-based estimates. If $D_i(\theta)$ is as simple as a quadratic, then simple iteration on the weights and explicit solution of $\sum w_{i-1}^*(\hat{\theta})D_i(\theta) = 0$ usually leads quickly to the estimate. This is typical of our martingale-based estimates. Approximate confidence intervals may be set in one of two ways. Firstly, under mild regularity conditions

$$\frac{\mathcal{M}_t(\theta)}{\sqrt{\left(\sum_{i=1}^{t} w_{i-1}^2 \, \mathrm{Var}(D_i|\mathcal{F}_{i-1})\right)}} \to N(0, 1) \qquad \text{as } t \to \infty. \tag{5.5}$$

The denominator is itself a random variable, interpreted as the sum of conditional variances of the weighted martingale differences. It is a feature of stochastic process inferences that random norming be employed. One can think of this as essentially conditioning on the information which was available for this particular experimental outcome. Bounding the left side of (5.5) by a normal quartile and solving for θ leads to approximate confidence intervals.

A second procedure is to directly use (5.2). It is simple to see that a linear approximation to $\mathcal{M}_t(\theta)$ in (5.5) leads to (5.2) so the two methods described are asymptotically equivalent and are exactly equivalent in the case $\mathcal{M}_t(\theta)$ is linear in θ. It is always simpler to use (5.2) than (5.5) in practice. An important theoretical objection to using (5.2) is that it is not parametrisation invariant; however, when the parameter θ is a population size N it is hard to see the meaning of a transformed parameter. We will mainly be using (5.2) in this paper for no other reason than it is simpler to calculate.

5.2 Homogeneous populations

In this section we will assume that the capture probabilities, P_j, are the same from fish to fish but may vary from occasion to occasion. In fact, the latter property of time heterogeneity has no effect on the estimation problem since it is the proportion of marked fish caught at each occasion which carries the information about N and not the total fish caught at each occasion. Hence the estimation procedure given by the martingale methods for homogeneous populations is also applicable for the time heterogeneous case (see Yip (1990a)).

For this section recall the notation m_i, M_i for the number of marked fish in the sample and population respectively and u_i, U_i for the number of unmarked fish in the sample and population respectively. Only when differences in the capture probabilities P_j are allowed need multiple marking information enter the inference. Note that conditional on $\mathscr{F}_{i-1}$ we have

$$u_i | \mathscr{F}_{i-1} \overset{d}{=} \mathrm{Bin}(U_{i-1}, \pi_i), \quad m_i | \mathscr{F}_{i-1} \overset{d}{=} \mathrm{Bin}(M_{i-1}, \pi_i),$$

which gives the distributional basis for the estimators developed in the next section.

5.2.1 A CONDITIONAL LIKELIHOOD APPROACH

For this simple model Darroch (1958) derived a likelihood for N based on the sample sizes $n_1, \ldots, n_t$ and the number of fish $\mathbf{a}_w$ with capture history w. This was

$$L(N; \mathbf{a}_w) = \frac{N!}{(N - M_t)! \prod_w a_w} \prod_{i=1}^{t} \pi_i^{n_i} (1 - \pi_i)^{N - n_i},$$

where the first product is over all distinct capture histories w. This depends on the nuisance parameters π_i which may be eliminated by conditioning on independent binomial variables $n_1, \ldots, n_t$ leading to a conditional likelihood

$$L(N; \mathbf{a}_w | n) = \frac{N!}{\prod_w (N - M_t)!} \prod_{i=1}^{t} \binom{N}{n_i}^{-1} \tag{5.6}$$

with solution the unique root greater than M_t of

$$\left(1 - \frac{M_t}{N}\right) = \prod_{i=1}^{t} \left(1 - \frac{n_i}{N}\right). \tag{5.7}$$

Solution of this equation requires Newton–Raphson iteration. The equation cited is usually credited to Darroch (1958) who also gave estimates of asymptotic bias and variance, but was first obtained by Chapman (1952) under a slightly different model. Darroch showed that for point and interval estimation of N, it makes no difference whether or not we condition on the n_i. Because of the form of the marginal likelihood for $n_1, \ldots, n_t$, results of Godambe (1976) identify the differentiated logarithm of (5.6) as the optimal estimating function among unbiased estimating functions depending on N

only. This leads directly to the estimator (5.7). However, it is not necessarily true that the best estimating equation is unbiased. For example, it may be possible to show that the best equation depends on certain nuisance parameters which when estimated produce a slightly biased but still optimum equation.

The main disadvantages of the conditional likelihood approach are computational difficulty (though this is not too great for the homogeneous model) and that extensions to more complex models require us to possibly completely rebuild the likelihood. As an illustration of our martingale approach we will derive a class of alternative estimators of N for this problem, which includes estimators that are almost surely very close to Darroch's conditional ML estimator. The general advantages of this alternative approach will be detailed in Section 5.4 after we have studied in some detail how to derive useful classes of estimators from it.

5.2.2　MARTINGALES IN THE HOMOGENEOUS MODEL

Consider occasion i. Just before this occasion we have exactly $C_{i-1}(k-1)$ fish in the population that are marked $k-1$ times and an unknown quantity $U_{i-1} = C_{i-1}(0) = N - M_{i-1}$ unmarked fish in the population. This partitions the population into i distinct groups and, since all fish are assumed to behave independently, any martingales based on different groups will be statistically independent. The distributional basis of our martingale approach is summarised by the $\mathscr{F}_{i-1}$-conditional distributions of the elements of the matrix $m_i(k)$, namely

$$m_i(k)|\mathscr{F}_{i-1} \stackrel{d}{=} \text{Bin}(C_{i-1}(k-1), \pi_i) \qquad k = 1, \ldots, i \qquad (5.8)$$

all independently. Summing over all occasions and subtracting the conditional expectation of $m_i(k)$ gives t independent ZMMs

$$\mathscr{M}_t^{(k)} = \sum_{i=k}^{t} \{m_i(k) - C_{i-1}(k-1)\pi_i\} \qquad k = 1, \ldots, t.$$

Note that apparently for $k = 2, \ldots, t$ these martingales do not involve N; only $\mathscr{M}_t^{(1)}$ carries information about N through the term $C_{i-1}(0) = N - M_{i-1}$. Consider the random variable

$$D_i = \{M_{i-1}u_i - U_{i-1}m_i\},$$

with conditional expectation zero. The last term involves m_i which is the sum of random variables in (5.8) excluding the first. Its function here is simply as a random variable which we can subtract from the informative statistic u_i to make the expectation conditionally zero. Summing over occasions and weighting these martingale differences gives us a class of martingale estimating functions

$$\sum_{i=1}^{t} w_{i-1}\{M_{i-1}u_i - U_{i-1}m_i\} = 0 \qquad (5.9)$$

which, provided the weights w_{i-1} are observable, leads explicitly to the estimator

$$\hat{N} = \frac{\sum_{i=1}^{t} w_{i-1}n_i M_{i-1}}{\sum_{i=1}^{t} w_{i-1}m_i}. \qquad (5.10)$$

For the particular choice $w_i = 1$ we obtain the estimator of Schnabel (1938) and Yip (1989) which may be thought of as an m_i-weighted average of natural 'Peterson' estimators, $n_i M_{i-1}/m_i$, from each occasion. An estimate of the asymptotic 'standard error' derived from (1.3) is given by

$$\sqrt{\left(\hat{N} \sum_{i=1}^{t} \hat{p}_i(1 - \hat{p}_i)M_{i-1}(\hat{N} - M_{i-1})\right) \bigg/ \sum_{i=1}^{t} m_i}.$$

If we choose weights which themselves depend on N then to solve (5.9) it is necessary to iteratively apply (5.10). Notice that the first terms in all of these sums are zero since $m_1 = M_o = 0$.

5.2.3   OPTIMAL MARTINGALE ESTIMATING FUNCTIONS

An optimal weight for (5.9) is given in Yip (1990a) and is found by differentiating the differences D_i, using the independent binomial distributions of u_i and m_i and invoking (5.3). This gives optimal weights

$$w_{i-1}^* = \frac{1}{(N - M_{i-1})(1 - \pi_i)}$$

and when π_i is estimated by n_i/N we obtain the estimated optimally weighted equation

$$\sum_{i=1}^{t} \frac{M_{i-1}u_i - (N - M_{i-1})m_i}{(N - M_{i-1})(N - n_i)} = 0, \qquad (5.11)$$

which must be solved iteratively. (The $i = 1$ term is always zero.) Nevertheless, because of the linearity of the numerator in N, computation of this estimator is easier than the ML estimator. Note that w_{i-1}^* is a non-decreasing function of i provided that π_i is non-decreasing, i.e. the catching effort does not decrease substantially later in the experiment. Hence, w_{i-1}^* gives more weight to later parts of the experiment as there are more marked individuals. An estimator of standard error and confidence intervals follow from (5.2), (5.4), which is given explicitly as

$$\sqrt{\left(\sum_{i=1}^{t} \frac{n_i M_{i-1}}{(\hat{N}^* - M_{i-1})(\hat{N}^* - n_i)}\right)\Bigg/ \sum_{i=1}^{t} \frac{m_i}{(\hat{N} - M_{i-1})(\hat{N}^* - n_i)}}.$$

A short simulation study was conducted to compare the solution of (5.11), $\hat{N}_{MG}^*$, with Darroch's conditional ML estimator, $\hat{N}_{ML}$. Two cases were considered. The first had parameters ($N = 100$, $t = 5$, $\pi_i = 0.1 \ \forall i$) and the second ($N = 500$, $t = 10$, $\pi_i = 0.05 \ \forall i$). Data was simulated 300 times for each of these models and for each simulation the two estimators $\hat{N}_{ML}$, $\hat{N}_{MG}^*$ were calculated together with the standardized quantity $(\hat{N} - N)/\hat{\sigma}$, where $\hat{\sigma}$ is an estimated normaliser given by (5.4) for $\hat{N}_{MG}^*$ and explicitly by Darroch (1958) and Seber (1982, p. 135) for $\hat{N}_{ML}$. The results are described below.

For the $N = 100$ simulation, about 40 of the entire population of 100 is typically sampled. The maximum difference between the two estimators is 4 per cent. The respective means and RMSE are (112.3, 29.4) for $\hat{N}_{ML}$ and (112.0, 29.4) for $\hat{N}_{MG}^*$. The standardized quantities have respective mean and variance $(-0.18, 1.12)$ and $(0.03, 1.05)$. For the MG estimator the simulated distribution of the standardized quantities was markedly more normal than for the ML estimator.

For the $N = 500$ simulation, about 200 of the entire population of 500 is typically sampled. The maximum difference between the two estimators is 1 per cent. The respective means and RMSE are (506.8, 47.7) for $\hat{N}_{ML}$ and (506.4, 47.7) for $\hat{N}_{MG}^*$. The standardized quantities have respective mean and variance $(-0.08, 1.04)$ and $(-0.01, 1.00)$. For the MG estimator the simulated distribution of the standardized quantities was markedly more normal than for the ML estimator.

There is no particular reason why the martingale estimator should be more normally distributed than the Darroch estimator. Nor is there any reason in general why the standardized martingale estimator should have variance closer to 1 than the Darroch estimator. The closeness of the estimators, however, can be anticipated since the optimal weights employed in (5.11) are precisely those which maximize the correlation of (5.11) with the true score function.

5.2.4   ALLOWING FOR REMOVALS

We can illustrate the conceptual simplicity of our approach by allowing for removals. It is not uncommon for some of the fish to be removed from the population during the course of the study. This might occur for instance if some fish died during the capture procedure or, even more commonly, during the marking procedure. For scientific reasons, fish may also be deliberately removed from the population for individual study. Let us denote the number of removed individuals up to time $i - 1$ by R_{i-1}. Then conditional on $\mathscr{F}_{i-1}$ we still have

$$u_i|\mathscr{F}_{i-1} \overset{d}{=} \text{Bin}(U_{i-1}, \pi_i) \qquad \text{and} \qquad m_i|\mathscr{F}_{i-1} \overset{d}{=} \text{Bin}(M_{i-1}, \pi_i),$$

all independently with $M_{i-1} + R_{i-1} + U_{i-1} = N$, where we must redefine M_{i-1} as the number of marked fish in the population before the ith occasion. There is no necessity to model the process of removals itself. Following Section 5.2.2 suggests the martingale differences

$$D_i = (N - M_{i-1} - R_{i-1})m_i - M_{i-1}u_i$$

be used to form the martingale estimating function

$$\sum_{t=1}^{t} w_{i-1}\{(N - M_{i-1} - R_{i-1})m_i - M_{i-1}u_i\}, \tag{5.12}$$

generating the estimator

$$\hat{N} = \frac{\sum_{i=1}^{t} w_{i-1}\{M_{i-1}n_i + R_{i-1}m_i\}}{\sum_{i=1}^{t} w_{i-1}m_i}, \tag{5.13}$$

provided w_{i-1} is observable. In case of no removals, i.e. $R_i = 0$, (5.13) reduces to (5.10). An optimal weight according to (5.3) is

$$w_{i-1}^* = \frac{1}{(1 - \pi_i)(N - M_{i-1} - R_{i-1})(N - R_{i-1})},$$

where π_i is again estimated by n_i/N. As an example consider the data of Overton (1965) which is analysed by Yip (1990b). A common cause of removal in capture–recapture studies is fatality during the marking process. This is the case with Overton's data, which comprises capture records for 10 capture occasions with removals occurring only for previously uncaught fish. Relevant statistics are listed in Table 5.1, together with the necessary calculations for computing (5.13) with $w_{i-1} = 1$. Over the course of the experiment 163 distinct aimals were caught, 92 of whom did not survive marking. For the optimal estimator $\hat{N}^*$, two iterations on the weights were necessary. The results for the data are

$$\hat{N}_1 = \frac{6201 + 986}{16} = 449, \qquad \hat{\sigma}\{\hat{N}_1\} = 83.9$$

and $\hat{N}^* = 443$, $\hat{\sigma}\{\hat{N}^*\} = 79.9$. The ML estimator is more difficult to compute and is $\hat{N}_{\text{ML}} = 445$, $\hat{\sigma}\{\hat{N}_{\text{ML}}\} = 88.0$.

5.3   Heterogeneous population

The methods of Section 5.2 are all based on the idea that as more marked

Table 5.1 Computations required for the estimates $\hat{N}_1$, $\hat{N}^*$ of the capture–recapture data with removals (From Overton (1965; Table 1)).

Ocassion i	n_i	m_i	r_i	M_{i-1}	R_{i-1}	$M_{i-1}n_i$	$R_{i-1}m_i$	w^*_{i-1}
1	20	0	10					
2	22	1	11	10	10	220	10	0.73
3	18	0	10	20	21	360	0	0.79
4	21	1	11	28	31	558	31	0.84
5	16	2	7	37	42	592	84	0.96
6	18	1	11	44	49	792	49	1.04
7	17	3	9	50	60	850	180	1.11
8	14	2	8	55	69	770	138	1.17
9	17	2	8	59	77	1003	154	1.27
10	16	4	7	66	85	1056	340	1.35
Total	179	16	92			6201	986	

animals appear in our samples we infer that more of the population have been marked. Apparently this reasoning depends critically on our assuming that fish are equally catchable at each sampling occasion; otherwise a preponderance of marked fish in our sample could be due to our repeatedly capturing the easily caught fish rather than the bulk of the fish population having been marked. According to Carothers (1973)

Equal catchability is an unattainable ideal in natural populations ...

In this section we therefore relax the assumption of homogeneity among fish and allow the capture probabilities P_j to differ from fish to fish but not from occasion to occasion. We will take $\pi_i = \pi \, \forall \, i$ and think of the P_j as realizations of some underlying random variable. Clearly, if too many of the P_j are very small it will be difficult to estimate N. We assume for this section that

$$P_j \overset{d}{=} \text{Beta}(\alpha, \beta),$$

since this provides a wide range of distributions and, for each mean value $\alpha/(\alpha + \beta)$, any variance is possible. The homogeneous case is approached in the limit as $\alpha, \beta \to \infty$ with convergent ratio.

With a heterogeneous population certain fish tend to be caught more often than certain others. This implies that fish who have been caught a certain number of times in a certain number of occasions will have capture probabilities P_j with a posterior distribution differing from the prior Beta(α, β) distribution. For example, the fish caught 5 times after 5 sampling occasions are more likely to be those individuals with a relative high capture probability. For a general density function f for P_j, the distribution conditional on being caught $k - 1$ times up to occasion $i - 1$ is just

$$\frac{p^{k-1}(1 - p)^{i-k}f(p)}{\int_0^1 p^{k-1}(1 - p)^{i-k}f(p)\,\mathrm{d}p},$$

which for the Beta case gives the distribution Beta($\alpha + k - 1, \beta + i - k$). It therefore follows that $m_i(k)$ has a Beta–Binomial distribution with parameters $(C_{i-k}(k - 1), \alpha + k - 1, \beta + i - k)$ conditional on $\mathscr{F}_{i-1}$.

5.3.1 SUFFICIENCY AND LIKELIHOOD

The likelihood for the heterogeneous model is not entirely straightforward as the important quantities P_j are unobservable but estimable. According to Overton and Burnham (1978) the likelihood is

$$L(N, f) = \frac{N!}{(N - M_t)!} \prod_{i=0}^{t} p_i^{C_t(i)}, \tag{5.14}$$

where

$$p_i = \int_0^1 \binom{t}{i} p^i (1 - p)^{t-i} f(p)\,\mathrm{d}p \qquad \text{and} \qquad C_t(O) = N - M_t.$$

Apparently then only the frequency of fish captured once, twice, ... at the conclusion of the experiment need be considered. In statistical parlance we say that $C_t(1), \ldots, C_t(t)$ are sufficient for N, α, β.*

The authors disagree that the above likelihood is the appropriate one for this model. We would claim that the full likelihood involves the entire capture history of each fish. To see this let X_i be the indicator of the event that a particular fish is caught on occasion i, with chance of capture P a random variable. We will try to calculate the likelihood for capture history of this fish. Firstly

$$\Pr(X_{i+1} = 1 | \mathscr{F}_i) = E(P | \mathscr{F}_i)$$

and this conditional expectation is given by integrating the distribution given at the end of the previous section. We note that the result depends on f and also on $k(X_i, \ldots, X_0)$, being the number of times the fish has been caught so far. The likelihood corresponding to the entire history of the fish thus depends on the number of times the fish has been caught on each occasion or, in simpler words, on the entire capture history of the fish. This is the sufficient statistic for N, α, β and a subset of this is the pair of matrices $m_i(k), C_i(k)$. The conditional distribution of $m_i(k)$ is Beta-Binomial and

---

* Burnham (1972) investigated this likelihood extensively for the Beta mixing density and found the ML approach to be unsatisfactory, if both α and β are unknown.

Table 5.2 Comparison of observed and model relative frequencies for simulated data

	Capture occasion k for fish								
	2	3	4	5	6	7	8	9	10
2	0.32 (104) 0.33								
3	0.32 (138) 0.29	0.34 (91) 0.42							
4	0.23 (142) 0.25	0.28 (65) 0.37	0.27 (11) 0.50						
5	0.22 (139) 0.22	0.42 (79) 0.33	0.50 (26) 0.44	1.00 (3) 0.56					
6	0.18 (133) 0.20	0.36 (76) 0.30	0.28 (46) 0.40	0.62 (13) 0.50	0.33 (3) 0.60				
7	0.20 (132) 0.18	0.32 (73) 0.27	0.37 (60) 0.36	0.56 (18) 0.45	0.40 (10) 0.54	1.00 (1) 0.64			
8	0.17 (122) 0.17	0.22 (68) 0.25	0.24 (68) 0.33	0.37 (30) 0.42	0.56 (16) 0.50	0.75 (4) 0.58	0.00 (1) 0.67		
9	0.16 (128) 0.15	0.20 (74) 0.23	0.24 (62) 0.31	0.39 (41) 0.38	0.44 (18) 0.46	0.30 (10) 0.54	0.50 (4) 0.62	0.00 (0) 0.69	
10	0.21 (123) 0.14	0.11 (80) 0.21	0.19 (62) 0.29	0.38 (40) 0.36	0.46 (26) 0.43	0.67 (15) 0.50	0.80 (5) 0.57	1.00 (2) 0.64	0.00 (0) 0.71

Capture occasion i

* Upper value observed, lower value expected.
† Figure in brackets is $C_{k-1}(i-1)$.

depends explicity on the parameters α, β. Indeed,there is ample empirical evidence that the partial experimental data displays a clear dependence on these parameters (see Table 5.2) demonstrating beyond doubt that the counts at the conclusion of the experiments are not sufficient. The only existing procedure for the heterogeneous model is the 'jackknife' procedure of Burnham and Overton (1978) which assumes nothing about the mixing distribution f. Starting with M_t, the MLE, as an initial estimate of N and assuming

$$E(M_t) = N + \sum_{i=1} a_i t^{-i},$$

they jackknife the bias parameters a_i and also the variance of the bias-reduced estimator. Our sufficiency considerations would suggest that this procedure ignores information about f available throughout the experiment. Certainly, simulation results show it to perform badly for highly heterogeneous populations (Burnham and Overton, 1978).

5.3.2 MARTINGALE APPROACH

We develop in this section a systematic approach to estimating N, α and β for the heterogeneous model with Beta mixing distribution. In contrast to the jackknife approach the extension to time heterogeneous models is clear and the logic of the procedure quite transparent. For certain weight functions, estimates actually have a closed form. Let X_{ij} be the indicator of the event fish j is sampled on occasion i. The expectation of X_{ij} is P_j. Thus $X_{ij} - P_j$ is a martingale difference, albeit unobservable. However, we may also use the conditional expectation to make a martingale difference. Since conditional on $\mathscr{F}_{i-1}$, P_j has Beta($\alpha + k - 1$, $\beta + i - k$) distribution, where $k - 1$ is the number of times fish j caught up to occasion $i - 1$, we also have that

$$\left\{ X_{ij} - \frac{\alpha + k - 1}{\alpha + \beta + i - 1} \right\} \tag{5.15}$$

is a martingale difference. Summing over all the $C_{i-1}(k-1)$ fish who have been caught $k - 1$ times shows that

$$\left\{ m_i(k) - C_{i-1}(k-1) \frac{\alpha + k - 1}{\alpha + \beta + i - 1} \right\} \tag{5.16}$$

is a martingale difference for $k = 1, \ldots, i$ and the t martingales

$$\mathscr{M}_t^{(k)} = \sum_{i=k}^{t} w_{i-1} \left\{ (\alpha + \beta + i - 1) m_i(k) - (\alpha + k - 1) C_{i-1}(k-1) \right\} \tag{5.17}$$

are independent observable martingales except for the parameters N, α, β. The fact that these martingales are independent follows from the fact that on each occasion the population has been partitioned into individuals who have been caught a different number of times.

5.3.3   ESTIMATES OF α, β

If the model is correct, then we ought to observe the proportion of fish caught $k - 1$ times up to occasion $i - 1$ being caught on the next occasion follow the law $(\alpha + k - 1)/(\alpha + \beta + i - 1)$ with random binomial variations. Now the martingales $\mathcal{M}_t^{(2)}, \ldots, \mathcal{M}_t^{(t)}$ depend on α, β only. We have $t - 1$ independent martingales and only 2 parameters to estimate. In general, we may obtain two weighted martingales for the two unknowns α, β by weighting the summed MG.

$$\sum_{k=2}^{t} \sum_{i=k}^{t} \{(\alpha + \beta + i - 1)m_i(k) - (\alpha + k - 1)C_i^{-k}(k - 1)\}, \qquad (5.18)$$

with two different sets of weights, say $w_{i-1,k1}$ and $w_{i-1,k2}$. When these weights are not functions of α or β we may, by equating both martingales to zero, obtain explicit estimates of α and β. If we let

$$S_1 = \sum_{i,k} w_{i-1,k1}\{m_i(k) - C_{i-1}(k - 1)\},$$

$$S_2 = \sum_{i,k} w_{i-1,k1} m_i(k),$$

$$S_3 = \sum_{i,k} w_{i-1,k1}(i - 1)m_i(k),$$

$$S_4 = \sum_{i,k} w_{i-1,k1}(k - 1)C_{i-k}(k - 1)$$

and Q_1, Q_2, Q_3, Q_4 be the same quantities for the second weight sequence $w_{i-1,k2}$, then explicit expressions for $\hat{\alpha}$, $\hat{\beta}$ are

$$\hat{\alpha} = \frac{Q_2(S_4 - S_3) - S_2(Q_4 - Q_3)}{Q_2 S_1 - S_2 Q_1}$$

and

$$\hat{\beta} = \frac{Q_1(S_4 - S_3) - S_1(Q_4 - Q_3)}{Q_1 S_2 - S_1 Q_2}. \qquad (5.19)$$

If the weights depend on α, β then a recursive procedure must be employed.

What weights should we use? If we consider linear combinations of weighted version of arbitrary martingales $\mathcal{M}_t^{(2)}, \ldots, \mathcal{M}_t^{(t)}$ depending on a vector parameter θ (i.e. more martingales than parameters), then it can be simply shown that an optimal set of equations for θ is obtained by simply using the optimal weights of (5.3) with respect to θ and then simply adding the resulting martingales, see Heyde (1987). In the present case, $\theta = (\alpha, \beta)$ and optimal weights are

$$w_{i-1,k1} = E\left(\frac{\partial D_{i,k}}{\partial \alpha}\bigg| \mathscr{F}_{i-1}\right)\bigg/ \mathrm{Var}(D_{i,k}|\mathscr{F}_{i-1})$$

and

$$w_{i-1,k2} = E\left(\frac{\partial D_{i,k}}{\partial \beta}\bigg| \mathscr{F}_{i-1}\right)\bigg/ \mathrm{Var}(D_{i,k}|\mathscr{F}_{i-1}),$$

where $D_{i,k}$ are the zero mean variables summed in (5.18). After a little calculation from the binomial distributions involved this yields

$$w^*_{i-1,k1} = \frac{1}{\alpha + k - 1} \quad \text{and} \quad w^*_{i-1,k2} = \frac{1}{\beta + i - k}.$$

An initial pair of solutions $\hat{\alpha}$, $\hat{\beta}$ for these weights is provided by (5.19) for a set of known weights. We found it convenient to use $w_{i-1,k1} = 1$ and $w_{i-1,k2} = 1/C_{i-1}(k-1)$.

We simulated the heterogeneous model for various values of α, β, N, t, the results of which we report a little later. Table 5.2 shows one typical outcome for $\alpha = 1$, $\beta = 4$, $N = 500$ and $t = 10$ showing the matrix of proportions $m_i(k)/C_{i-1}(k-1)$. These should follow the rule $(\alpha + k - 1)/(\alpha + \beta + i - 1)$ and indeed increases to the right along each line and decreases down each column are quite apparent. Our estimation procedure is nothing more than a very special binomial regression which fits appropriate values of α and β to the observed proportions. For this data set the estimates were 1.12, 4.73.

5.3.4 MARTINGALE ESTIMATING FUNCTIONS FOR N

In order to estimate the population size N we equate the first martingale $\mathscr{M}_t^{(1)}$ to zero to obtain the equation

$$\mathscr{M}_t^{(1)} = \sum_{i=1}^{t} w_{i-1}\{m_i(1)(\alpha + \beta + i - 1) - (N - M_{i-1})\alpha\} = 0, \quad (5.20)$$

for observable weights w_{i-1}. Since this equation is linear in N an explicit solution can be given if w_{i-1}, α, β are specified. In practice we will replace them by estimates. Applying (5.3) to the differences of this martingale leads to optimal weights

$$w^*_{i-1} = \frac{\alpha + \beta + i - 1}{(N - M_{i-1})(\beta + i - 1)}.$$

This attributes more weight to data late in the experiment as then M_{i-1} is closer to N. For constant weights $w_{i-1} = 1 \; \forall \; i$ we obtain an explicit solution of (5.20) namely

$$\hat{N} = M_t \frac{t\alpha + \alpha + \beta - 1}{t\alpha} + \frac{(1-\alpha)\sum_{i=1}^{t} m_i(1)i}{t\alpha}.$$

The optimal estimator $\hat{N}^*_{MG}$ requires only a few iterations to converge if $\hat{N}_1$ is used as the starting value. Estimates of α, β from Section 5.3.3 are substituted into (5.20) at the beginning of the process.

Below we present the results of a small simulation study. Each line summarizes the results of 100 simulations of the heterogeneous capture–recapture model for the specific values of N, t, α, β. The fish capture probabilities P_j were taken as the N evenly spaced quantiles of the Beta distribution. The average of $\hat{N}_{MG}$ together with the lower and upper quartiles are the main summary statistics of its performance. The estimator of the right-hand column was the homogeneous Schnabel estimator described in (5.10). Several comments are in order.

Firstly, estimation of N seems easier when N is large. For a given value of N, it seems that the performance is slightly better for more homogeneous populations than for more heterogeneous ones. (Compare lines 4 and 6 of Table 5.3.) However, estimation is still possible under *extreme* heterogeneity as evidenced in lines 3 and 9. The good performance for homogeneous populations is despite the fact that the estimation of the mixing parameters α, β is apparently much more difficult. As expected, the larger the average fish capture probability the more fish will be caught and the easier N is to estimate (compare lines 7 and 8).

The quartiles listed suppress the occurrence of a small number of extremely poor estimates of N. To appreciate how this happens one must

Table 5.3 Simulated performance of $\hat{N}_{MG}$

N	α	β	t	med($\hat{\alpha}$)	med($\hat{\beta}$)	$E(\hat{N}_{MG})$	$q_{0.25}$	$q_{0.75}$	$E(\hat{N}_{HOMO})$
1000	1	19	10	1.05	19.98	1075	889	1140	678
1000	1	4	5	1.03	4.18	1016	908	1101	916
1000	0.5	4.5	10	0.53	4.60	1043	859	1130	494
500	20	80	10	12.54	50.50	500	490	510	569
500	20	80	5	2.71	14.16	509	464	552	490
500	1	4	10	1.08	4.18	496	408	672	414
300	2	3	5	2.26	3,26	304	277	325	369
300	2	18	5	0.64	9.80	323	173	360	221
250	1	9	10	1.17	10.59	260	200	286	188
200	0.5	4.5	10	0.56	5.08	223	142	270	99
100	2	3	5	2.19	3.32	103	94	108	86

realize that it is possible for the estimates of α, β to be negative or very close to zero, especially for homogeneous populations. This is explained in Section 5.4.1. A glance at the formula for $\hat{N}_1$ in this section reveals that this may have a drastic effect on $\hat{N}_{MG}^*$. A plot against $\hat{\alpha}$ for the simulated estimates shows that where the mixing parameters are accurately estimated then so is N. Unfortunately, we have no way of knowing whether or not we have a good estimate of α from a single realization unless it turns out to be negative. One would recommend against the use of $\hat{N}_{MG}^*$ if $\hat{\alpha}$ turns out to be very close to zero or even negative. In fact it is possible that $\hat{N}_{MG}$ turns out to be negative when $\hat{\alpha}$ is small and negative. Such absurdities can be removed by taking the estimate to be the greater of $\hat{N}_{MG}$ and M_r. This only occurred twice in our simulations, both times in line 8 where the number of capture occasions is small and the chance of capture also small. These are the conditions under which we would expect difficulty in estimating α. A possible alternative procedure for such cases is developed in the next section.

Finally we show a comparison with the homogeneous estimator of Schnabel (1938). Figure 5.1 shows histograms of the two estimators based on 100 simulations of the capture–recapture process with $N = 500$, $t = 5$, $\alpha = 20$ and $\beta = 80$. This population is quite homogeneous and so we would

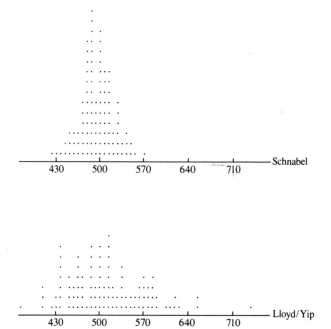

Fig. 5.1 Comparison of homogeneous and heterogeneous estimators for an almost homogeneous population ($N = 500$, $t = 5$).

expect the Schnabel estimator to perform better. Both estimators are seen to have small bias (around 10) but the variance of the heterogeneous estimator is significantly higher. This is only to be expected as a cost for its extra flexibility.

5.3.5 AN ESTIMATION EQUATION FREE OF α, β

The estimation equations described in the previous section may break down when α and β are very difficult to estimate. An alternative equation can be developed which does not explicitly depend on the mixing parameters and would therefore be expected to perform better under such conditions. Consider the martingale

$$\mathcal{M} = \sum_{i=1}^{t} \left\{ m_i(1) - \frac{C_{i-1}(0)\alpha}{\alpha + \beta + i - 1} \right\}.$$

In Section 5.3.4 we essentially reweighted this MG and substituted $\hat{\alpha}, \hat{\beta}$ for α, β as an estimate of the second term. This second term may actually be estimated without estimating α, β explicitly. Consider the quantity

$$\hat{e}_i^\kappa = \kappa \frac{C_{i-1}(0)}{C_{i-1}(\kappa - 1)} m_i(\kappa) - (\kappa - 1) \frac{C_{i-1}(0)}{C_{i-1}(\kappa)} m_i(\kappa + 1),$$

for $\kappa \in \{2, \ldots, i\}$. It is trivial to show that the $\mathcal{F}_{i-1}$-conditional expectation of this quantity is the second term of $\mathcal{M}$. Thus we may form a martingale free of α and β, namely

$$\mathcal{M} = \sum_{i=1}^{t} w_{i-1}\{m_i(1) - \hat{e}_i^\kappa\}. \tag{5.21}$$

The main advantage of this estimating function is that the necessity of directly estimating α, β is removed. It was suggested by Yip (1988) with $\kappa = 2$, though there is no particular reason why this choice should be made. Indeed, to make use of as much information as possible we could average $\hat{e}_i^2, \ldots, \hat{e}_i^i$. We have not investigated this estimator in the present paper.

The estimator $\hat{N}_{\text{YIP}}$ derived from (5.21) with unit weights gives an explicit solution for $\hat{N}$. Optimal weights may be chosen via (5.3), though these will depend on α, β which must then be estimated separately, say by the MG methods of Section 5.3.4. Error in their estimation, however, would have a much less profound effect than it does on $\hat{N}_{\text{MG}}$.

Table 5.4 presents the results of a very small simulation study of the relative performance of $\hat{N}_{\text{YIP}}$ and $\hat{N}_{\text{MG}}$. Apparently $\hat{N}_{\text{YIP}}$ substantially succeeds in divorcing the estimator of N from errors of estimation in α, β. With more capture occasions and therefore a better chance of estimating the heterogeneity parameters, the performance of $\hat{N}_{\text{MG}}$ is superior.

Table 5.4 $\hat{N}_{MG}$ and $\hat{N}_{YIP}$ compared

N	α	β	t	$E(\hat{N}_{MG})$	$E(\hat{N}_{YIP})$	RMSE($\hat{N}_{MG}$)	RMSE($\hat{N}_{YIP}$)
50	2	3	5	56.5	51.7	11.5	6.1
50	1	4	5	21.3	51.0	73.1	23.1
100	2	3	5	102.7	101.1	11.5	8.0
300	2	18	5	323	309	222.8	137.7
500	1	9	10	506	458	76.7	137.4
1000	0.5	4.5	10	1043	728	200.0	282.1

Note that for $N = 50$, $\alpha = 1$, $\beta = 4$, $t = 5$ only around 30 individuals are sampled in total. It is extremely difficult to estimate α, β from such scanty information. Larger numbers of sampled fish facilitate the estimation of α, β and we might expect $\hat{N}_{MG}$ to perform better. This is apparent in the last line. It should be noted however that improvements to $\hat{N}_{YIP}$ are possible firstly by optimal weighting and secondly by using the average of $\hat{e}_i^2, \ldots, \hat{e}_i^i$ as an estimate of $E(m_i(1)|\mathcal{F}_{i-1})$. Further work is required to determine whether the approach offers any advantages in general.

5.4 Advantages of the MG approach

In Sections 5.2 and 5.3 we have detailed how inference may be made from capture–recapture experiments by building up sets of martingales involving the unknown parameters, weighting and combining these in an optimal way, and equating to zero. In this section we discuss the advantages of this approach. It should first be noted, however, that for many common statistical models, for instance ARMA time-series models and GLM models for independent data, the likelihood-based methods are themselves martingale methods. For capture–recapture experiments the likelihood methods proposed by Darroch (1958) and Burnham and Overton (1978) are also based on complicated martingales. It should be mentioned that martingale estimating functions may themselves be regarded as score functions from 'quasi-likelihoods' which only incorporate the essential mean variance relations specified in the model. In this sense we may regard MG methods as simplified procedures which conform to the likelihood methods in spirit.

5.4.1 SIMPLICITY OF INTERPRETATION

The beauty of martingale methods is that they are simply based on equating a statistic to its (conditional) expectation. The parameters appearing in the model for the (conditional) expectation are chosen to produce as good agreement as possible across the entire data set.

An immediate advantage is that it becomes possible to give heuristic explanations of the performance of the procedure. For instance, in Section 5.3.4 we considered MG differences

$$m_i(1) - \frac{C_{i-1}(0)\alpha}{\alpha + \beta + i - 1},$$

and aimed to use these to estimate N with α, β substituted with estimates. Immediately we see that estimation errors in these mixing paramerers, particularly if they take small values, will severely effect the germane property of the martingale sequence, namely zero conditional expectation. Therefore, we concluded that errors of estimation in α and β may have a quite disastrous effect on estimation of N. This in turn led to alternative estimators which are resistant to these sort of errors. In Section 5.3.3 we considered martingales of the form

$$m_i(k) - \frac{C_{i-1}(k-1)(\alpha + k - 1)}{\alpha + \beta + i - 1}$$

and could immediately think of the estimation procedure as a special type of regression. We can then quite quickly see that if the population is quite homogeneous so that α, β are large, then the quantity $(\alpha + k - 1)/(\alpha + \beta + i - 1)$ will change hardly at all with i and k, and therefore trying to fit a model to such data will be essentially like trying to fit the slope parameter of a line to a cloud of points. We may thus expect estimates of α and β to be extremely erratic for relatively homogeneous populations. We indeed see this for large α, β in the simulation results (Table 5.3, lines 4, 5) although the estimates of N are mostly unaffected. We may further argue that in order to fit a successful regression of this type will require several capture occasions and a good spread of fish across each column of the matrix $m_i(k)$. This is analogous to requiring a good spread of 'x-values' in a simple linear regression.

Finally, the martingale differences are all independent and so the fitted differences may be regarded as 'residuals' and the sum of squares of these is suggested as a natural and simple goodness-of-fit statistic. This points the way to diagnostic plots and a more explicit regression approach to the analysis of capture–recapture experiments in general. A goodness-of-fit test for the homogeneous model is quite simply developed as follows. If all fish have the same probability of capture P then $m_i(k)$ will be binomially distributed with parameters $C_{i-1}(k-1)$ and P. Departures from homogeneity will be detected by observing systematic departures of the ratio $m_i(k)/C_{i-k}(k-1)$ from some constant P. An obvious estimator of this constant is

$$\hat{P} = \frac{\sum_{k=1}^{t} \sum_{i=k}^{t} m_i(k)}{\sum_{k=1}^{t} \sum_{i=k}^{t} C_{i-1}(k-1)} \tag{5.22}$$

being itself derived from a natural MG estimating equation for p. An approximately chi-squared distributed measure of departure is

$$\frac{\{m_i(k) - C_{i-k}(k-1)\hat{P}\}^2}{\hat{P}(1-\hat{P})C_{i-1}(k-1)}, \tag{5.23}$$

provided conditions for normality approximation of the Binomial hold. This will roughly hold when $C_{i-1}(k-1)\hat{P} \geq 10$. Thus we may obtain an approximate chi-squared test statistic of the homogeneity model (both with respect to time and fish) by summing these quantities over the lower-triangular matrix $m_i(k)$. It is clear how one could specifically test for time heterogeneity, assuming either homogeneity or heterogeneity of the fish. Indeed, once the likely patterns in the mean values of $m_i(k)$ have been identified under two competing models a test statistic follows almost routinely.

5.4.2 EXTENSIONS TO MORE COMPLEX MODELS

The martingale differences used in Section 5.3 were essentially

$$X_{ij} - E(X_{ij}|\mathcal{F}_{i-1}),$$

for each fish j and each occasion i. Under the heterogeneous model we had a simple expression for $E(X_{ij}|\mathcal{F}_{i-1})$ in terms of α, β and the fish's capture history, namely the number of times it had been caught so far (see equation (5.15)). This immediately opens the way for more general models so long as we can given an expression for $E(X_{ij}|\mathcal{F}_{i-1})$ in terms of parameters. For example, we might allow time dependence and model the probability of fish j being caught on occasion i by

$$P_{ij} = \pi_i P_j,$$

where P_j has some parametric density $f(\cdot; \theta)$ implying a model for the conditional expectation of X_{ij}, namely

$$E(X_{ij}|\mathcal{F}_{i-1}) = \Pi(z_i) \frac{\int_0^1 p^k(1-p)^{i-k} f(p; \theta) \, dp}{\int_0^1 p^{k-1}(1-p)^{i-k} f(p; \theta) \, dp}, \tag{5.24}$$

where $k - 1 = $ number of times fish j is caught and z_i is a factor or covariate which tells us something about the overall capture probabilities on occasion i, say the state of the weather or the duration of the capture effort. We could further model behaviour of the fish into the model. If fish become more wary of capture the more often they are caught we could multiply by a parametric function decreasing in k, say $1/(k + \gamma)$, for unknown 'learning' parameter γ. We could relax the beta mixing distribution to an arbitrary one simply by replacing $E(P_j|\mathscr{F}_{i-1})$ by an unknown constant p_j and estimate these quantities by observed fish-specific frequencies. In summary there seems to be no limit to the type of models we might consider within the single unifying framework, though there is no guarantee that the estimators will always perform well in practice unless enough data is available.

5.4.3 BIAS REDUCTION

The estimators produced by MG-estimating functions typically have bias of order n^{-1}, where n is the number of terms in the MG. In our application we have $n = \frac{1}{2}t^2$ for estimating α, β and $n = t$ for estimating N. The same order of bias is true of likelihood-based estimators. The standard derivation of the asymptotic distribution of the solution, $\hat{\theta}$, of the MG estimation equation

$$\mathscr{M}_n(\theta) = \sum_{i=1}^{n} D_i(\theta) = 0$$

is based on a first-order Taylor expansion. Taking a second-order expansion we may obtain an expression for the bias as

$$E(\hat{\theta} - \theta) = -\frac{\bar{\sigma}^2 \bar{\mu}''}{2n\bar{\mu}'^3}, \tag{5.25}$$

where

$$\bar{\sigma}^2 = \frac{1}{n} \sum_{i=1}^{n} \text{Var}(D_i(\theta)|\mathscr{F}_{i-1}),$$

$$\bar{\mu}' = \frac{1}{n} \sum_{i=1}^{n} E(D_i'(\theta)|\mathscr{F}_{i-1})$$

and

$$\bar{\mu}'' = \frac{1}{n} \sum_{i=1}^{n} E(D_i''(\theta)|\mathscr{F}_{i-1}).$$

This immediately implies that if the MG differences $D_i(\theta)$ are linear in θ, then the bias is zero to $O(n^{-1})$. This was the case for the MG-estimating functions considered, which may explain the good performance in mean when $t = 10$. For non-linear martingale differences (5.25) provides a correction to $\hat{\theta}$ provided enough is known about the conditional distribution of $D_i(\theta)$ to compute the quantities involved.

5.4.4 FURTHER WORK

There are two aspects of the MG approach which lack a proper theoretical foundation. The first issue is the effect of estimating nuisance parameters such as α, β in the MG difference

$$D_{ij} = X_{ij} - \frac{\alpha + k - 1}{\alpha + \beta + i - 1}$$

on the properties of the estimator. In particular, if $\hat{\alpha}, \hat{\beta}$ are not $\mathscr{F}_{i-1}$-measurable, as was the case with our procedure, then D_{ij} is not an MG difference. Resolution of this issue requires the consideration of the system of MG estimating functions together rather than separately.

The second issue is a related and more difficult one and concerns estimation efficiency. Under what circumstances is it preferable to use an MG-estimating function free of nuisance parameters and under what circumstances is it preferable to try and estimate all parameters simultaneously? We touched on this issue in Section 5.3.5, where $\hat{N}_{\text{YIP}}$ and $\hat{N}_{\text{MG}}$ were compared. One practical consideration is that we may not be able to optimally weight martingales free of nuisance parameters as nuisance parameters appear in the optimal weights. Indeed, this motivated the first author to study the more general approach detailed in this paper.

The equations for α, β derived in Section 5.3.3 do not involve $\mathscr{M}_t^{(1)}$. The stability of the estimates could presumably be improved by including this term but would require an extra level of recursion in the calculation of the estimates. Nevertheless, it seems an approach worth pursuing and is more in the spirit of estimating-function theory.

Finally, the procedures have not been applied to populations whose capture distribution does not look like a Beta–Binomial. In as much as fitting a Beta–Binomial distribution at least gives the correct mean and variance, the authors would expect the procedures to be quite robust to departures from this model assumption. It is likely that the accuracy of the results would be affected mostly by the proportion of very small capture probabilities, regardless of the overall shape of the distribution, though we have seen that even when this proportion is large estimation of N may still sometimes be achieved with reasonable accuracy. It is hoped that the generality of the theory of Section 5.4 makes it clear how a smooth distribution f could be fitted to the data and, in principle at least, be incorporated into the estimating equation for N as were the Beta–Binomial parameters. This is a promising direction for further development.

5.5 Acknowledgement

The authors are grateful to a referee whose meticulous reading of the manuscript and insightful suggestions greatly added to the clarity of the paper.

References

Becker, N. G. (1984). Estimating population size from capture–recapture experiments in continuous time. *Austral. J. Stat.*, **26**, 1–7.

Burnham, K. P. (1972). Estimation of population size in multiple capture–recapture studies when capture probabilities vary among animals. Ph.D. Dissertation, Ore. St. Univ. Corvallis.

Burnham, K. P. and Overton, W. S. (1978). Estimation of the size of a closed population when capture probabilities vary among animals. *Biometrika*, **65**, 625–33.

Carothers, A. D. (1973). Capture–recapture methods applied to a population with known parameters. *J. Anim. Ecol.*, **42**, 125–46.

Chapman, D. G. (1952). Inverse multiple and sequential sample census. *Biometrics*, **8**, 286–306.

Darroch, J. N. (1958). The multiple-recapture census: I. Estimation of a closed population. *Biometrika*, **45**, 343–59.

Godambe, V. P. (1976). Conditional likelihood and unconditional optimum estimating equations. *Biometrika*, **63**, 277–84.

Godambe, V. P. (1985). The foundations of finite sample estimation in a stochastic process. *Biometrika*, **72**, 419–28.

Heyde, C. C. (1987). On combining quasi-likelihood estimating functions. *Stoch. Proc. App.*, **25**, 281–7.

Lloyd, C. J. (1987). Optimal martingale estimating equations in a stochastic process. *Stat. Prob. Letters*, **5**, 381–7.

Otis, D. L., Burnham, K. P., White, G. C., and Anderson, D. R. (1978). Statistical inference from capture data on closed animal populations. *Wildlife Monograph*, **62**, 1–135.

Overton, W. S. (1965). A modification of the Schnabel estimator to account for removal of animals from the population. *Wildlife Monographs*, **29**, 392–5.

Schnabel, Z. E. (1938). The estimation of the total fish population of a lake. *Amer. Math. Monthly*, **39**, 348–52.

Seber, G. A. F. (1982). *Estimation of animal abundance and related parameters* (2nd edn.), Macmillan, New York.

Yip, P. (1988). Statistical inference via martingale theory in biomedical applications. Ph.D. Thesis, La Trobe University, Melbourne.

Yip, P. (1989). An inference procedure for a capture recapture experiment with time dependent capture probabilities. *Biometrics*, **45**, 471–9.

Yip, P. (1990a). A martingale estimating equation for a capture recapture experiment in discrete time. To appear in *Biometrics*.

Yip, P. (1990b). Estimation of population size from a capture–recapture experiment with known removals. To appear in *Theor. Pop. Biol.*

6
The role of unbiasedness in estimating equations

Takemi Yanagimoto and Eiji Yamamoto

ABSTRACT

It is appealing to impose unbiasedness on an estimating equation. However, it seems necessary to investigate further the adequacy of the requirement. We show that existing moment estimators can be improved by enforcing the corresponding estimating equation to be unbiased. We also note that the conditional maximum likelihood equations can be regarded as the profile likelihood estimating equations, corrected for their bias.

6.1 Introduction

Let $\mathbf{x} = (x_1, \ldots, x_n)'$ be a sample of size n from a population having the density (or probability) function $p(x; \theta, \mu)$. An estimator of θ, $\hat{\theta}$, is usually defined as the solution of an estimating equation,

$$g(\mathbf{x}; \theta) = 0. \tag{6.1}$$

The estimating equation is called unbiased, if the expectation $E(g(\mathbf{x}; \theta)) = 0$ for any θ and μ. For convenience we call also the estimating function $g(\mathbf{x}; \theta)$ unbiased.

The notion of unbiasedness of an estimator is familiar in the statistical theory. In fact a highly sophisticated theory of uniformly minimum variance unbiased estimator and best linear unbiased estimator has been constructed. The aim of this article is to emphasize that unbiasedness of an estimating equation is also natural and appealing as well. It is surprising that little attention has been paid to the notion in spite of its practical importance. In some cases the imposition of unbiasedness is regarded as a prerequisite (cf. Liang 1987). In other cases, however, unbiasedness is not regarded as an important criterion for yielding an estimator. As will be observed later, the estimating equations corresponding to conventional moment estimators are often not unbiased. The profile likelihood estimating equation for a dispersion parameter is generally not unbiased. Therefore, we need to discuss importance of the property of unbiasedness in relation to an estimating equation. There are definite advantages in imposing unbiasedness on an estimating equation.

In fact many conventional moment estimators can be easily improved. A more important aspect from a theoretical point of view is the simplicity of the resulting estimating-function theory. For example Godambe (1960) introduced a criterion for unbiased estimating equations,

$$M(g(\mathbf{x}; \theta)) = V(g(\mathbf{x};\theta))/\{E(dg(\mathbf{x}; \theta)/d\theta)\}^2, \qquad (6.2)$$

which represents the reciprocal of the sensitivity. He assumes unbiasedness of $g(\mathbf{x}; \theta) = 0$ as a prerequisite.

We start with two simple, illustrative examples, which will be examined more later.

Example 6.1 (Coefficient of variation) Let $\mathbf{x}$ be a sample vector from $p(x; \theta, \mu)$ on the support of the non-negative real values with mean μ and squared coefficient of variation θ. The conventional moment estimator of θ is $\hat{\theta}_M = s^2/\bar{x}^2$ with $s^2 = \sum (x_i - \bar{x})^2/(n - 1)$. A corresponding estimating function $g(\mathbf{x}; \theta)$ is $s^2 - \theta\bar{x}^2$. Then the expectation $E(g(\mathbf{x}; \theta))$ is $-\theta^2\mu^2/n$, which is not zero. An unbiased estimating function is $g(\mathbf{x}; \theta) = s^2 - \theta(\bar{x}^2 - s^2/n)$, which yields $\hat{\theta}_{UB} = s^2/(\bar{x}^2 - s^2/n)$. It seems that the introduction of the term $-s^2/n$ is reasonable. Note that $\bar{x}^2 - s^2/n$ takes a positive value, unless all the sample values are equivalently zero.

Example 6.2 (Ratio of two means) Let $\mathbf{x}$ be a sample vector of size n from $p(x; \lambda_1)$ with mean λ_1, and support, the positive real line. Similarly, let $\mathbf{y}$ be a positively supported sample vector of size m from $p(y; \lambda_2)$. A naïve moment estimator of $\theta = \lambda_1/\lambda_2$ is $\bar{x}/\bar{y}$, which is upward biased if the expectation exists. The corresponding estimating equation is

$$g(\mathbf{z}; \theta) = \bar{x} - \bar{y}\theta = 0 \qquad (6.3)$$

with $\mathbf{z} = (\mathbf{x}', \mathbf{y}')'$, which is unbiased.

If we specify an explicit form of $p(\mathbf{y}; \lambda_2)$, it may be possible to construct an unbiased estimator $c\bar{x}/\bar{y}$ for an appropriate $c < 1$. When $p(x; \lambda)$ is exponential, we may set $c = (m - 1)/m$. This looks, however, less satisfactory, if the quantities θ and $1/\theta$ denote discrepancy of two means in a dual way.

We will subsequently also discuss the role of unbiasedness of estimating functions in relation to conditional, marginal and more generally partial likelihood inference.

6.2 Basic properties

Suppose that an estimating function $g(\mathbf{x}; \theta)$ has a form,

$$g(\mathbf{x}; \theta) = g_1(\mathbf{x}) - \theta g_2(\mathbf{x}). \qquad (6.4)$$

This special form contains many important applications, and is convenient for studying the estimating function approach to inference (Durbin 1960). The notion of unbiasedness of an estimating equation is a generalization of that of an estimator. In fact both are equivalent when $g_2(\mathbf{x}) = 1$.

When a parameter θ, other than $h(\theta)$ or $h(\theta, \mu)$, is actually of our interest, it is reasonable to impose $\hat{\theta}$ to be unbiased. In practical situations, however, the assumption is not as realistic as the statistical theory often implies. Consider, for example, the problem of estimating the ratio of means in Example 6.2. The parameters $\theta = \mu_1/\mu_2$ and $1/\theta = \mu_2/\mu_2$ often represent a common measure in a practical sense.

Unbiasedness of an estimating equation does not imply that of the estimator induced from the estimating equation. They are, however, obviously related. In Example 6.1 a biased estimating equation was modified so as to be unbiased. Such a treatment is reasonably expected to reduce the bias of the induced estimator. To show this consider an asymptotic case when the sample size tends to infinity. Suppose $g_1(\mathbf{x})$ and $g_2(\mathbf{x})$ have expectations, $\gamma\theta$ and γ, respectively. Then the formal expansion is as follows:

$$\hat{\theta} = \frac{\gamma\theta + g_1(\mathbf{x}) - \gamma\theta}{\gamma + g_2(\mathbf{x}) - \gamma} \fallingdotseq \theta + (g_1(\mathbf{x}) - \theta g_2(\mathbf{x}))/\gamma, \qquad (6.5)$$

where $\hat{\theta}$ is computed using (6.4). This illustrates how unbiasedness of an estimating equation can eliminate the first term of the bias of the induced estimator.

The imposition of unbiasedness on an estimating equation is essential in the sparsely asymptotic case. Suppose that there exist K strata; the kth stratum has the density function $p(\mathbf{x}; \theta, \mu_k)$ with μ_k depending on the stratum. The sample $\mathbf{x}$ is written as $(\mathbf{x}_1', \ldots, \mathbf{x}_k')'$, where $\mathbf{x}_k$ is the n_k-dimensional sample vector. We call a sample sparse when K is large and each sample size n_k is small. It is known that the conventional estimator is often inconsistent in the sparsely asymptotic case. On the other hand the imposition of unbiasedness obviously implies consistency of an induced estimator even in the sparsely asymptotic case.

A typical example of inconsistency in the sparsely asymptotic case is found in the unconditional maximum likelihood estimator. The familiar distributions under the study are listed in Table 6.1. The parametrizations in Table 6.1 may be slightly different from the usual ones. The parameter θ is the dispersion parameter (Jorgensen 1987). In these distributions, the density function is factored into

$$p(\mathbf{x}; \theta, \mu) = pc(\mathbf{x}; \theta|t)pr(t; \theta, \mu) \qquad (6.6)$$

for a statistic t, where $pr(t; \theta, \mu)$ is the marginal density function of t. In actual examples t is usually the maximum likelihood estimator of μ, and is

Table 6.1   Familiar models satisfying the factorization property

Model	Density (probability) function	t
Normal	$\dfrac{1}{\sqrt{(2\pi\theta)}}\,e^{-(1/2\theta)(x-\mu)^2}$	$\bar{x}$
Inverse Gauss	$\sqrt{(1/2\pi\theta x^3)}\,e^{-(x-\mu)^2/2\theta\mu^2 x}$	$\bar{x}$
Gamma	$\dfrac{x^{1/\theta-1}}{\Gamma(1/\theta)(\theta\mu)^{1/\theta}}\,e^{-x/\theta\mu}$	$\bar{x}$
Two-parameter Exponential	$\dfrac{1}{\theta}\exp{-(x-\mu)/\theta}$	$x_{(1)}$*
Negative Binomial	$\dbinom{1/\theta+x-1}{x}\left(\dfrac{1}{1+\theta\mu}\right)\left(\dfrac{\theta\mu}{1+\theta\mu}\right)$	$\bar{x}$
2 × 2 table	$\dbinom{n}{x}p^x(1-p)^{n-x}\dbinom{m}{y}q^y(1-q)^y$	$x+y$
	with $\mu=np+mq$ $\theta=p(1-q)/q(1-p)$	

* The minimum order statistic.

sufficient for μ for a fixed θ. Write $l(\mathbf{x};\theta,\mu)=\log(p(\mathbf{x};\theta,\mu))$ and so forth. Then the profile likelihood estimating equation for θ is written as

$$l_\theta(\mathbf{x};\theta,\hat{\mu}(\theta))=lc_\theta(\mathbf{x};\theta|t)+lr_\theta(t;\theta,\hat{\mu}(\theta))$$

$$=\frac{pc_\theta(\mathbf{x};\theta|t)}{pc(\mathbf{x};\theta|t)}+\frac{pr_\theta(t;\theta,\hat{\mu}(\theta))}{pr(t;\theta,\hat{\mu}(\theta))}=0, \tag{6.7}$$

with $\hat{\mu}(\theta)$ being the solution of $pr_\mu(t;\theta,\mu)=0$. The expectation of $lc_\theta(\mathbf{x};\theta|t)$ is zero, but that of $lr_\theta(t;\theta,\hat{\mu}(\theta))$ is often not zero (Anderson 1970). Yanagimoto (1987) notes that the behaviour of these two terms with respect to θ are quite different in many familiar distributions. These imply inconsistency of the unconditional maximum likelihood estimator in the sparsely asymptotic case for all the distributions in Table 6.1. A simple way to avoid inconsistency is to disregard the second term, that is, to use only the first term, which is the conditional likelihood.

6.3   Method of moments

The method of moments for obtaining an estimator is intuitively appealing. It seems that a reason why the moment estimator has attracted less attention is its lack of theoretical developments. In fact the moment estimator is conventionally obtained by equating the first L sample moments and the corresponding population ones, with L being the number of unknown parameters. There are two sample moments systems: the raw one, $\sum x_i^l/n$, and the central one, $\bar{x}$, $\sum (x_i - \bar{x})^l/n$. There is no rule for the choice of the sample moments system, though the latter looks appealing. Earlier, when computational facilities were not easily available, ease of computation was a major consideration. Now, a more important criterion for selecting a method is its frequency performance.

The fundamental characteristic of the moment estimator is its simple form. This is expected to be associated with robustness compared with the maximum likelihood estimator, which depends heavily on the form of the density function. As is shown in Section 6.1, the estimating equation corresponding to the conventional moment estimator is not always unbiased. The improved moment estimator is only slightly more complicated. It is curious that the performance of moment estimators has been virtually unexplored. We add the examples.

Example 6.3 (Negative binomial distribution)   The conventional moment estimator is the solution of

$$g(\mathbf{x}; \theta) = s^2 - \bar{x} - \theta \bar{x}^2 = 0.$$

Since this estimating equation is not unbiased, we recommend an estimator induced from the unbiased estimating equation,

$$s^2 - \bar{x} - \theta(\bar{x}^2 - s^2/n) = 0. \tag{6.8}$$

Bliss and Owen (1958) discussed the moment estimator based on (6.8). The estimator, however, is not employed in recent standard monographs.

Example 6.4 (Beta-binomial distribution)   The Beta-Binomial distribution $BB(m; \mu, \theta)$ has the probability function

$$p(x; m, \mu, \theta) = \binom{m}{x} \prod_{r=0}^{x-1} \left(\mu + r \frac{\theta}{1-\theta} \right)$$

$$\times \prod_{r=0}^{m-x-1} \left(1 - \mu + r \frac{\theta}{1-\theta} \right) \bigg/ \prod_{r=0}^{m-1} \left(1 + r \frac{\theta}{1-\theta} \right).$$

The distribution is widely employed as an overdispersed one compared with the binomial. It has mean μ and variance $m\mu(1 - \mu)(1 + (m - 1)\theta)$. Suppose that $x_1, \ldots, x_n$ be a sample of size n from $BB(m; \mu, \theta)$. Yamamoto and Yanagimoto (1989) introduced an unbiased estimating equation for $BB(m; \mu, \theta)$

$$(mn - 1)s^2 - n\bar{x}(m - \bar{x}) - \theta(m - 1)\{n\bar{x}(m - \bar{x}) + s^2\} = 0.$$

A naïve way to obtain an estimator is given by solving

$$ms^2 - \bar{x}(m - \bar{x}) - \theta(m - 1)\bar{x}(m - \bar{x}) = 0.$$

In the above examples, the unbiased moment-estimating equations were introduced in a heuristic way for their intuitive appeal. Under the assumption of familiar distributions, these functions except the case of the Beta–Binomial distribution will be characterized as unbiasedized functions in terms of the conditional likelihood in the final section.

Modification of a moment estimator in terms of unbiasedness of an estimating equation provides an improvement as discussed in the previous section. To confirm this assertion, a case-by-case study for a finite sample size is necessary. Yamamoto and Yanagimoto (1989) studied the case of the Beta–Binomial distribution in depth.

The following example shows better performance of the modified moment estimator, $\hat{\theta}_{ub}$, in the gamma distribution, which was defined Example 6.1 (Section 6.1).

Example 6.5 (Gamma distribution)   The Gamma distribution $Ga(\mu, \theta)$ is one of the most familiar positive distributions. The estimator for μ is $\hat{\mu} = \bar{x}$, which is employed without any dispute. The choice of an estimator θ is the subject of the present study. We compare $\hat{\theta}_{ub}$ with $\hat{\theta}_m$, and also briefly with the unconditional and the conditional maximum likelihood estimators. Note that the parameter $1/\theta$ is also of interest.

First, it holds for any sample $\mathbf{x}$ that $0 \leq \hat{\theta}_m \leq n$. This comes from the fact that $\bar{x}^2 \geq s^2/n$. This restriction of the range of the estimator depending on the sample size looks undesirable.

Next we show that $E(1/\hat{\theta}_m) > E(1/\hat{\theta}_{ub}) > 1/\theta$ for any θ and μ. This means that the bias of $1/\hat{\theta}_m$ is greater than that of $1/\hat{\theta}_{ub}$. To prove this recall that the vector variable, $\mathbf{V} = (V_1, \ldots, V_n) = (x_1/n\bar{x}, \ldots, x_n/n\bar{x})$, has the Dirichlet distribution with the common parameter $1/\theta$. The first and second moments of V_i are $1/n$ and $(1 + \theta)/n(n + \theta)$. Then it follows that

$$E(\bar{x}^2/s^2) = \frac{n - 1}{n^2} E(1/(\textstyle\sum V_i^2 - 1/n))$$

$$\geq \frac{n - 1}{n^2} \frac{1}{E(\sum V_i^2 - 1/n)}$$

$$= 1/\theta + 1/n.$$

This means $1/\hat{\theta}_m$ has the bias greater than $1/n$. Note that $1/\hat{\theta}_{ub} = 1/\hat{\theta}_m - 1/n$; the correction term is rightly $1/n$. In other words $1/\hat{\theta}_m$ adds spuriously a constant $1/n$ to $1/\hat{\theta}_{ub}$ for any sample, which is still an upward biased estimator.

Similarly, we have that $E(\hat{\theta}_m) = n\theta/(n + \theta)$, which is downward biased. The evaluation of $E(\hat{\theta}_{ub})$ looks complicated, and we conduct a small simulation study

Table 6.2 Empirical biases by 10 000 replications of the four estimators of the shape parameter in $Ga(\mu, \theta)$ (the upper case) and its reciprocal (the lower) for the sample size 20

θ	Proposed moment est.	Conventional moment est.	Unconditional MLE	Conditional MLE
0.5	−0.0094	−0.0122*	−0.0277	−0.0064
	0.2996	0.3496	0.3240	0.2216
1	−0.0077	−0.0476*	+0.0440	−0.0053
	0.1658	0.2158	0.1399	0.0943
2	0.0744	−0.1818*	−0.0727	−0.0035
	0.0959	0.1459	0.0604	0.0404

* The number is theoretical.

for comparing biases of the four estimators. The results are summarized in Table 6.2. We observe that the conventional moment estimator has the largest bias in all but one case, and in contrast the conditional maximum likelihood estimator has the smallest. The proposed estimator and the unconditional maximum likelihood estimator are comparable. Recall that the proposed estimator does not depend heavily on the form of the gamma density function. Therefore, the estimator looks useful, when robustness of an estimator is desired.

6.4 Advantages of unbiasedness

In the two previous sections we show that the prerequisite of unbiasedness of an estimating equation is reasonable and actually useful. Unbiasedness is more important than we showed. It permits us a type of likelihood inference. An attractive approach to extending likelihood inference is the quasi-likelihood one (McCullagh and Nelder 1983).

Godambe (1960) introduced a criterion for unbiased estimating equations, $M(g(\mathbf{x}, \theta))$ in (6.2). He pointed out that the likelihood estimating equation attains the minimum under the regularity conditions when $p(x; \theta, \mu) = p(x; \theta)$. Godambe (1976) extended the result to the case of the conditional likelihood estimating equation. This measure gives some insight into the quasi-likelihood approach (Godambe and Hyde 1987).

To show usefulness of his criterion, we compare two estimating equations for the ratio of means in Example 6.2 (Section 6.1).

Example 6.2 (Continued)   Here we consider the estimation problem of the ratio of the means under the assumption that both the density functions are exponential. Write the estimating function in (6.3) as $g_1(z; \theta)$, and set $g_2(\mathbf{z}; \theta) = \{(m - 1)/m\}\bar{x}/\bar{y} - \theta$. It follows that $M(g_1(\mathbf{z}; \theta)) = (1/n + 1/m)\theta^2 < \{(n + m - 1)/n(m - 2)\}\theta^2 = M(g_2(\mathbf{z}; \theta))$. Obviously, the same inequality holds between the estimating functions, $\bar{y} - \bar{x}/\theta$ and $\{(n - 1)/n\}\bar{y}/\bar{x} - 1/\theta$. These present theoretical evidence supporting our assertion that $g_1(\mathbf{z}; \theta)$ is more appealing.

Consider the estimation problem of a parameter common through multiple strata. Let $\mathbf{x}_k, k = 1, \ldots, K$ be a sample from a population with the density function $p(x; \theta, \mu_k)$, when μ_k depends on the kth population. Suppose we have an estimating equation $g_k(\mathbf{x}; \theta) = 0$ to each sample. Then our problem becomes how to combine these K equations. This problem can be solved, when estimating equations are simple ones such as those derived from the moment estimators.

Yanagimoto (1989a) showed Godambe's criterion is useful in determining the optimum weights of the combined estimating equation

$$g(\mathbf{x}; \theta) = \sum w_k g_k(\mathbf{x}_k; \theta)$$

with $\mathbf{x} = (\mathbf{x}_1', \ldots, \mathbf{x}_K')'$. The weights are optimum, when $M(g(\mathbf{x}, \theta))$ is uniformly minimized for θ and μ with respect to the w_k. The minimum of $M(g(\mathbf{x}; \theta))$ is attained at $w_k = E(dg_k(\mathbf{x}_k; \theta)/d\theta)/V(g_k(\mathbf{x}_k; \theta))$, if it is free from the parameters, θ and μ_k. If all the estimating equations are the likelihood estimating equations, then the summation of them is the optimum combined likelihood estimating equation. A similar situation is discussed in Godambe (1985). Since the condition is too restrictive in actual examples, Yanagimoto introduced the locally optimum weights at θ_0 for a suitable point θ_0. The well-known Mantel–Haenszel estimator is characterized in terms of this measure.

Let x_k and y_k be samples from the binomial distributions $Bi(n_k; p_k)$ and $Bi(m_k; q_k)$. The problem is to estimate the common odds ratio $\theta = p_k(1 - q_k)/q_k(1 - p_k)$. Write $\mathbf{z}_k = (x_k, y_k)'$ and $\mathbf{z} = (\mathbf{z}_1', \ldots, \mathbf{z}_k')'$. The moment estimator of θ from the kth stratum is the solution of the unbiased estimating equation

$$g_k(\mathbf{z}_k; \theta) = x_k(m_k - y_k) - \theta y_k(n_k - x_k) = 0.$$

Then for $\theta = 1$,

$$E\left(\frac{d}{d\theta} g_k(\mathbf{z}_k; \theta)\right)\Big/ V(g_k(\mathbf{z}_k; \theta)) = \frac{n_k m_k p_k(1 - p_k)}{n_k m_k^2 p_k(1 - p_k) + m_k n_k^2 p_k(1 - p_k)}$$

$$= \frac{1}{n_k + m_k}.$$

Consequently, the locally optimum combined estimating equation at $\theta_0 = 1$ becomes

$$g(z; \theta) = \sum \frac{1}{n_k + m_k} \{x_k(m - y_k) - \theta y_k(n_k - x_k)\} = 0,$$

which implies the Mantel–Haenszel estimator.

Next we discuss the test problem of testing $\theta = \theta_0$. To construct it, we need an estimator of $V(g(\mathbf{x}, \theta))$, $\hat{V}(g(\mathbf{x}; \theta))$. Then a naïve test statistic is expressed as

$$T = g(\mathbf{x}; \theta)/\sqrt{\hat{V}(g(\mathbf{x}; \theta))}.$$

The test statistic is expected to have approximately the standard normal distribution or the t-statistic in the small sample case. The role of unbiasedness of an estimating equation is important. It makes the approximation accurate. The well-known Mantel–Haenszel test statistic is given as T^2 for an unbiased estimator $\hat{V}(g(\mathbf{x}; \theta))$. A similar procedure for testing is adopted by Mantel and Godambe (1989).

The above test statistic is useful, when an estimating equation is expressed as $g_1(\mathbf{x}) - \theta g_2(\mathbf{x}) = 0$ as in (6.4). We observed such examples in moment estimators. In these examples $g_1(\mathbf{x})$ and $g_2(\mathbf{x})$ have asymptotically normal distributions $N(\gamma(\mu, \theta), \sigma_1^2)$ and $N(\gamma(\mu, \theta), \sigma_2^2)$. Then the estimating function, $g_1(\mathbf{x}) - \theta g_2(\mathbf{x})$, has the asymptotic normal distribution $N(0, \sigma_3^2)$ for a suitable σ_3^2. An alternative, more familiar, method for constructing a test statistic is to consider the asymptotic distribution of $\hat{\theta} = g_1(\mathbf{x})/g_2(\mathbf{x})$. It is anticipated that the accuracy of the asymptotic normal distribution of $g(\mathbf{x}, \theta)$ is better than $\hat{\theta}$ in most cases.

Yanagimoto (1989b) gave an accurate confidence interval of the odds ratio in the negative binomial distribution, which is a generalization of the usual odds ratio.

6.5 Conditional inference

When the density function is factored as in (6.6), it is recommended that the estimation and the test for θ be based only on the factor $pc(\mathbf{x}; \theta|t)$. It is a challenging problem to pursue why and in what situations conditional inference is employed. As Cox and Reid (1987) reviewed, an original motivation of the use of conditional inference is rather philosophical. It was hoped to 'induce relevance of the probability calculations to the particular data under study'. A more practical motivation is, however, perhaps to obtain a naïve method for yielding an estimator or a test having better performance. Yanagimoto and Anraku (1989) and references therein claimed superiority of the conditional maximum likelihood estimator of θ over the unconditional one in all the models in Table 6.1.

Yanagimoto (1987) recommended the conditional MLE, since the residual profile likelihood behaves unfavourably. The unfavourable behaviour implies biasedness of an profile likelihood estimating equation of θ. This suggests that a key reason why the conditional maximum likelihood estimator is recommended is that with such a modification the likelihood estimating equation is 'unbiasedized' by disregarding the unfavourable residual likelihood. The notion of obstructiveness of the residual likelihood in Yanagimoto (1987) is a candidate for defining such an unfavourable residual likelihood.

When the factorization property (6.6) holds, we can define 'unbiasedization' of an estimating function, $g(\mathbf{x}; \theta)$, as

$$g_{ub}(\mathbf{x}; \theta) = g(\mathbf{x}; \theta) - E(g(\mathbf{x}; \theta)|t). \tag{6.9}$$

Note that the expectation of $g(\mathbf{x}; \theta)$ with respect to the conditional density function $pc(\mathbf{x}; \theta|t)$ is free from μ. It follows that the estimating equation, $g_{ub}(\mathbf{x}; \theta)$, is unbiased. This is a generalization of the use of the conditional likelihood equation. In fact set $g(\mathbf{x}; \theta) = l(\mathbf{x}; \theta)$. Then the equality (6.7) is rewritten as $lc(\mathbf{x}; \theta|t) = l(\mathbf{x}; \theta) - lr(t; \theta, \hat{\mu}(\theta))$, which corresponds with (6.9). An approximated method for unbiasedization in a more general manner is discussed in McCullagh and Tibshirani (1990).

We assume further that the statistic t is complete for μ for an arbitrary fixed θ. Write the second term in the right-hand side of (6.9) as $-g_e(t; \theta)$. The assumption of completeness implies $g_e(t; \theta) \equiv 0$, if $g(\mathbf{x}; \theta)$ is unbiased. Note that $\mathrm{Cov}(g_{ub}(\mathbf{x}; \theta), g_e(t; \theta)) = 0$, if it exists. Therefore the decomposition in (6.9) is orthogonal in this sense. Using these facts, we obtain the following proposition.

Proposition *Suppose that (6.6) holds, and that t is complete for μ for any θ. Then for any unbiased estimating function $h(\mathbf{x}; \theta)$*

$$\mathrm{Cor}^2(g(\mathbf{x}; \theta), h(\mathbf{x}; \theta)) \leqslant \mathrm{Cor}^2(g(\mathbf{x}; \theta), g_{ub}(\mathbf{x}; \theta)),$$

if both the correlation coefficients exist.

The above proposition provides some kind of optimality for the unbiased estimating function g_{ub} defined by (6.9).

Example 6.6 Consider the problem of estimation of variance in $N(\mu, \theta)$. The Normal distribution satisfies the assumptions in the proposition. Let an estimating function $g(\mathbf{x}; \theta) = \sum (x_i - \bar{x})^2/n - \theta$. Setting $t = \bar{x}$, we get $g_{ub}(\mathbf{x}; \theta) = \sum (x_i - \bar{x})^2/n - (n-1)\theta/n = \{(n-1)/n\}(s^2 - \theta)$.

Example 6.5 (Continued) We return to the estimation problem of θ in $Ga(\mu, \theta)$. The Gamma distribution also satisfies the two assumptions. A conventional moment estimator is $\hat{\theta} = s^2/\bar{x}^2$. Therefore we consider $g(\mathbf{x}; \theta) = s^2 - \theta\bar{x}^2$. Setting $t = \bar{x}$, we get

$$g_{ub}(\mathbf{x}; \theta) = s^2 - E(s^2|\bar{x}) = s^2 - \theta\bar{x}^2/(n + \theta)$$

$$= \left\{\frac{n}{(n + \theta)}\right\}\left\{s^2 - \theta\left(\frac{\bar{x}^2 - s^2}{n}\right)\right\}.$$

A similar result is valid for the shape parameter in the Negative Binomial distribution.

In familiar distributions such as the Normal, the Inverse Gaussian and the Gamma, conditional inference can be regarded also as inference based on the marginal likelihood. In fact the factorization form (6.6) is expressed more specifically as

$$p(\mathbf{x}; \theta, \mu) = pc(w; \theta|t)pr(t; \theta, \mu)$$

$$= pc(w; \theta)pr(t; \theta, \mu)$$

with a suitable statistic w independent of t. Inference based on the marginal likelihood has also attracted considerable attention (Kalbfleish and Sprott 1970). In come cases it looks more appealing than conditional inference. In fact the estimate s^2 of σ^2 in the normal distribution is recognized as the unbiased estimator based on the marginal distribution of s^2. In some examples the density function is factored into

$$p(\mathbf{x}; \theta, \mu) = pm(w; \theta)pc(t; \theta, \mu|w). \tag{6.10}$$

Such examples include the von Mises distribution, whose density function is written as

$$\prod_{i=1}^{n} \frac{1}{2\pi I_0(1/\theta)} \exp(\cos(x_i - \mu)/\theta)$$

$$= \frac{I_0(R/\theta)}{(2\pi)^{n-1} I_0(1/\theta)^n} \frac{1}{2\pi I_0(R/\theta)} \exp\{(R/\theta)\cos(\bar{x} - \mu)\},$$

where $I_t(x)$ is the modified Bessel function of the first kind, the support is the unit circle, and R and $\bar{x}$ are defined by

$$R \cos \bar{x} = \sum \cos x_i$$

$$R \sin \bar{x} = \sum \sin x_i.$$

Suppose the marginal density function in (6.10) satisfies usual regularity conditions. Then

$$g_m(w; \theta) = \frac{d}{d\theta} \log pm(w; \theta) = 0$$

is unbiased. Here we can find a common interpretation for conditional and marginal inference. They can be regarded as unbiasedization of estimating equations without detracting essential information from the full likelihood. Recall that the estimating equation derived from the partial likelihood in the well-known proportional hazard model (Cox 1972) is also unbiased. We believe unbiasedness of the resulting estimating equation is a key to directing future researches on useful, promising likelihood inference, such as conditional, marginal and partial likehood inference.

6.6   Acknowledgement

The authors thank the referee for his helpful comments and for calling their attention to the paper by McCullagh and Tibshirani. Thanks are also due to Drs M. Lesperance and H. Mantel for their detailed comments on the earlier version of the manuscript.

References

Anderson, E. B. (1970). Asymptotic properties of conditional maximum-likelihood estimators. *J. Roy. Stat. Soc.* B, **32**, 283–301.

Bliss, C. I. and Owen, A. R. G. (1958). Negative binomial distributions with a common k. *Biometrika*, **45**, 37–58.

Cox, D. R. (1972). Regression models and life tables (with discussion), *J. Roy. Stat. Soc.* B, **34**, 187–220.

Cox, D. R. and Reid, N. (1987). Parameter orthogonality and approximate conditional inference (with discussion), *J. Roy. Stat. Soc.* B, **49**, 1–39.

Durbin, J. (1960). Estimation of parameters in time-series regression models. *J. Roy. Stat. Soc.* B, **22**, 139–153.

Godambe, V. P. (1960). An optimum property of regular maximum likelihood estimation. *Ann. Math. Stat.*, **31**, 1208–11.

Godambe, V. P. (1976). Conditional likelihood and unconditional optimum estimation equations. *Biometrika*, **63**, 277–84.

Godambe, V. P. (1985). The foundations of finite sample estimation in stochastic process. *Biometrika*, **72**, 419–28.

Godambe, V. P. and Heyde, C. C. (1987). Quasi-likelihood and optimal estimation. *Int. Stat. Rev.*, **55**, 231–44.

Jorgensen, B. (1987). Exponential dispersion models (with discussion). *J. Roy. Stat. Soc.* B, **49**, 127–62.

Kalbfleish, J. D. and Sprott, D. A. (1970). Application of likelihood methods to models involving large numbers of parameters (with discussion). *J. Roy. Stat. Soc.* B, **32**, 175–208.

Liang, K.-Y. (1987). Estimating functions and approximate conditional likelihood. *Biometrika*, **74**, 695–702.

McCullagh, P. and Nelder, J. A. (1983). *Generalized linear model*. Chapman and Hall, London.

McCullagh, P. and Tibshirani, R. (1990). A simple method for the adjustment for profile likelihoods. *J. Roy. Stat. Soc.* B, **52**, 325–44.

Mantel, H. J. and Godambe, V. P. (1989). Conditional inference in a stratified model with nuisance parameter. Submitted for publication.

Yamamoto, E. and Yanagimoto, T. (1989). Moment estimators for the beta-binomial distribution, *J. Appl. Stat.* (In press.)

Yanagimoto, T. (1987). A notion of an obstructive residual likelihood. *Ann. Inst. Stat. Math.*, **39**, 247–61.

Yanagimoto, T. (1989a). Combining moment estimates of a parameter common through strats. *J. Stat. Plan. Inf.*, **25**, 187–98.

Yanagimoto, T. (1989b). The Mantel–Haenszel statistics for the extended odds ratio in the negative binomial distribution. Manuscript.

Yanagimoto, T. and Anraku, K. (1989). Possible superiority of the conditional MLE over the unconditional MLE. *Ann. Inst. Stat. Math.*, **41**, 269–78.

7
Use of a quadratic exponential model to generate estimating equations for means, variances, and covariances

Lue Ping Zhao and Ross L. Prentice

ABSTRACT

A quadratic exponential family model has been shown (Gourieroux *et al.*, 1984) to be unique in quite generally yielding consistent 'pseudo-maximum likelihood' estimates of parameters in the marginal means, variances and covariances even if sampling takes place from outside this family. Furthermore, this class of models can be shown to generate simple score-estimating equations for mean, variance and covariance parameters that extend standard equations for mean parameters arising as solution to score equations from generalized linear models. In fact, these equations are quite similar to those that have been proposed on *ad hoc* grounds for marginal means and correlations.

Some aspects of applying the quadratic exponential family and its associated estimating equations to discrete and to mixed continuous and discrete data are presented. A special member of the family, termed the generalized normal model, is also discussed in relation to these same types of response variables and in respect to practical aspects of model fitting. Mean, variance and covariance models, and corresponding estimating equations that yield consistent estimates of mean parameters under variance misspecification, or consistent estimates of mean and variance parameters under covariance misspecification, is also identified. The importance of simultaneous mean, variance and covariance estimation, in a variety of applications areas, is noted.

7.1 Introduction

Estimating functions can provide convenient estimation procedures for parameters in the means, variances and covariances of multivariate data even if a convenient joint distribution is unavailable. Specifically, based on K independent random observations $Y_k^T = (Y_{k1}, \ldots, Y_{kn_k})$, $k = 1, \ldots, K$, Liang and Zeger (1986) and Zeger and Liang (1986) proposed that the mean parameter vector α in $\mu_k = E(Y_k) = \mu_k(\alpha)$ be estimated by solving

$$\sum_k \left(\frac{\partial \mu_k}{\partial \alpha} \right)^T \text{var}(Y_k)^{-1}(y_k - \mu_k) = 0,$$

where var(Y_k) is the variance matrix of Y_k with variance vector $\sigma_k^T = (\sigma_{k1}, \ldots, \sigma_{kn_k})$ and covariance vector $\zeta_k^T = (\zeta_{k12}, \zeta_{k13}, \ldots,)$, y_k is the observed value of Y_k, summation is over all independent observations and 0 is a vector (or matrix) of zeros. The variance matrix var(Y_k) might involve parameter vectors δ and β in variances and covariances, respectively, which are estimated by simple functions of residuals. Prentice (1988) formalized this idea by introducing additional estimating equations, for example, in the current context,

$$\sum_k \left(\frac{\partial \sigma_k}{\partial \delta}\right)^T \text{var}(\tilde{Z}_k)^{-1} (\tilde{z}_k - \sigma_k) = 0,$$

$$\sum_k \left(\frac{\partial \zeta_k}{\partial \beta}\right)^T \text{var}(\tilde{W}_k)^{-1} (\tilde{w}_k - \zeta_k) = 0,$$

where $\tilde{Z}_k^T = [(Y_{k1} - \mu_{k1})^2, \ldots]$ with observed value $\tilde{z}_k$, $\tilde{W}_k^T = (\tilde{W}_{k12}, \tilde{W}_{k13}, \ldots)$ with observed value $\tilde{w}_k$ and $\tilde{W}_{kij} = (Y_{ki} - \mu_{ki})(Y_{kj} - \mu_{kj})$. The joint estimators of (α, δ, β) have been shown to be quite generally $K^{1/2}$-consistent and to have an asymptotic normal distribution. Since the additional two sets of estimating equations are introduced in an *ad hoc* basis, the efficiency of estimating δ and β is of particular concern, especially since adequate specification of var(Y_k) is important in the joint estimation of means, variances and covariances. Hence there is interest is seeking a class of parametric models that generate estimating equations more systematically, and that yield consistent and asymptotically normally distributed parameter estimates.

One such parametric model for a response vector $Y^T = (Y_1, \ldots, Y_n)$ is the generalized quadratic model (GQM) with density function with respect to a Legesque–Stieltjes measures $v(y)$

$$f(y) = \Delta^{-1} \exp[y^T \theta + z^T \rho + w^T \omega + c(y)], \tag{7.1}$$

where $\Delta = \int_y \exp(y^T \theta + z^T \rho + w^T \omega + c(y)] \, dv(y)$ is a normalizing constant, $\theta^T = (\theta_1, \ldots, \theta_n)$, $\rho^T = (\rho_1, \ldots, \rho_n)$ and $\omega^T = (\omega_{12}, \omega_{13}, \ldots,)$ are canonical parameter vectors, $z^T = (y_1^2, \ldots)$ is the observed value of Z, $w^T = (y_1 y_2, y_1 y_3, \ldots,)$ is the observed value of W, and $c(y)$ is a specified function of y which could be indexed by other nuisance parameters. Also, let $\tau = E(Z)$ and $\eta = E(W)$. The GQM (7.1) admits discrete data and mixtures of discrete and continuous data with appropriate definition of the measure $v(y)$. Specifically, $v(y_i)$ for a continuous or discrete random variable Y_i is a Lebesgue or counting measure. Correspondingly, $\int_{y_i}$ denotes Lebesgue integral or summation.

The canonical parameter vector typically has a one-to-one functional relationship with the marginal moment (μ, τ, η) (Prentice and Zhao 1991).

Under GQM, the marginal moments satisfy equations

$$\mu = \Delta^{-1} \int_y y \exp[y^T\theta + z^T\rho + w^T\omega + c(y)] \, dv(y),$$

$$\tau = \Delta^{-1} \int_y z \exp[y^T\theta + z^T\rho + w^T\omega + c(y)] \, dv(y),$$

$$\eta = \Delta^{-1} \int_y w \exp[y^T\theta + z^T\rho + w^T\omega + c(y)] \, dv(y).$$

The Jacobian matrix of partial derivatives of (μ, τ, η) with respect to (θ, ρ, ω) is the variance–covariance matrix for (Y, Z, W). Thus canonical parameters are uniquely determined by specified marginal moments, unless the joint distribution of pairwise products $Y_i Y_j$, all (i, j), degenerates. For specified (μ, τ, η), the canonical parameters satisfy the equations

$$g = \int_y \begin{pmatrix} y - \mu \\ z - \tau \\ w - \eta \end{pmatrix} \exp[y^T\theta + z^T\rho + w^T\omega + c(y)] \, dv(y) = 0. \qquad (7.2)$$

An iterative procedure is generally necessary in order to solve equations (7.2) for (θ, ρ, ω). Starting from an initial value $(\theta_0, \rho_0, \omega_0)$, one updates to a new value $(\theta_1, \rho_1, \omega_1)$ via

$$\begin{pmatrix} \theta_1 \\ \rho_1 \\ \omega_1 \end{pmatrix} = \begin{pmatrix} \theta_0 \\ \rho_0 \\ \omega_0 \end{pmatrix} - G_0^{-1} g_0,$$

where g_0, G_0 are values of function G, g evaluated at the initial value and

$$G = \int_y \begin{pmatrix} y - \mu \\ z - \tau \\ w - \eta \end{pmatrix} (y^T z^T w^T) \exp[y^T\theta + z^T\rho + w^T\omega + c(y)] \, dv(y)$$

is the first derivative of g with respect to the (θ, ρ, ω).

Given specified means, variances and covariances, Gourieroux et al. (1984) have shown that the GQM is unique in that maximum likelihood estimates of mean, variance and covariance parameters are consistent and asymptotically normal under mild regularity, even if (7.1) does not obtain. We refer to the distribution (7.1) as a generalized quadratic model to stress the relationship with generalized linear models, although the quadratic exponential family terminology used by Gourieroux et al. (1984) seems equally appropriate.

The objective of this paper is to review the use of the GQM to generate score equations for mean, variance, and covariance parameters. The score-estimating equations turn out to be quite similar to the previously proposed estimating equations. Also, the GQM provides a framework for generating a range of estimating equations about means, variances, and covariances that can accommodate certain model misspecifications. A procedure for estimating mean, variance, and covariance parameters based on a subclass model of GQM, termed the generalized normal model (GNM), is also reviewed here. The importance of simultaneously estimating mean, variance and covariance paameters in longitudinal studies, in studies with multiple responses, and in genetic studies will also be mentioned. Most of the results on the GQM and GNM have been given previously by Zhao and Prentice (1990), Prentice and Zhao (1991), and Zhao and Prentice (1991). The proposed estimating procedure has been applied to a clinical trial with bivariate responses and to a cross-over-designed clinical trial in these papers.

7.2   Score equation under the generalized quadratic model

Consider K independent observations $Y_1, Y_2, \ldots, Y_K$, where each Y_k is assumed to arise from a GQM (7.1) with specified means, variances, and covariances $[\mu_k(\alpha), \sigma_k(\delta, \alpha), \zeta_k(\beta, \alpha, \delta)]$, which in the context of regression analysis could be, for example, given by

$$\mu_{ki} = \mu_i(x_{ki}^{\mathrm{T}}\alpha),$$

$$\sigma_{ki} = \phi_i(\dot{x}_{ki}^{\mathrm{T}}\delta)v(\mu_{ki}),$$

$$\zeta_{kij} = \gamma_{kij}(\ddot{x}_{kij}^{\mathrm{T}}\beta)(\sigma_{ki}\sigma_{kj})^{1/2}, \qquad i \neq j = 1, \ldots, n, \qquad (7.3)$$

where $x_{ki}, \dot{x}_{ki}, \ddot{x}_{kij}$ are covariate vectors in mean, variance scalar ϕ_{ki} and correlation coefficient γ_{ij}, respectively, and $v(\mu_{ki})$ is a specified function. The marginal moments (μ_k, ζ_k, η_k) are specified by (α, δ, β) owing to the relationships $\tau_{ki} = \sigma_{ki} + \mu_{ki}^2$ and $\eta_{kij} = \zeta_{kij} + \mu_{ki}\mu_{kj}, i, j = 1, \ldots, n_k$. Further, the canonical parameters $(\theta_k, \rho_k, \omega_k)$ are also specified by (α, δ, β) and are obtainable by solving equation (7.2).

The kth log-likelihood contribution is given by

$$l_k = y_k^{\mathrm{T}}\theta_k + z_k^{\mathrm{T}}\rho_k + w_k^{\mathrm{T}}\omega_k + c(y_k) - \ln(\Delta_k).$$

Upon noting that the derivative of the l_k with respect to the $(\theta_k, \rho_k, \omega_k)$ is given by

$$f_k = \begin{pmatrix} y_k - \mu_k \\ z_k - \tau_k \\ w_k - \eta_k \end{pmatrix},$$

one applies the chain rule to obtain score equations with respect to the (α, δ, β), giving

$$u = \frac{1}{\sqrt{K}} \sum_k D_k^\mathrm{T} V_k^{-1} f_k = 0, \tag{7.4}$$

where V_k is a variance matrix for $(Y_k^\mathrm{T}, Z_k^\mathrm{T}, W_k^\mathrm{T})$ and

$$D_k = \begin{pmatrix} \dfrac{\partial \mu_k}{\partial \alpha} & \dfrac{\partial \mu_k}{\partial \delta} & \dfrac{\partial \mu_k}{\partial \beta} \\[2mm] \dfrac{\partial \tau_k}{\partial \alpha} & \dfrac{\partial \tau_k}{\partial \delta} & \dfrac{\partial \tau_k}{\partial \beta} \\[2mm] \dfrac{\partial \eta_k}{\partial \alpha} & \dfrac{\partial \eta_k}{\partial \delta} & \dfrac{\partial \eta_k}{\partial \beta} \end{pmatrix}$$

is a derivative matrix in which, for example, $\partial \mu_k / \partial \delta = 0$, $\partial \mu_k / \partial \beta = 0$, and $\partial \tau_k / \partial \beta = 0$ under specification (7.3). Also, the Fisher information matrix is given by

$$I = \frac{1}{K} \sum_k D_k^T V_k^{-1} D_k. \tag{7.5}$$

The maximum likelihood estimators of (α, δ, β) are obtained by solving the score equations (7.4). In general, there is no explicit solution. A Newton–Raphson algorithm based on u and I may be used to solve iteratively the score equations (7.4). Let $(\hat\alpha, \hat\delta, \hat\beta)$ be maximum likelihood estimators. As $K \to \infty$, the estimators are quite generally $K^{1/2}$-consistent estimators of (α, δ, β) and, under (7.1), have an asymptotic normal distribution

$$\sqrt{K} \begin{pmatrix} \hat\alpha - \alpha \\ \hat\delta - \delta \\ \hat\beta - \beta \end{pmatrix} \to N(0, I^{-1}). \tag{7.6}$$

Moreover, the $(\hat\alpha, \hat\delta, \hat\beta)$, provided that link functions are correctly specified, are still consistent, even if the GQM is misspecified, in which case $(\hat\alpha, \hat\delta, \hat\beta)$ can be referred to as pseudo-maximum likelihood estimates. Under such misspecification $K^{1/2}[(\hat\alpha - \alpha)^\mathrm{T}, (\hat\delta - \delta)^\mathrm{T}, (\hat\beta - \beta)^\mathrm{T}]$ will quite generally have an asymptotic normal distribution with mean 0 but with variance consistently estimated by

$$\frac{1}{K} I^{-1} \left[\sum_k D_k^\mathrm{T} V_k^{-1} f_k f_k^\mathrm{T} V_k^{-1} D_k \right] I^{-1}, \tag{7.7}$$

where $(y_k - \mu_k)(y_k - \mu_k)^\mathrm{T}$ in $f_k f_k^\mathrm{T}$ can also be replaced by the variance matrix of Y_k. This asymptotic result can be derived in the manner of Liang and Zeger (1986), or can be asserted on the basis of Theorem 3.2 in White (1982).

The quantities f_k and D_k are readily available from specified means, variances and covariances. The computation of the V_k, however, is not straightforward, since it involves the third and fourth moments. Provided that the empirical density function $\hat{f}(y_k)$, obtained by solving the parameters in (7.1), is available, then V_k is estimated by

$$\int_{y_k} f_k f_k^{\mathrm{T}} \hat{f}(y_k)\, \mathrm{d}v(y_k).\qquad(7.8)$$

But $\hat{f}(y_k)$ is indexed by canonical parameters $(\hat{\theta}_k, \hat{\rho}_k, \hat{\omega}_k)$ which are functions of (μ_k, ζ_k, η_k). Hence these canonical parameters estimates need to be computed iteratively by solving (7.2) for each $k = 1, \ldots, K$.

As an alternative to this fully parametric approach, we may, following Liang and Zeger (1986), introduce a 'working' variance matrix V_k and estimate (α, δ, β) by solving equation (7.4). Such an approach is simplified by re-expressing all quantities in terms of centralized moments of Y_k, $k = 1, \ldots, K$, in which case (7.4), (7.5), and (7.7) can be rewritten as

$$\frac{1}{\sqrt{K}}\sum_k \tilde{D}^{\mathrm{T}} \tilde{V}_k^{-1} \tilde{f}_k = 0,$$

$$\tilde{I} = \frac{1}{K}\sum_k \tilde{D}_k^{\mathrm{T}} \tilde{V}_k^{-1} \tilde{D}_k,$$

$$\frac{1}{K}\tilde{I}^{-1}\left[\sum_k \tilde{D}_k^{\mathrm{T}} \tilde{V}_k^{-1} \tilde{f}_k \tilde{f}_k^{\mathrm{T}} \tilde{V}_k^{-1} \tilde{D}_k\right] \tilde{I}^{-1},\qquad(7.9)$$

where

$$\tilde{D}_k = \begin{pmatrix} \dfrac{\partial \mu_k}{\partial \alpha} & \dfrac{\partial \mu_k}{\partial \delta} & \dfrac{\partial \mu_k}{\partial \beta} \\[2mm] \dfrac{\partial \sigma_k}{\partial \alpha} & \dfrac{\partial \sigma_k}{\partial \delta} & \dfrac{\partial \sigma_k}{\partial \beta} \\[2mm] \dfrac{\partial \zeta_k}{\partial \alpha} & \dfrac{\partial \zeta_k}{\partial \delta} & \dfrac{\partial \zeta_k}{\partial \beta} \end{pmatrix}, \qquad \tilde{f}_k = \begin{pmatrix} y_k - \mu_k \\ z_k - \sigma_k \\ w_k - \zeta_k \end{pmatrix},$$

and $\tilde{V}_k$ is a variance matrix for $(Y_k^{\mathrm{T}}, \tilde{Z}_k^{\mathrm{T}}, \tilde{W}_k^{\mathrm{T}})$. Note that

$$\zeta_{kijl} = E[(y_{ki} - \mu_{ki})(y_{kj} - \mu_{kj})(y_{kl} - \mu_{kl})],$$

and

$$\zeta_{kijlm} = E[(y_{ki} - \mu_{ki})(y_{kj} - \mu_{kj})(y_{kl} - \mu_{kl})(y_{km} - \mu_{km})] - \zeta_{kij}\zeta_{klm}$$
$$= \zeta_{kijlm} - \zeta_{kij}\zeta_{klm}, \qquad i, j, l, m = 1, \ldots, n_k,$$

where $\zeta_{kii} = \sigma_{ki}$ are elements of the covariance matrices of $(Y_k, \tilde{Z}_k)$ and $(Y_k, \tilde{W}_k)$, and of the variance and covariance matrixes of $(\tilde{Z}_k, \tilde{W}_k)$. Working values for ζ_{kijl} and ζ_{kijlm} can be assigned by the data analyst. For example, the assumption of

$$\zeta_{ijl} = 0 \qquad \text{and} \qquad \zeta_{kijlm} = \zeta_{kij}\zeta_{klm} + \zeta_{kil}\zeta_{kjm} + \zeta_{kim}\zeta_{kjl},$$

suggested by a multivariate normal distribution for Y_k, leads to a convenient estimation for (α, δ, β). Also, one could introduce additional parameters into the working variance matrices V_k, following Liang and Zeger (1986) and Prentice and Zhao (1991).

The property of $K^{1/2}$-consistent estimation of (α, δ, β) is drawn under the assumption that the mean, variance and covariance models are correctly specified. In general, estimators of α, δ, or β may be inconsistent if any of these models are incorrect. In contrast, the simple estimating equations discussed in the previous section have the advantage of yielding consistent estimation of α even under misspecification of the variance and covariance components (Liang and Zeger 1986; Zeger and Liang 1986) and consistent estimation of (α, δ), or (α, β) regardless of the misspecification of variance and covariance components, respectively.

We should point out, however, that it is possible to construct, based on a model closely related to GQM, a parametric score-estimating equation yielding consistent estimation of (α, δ), (α, β) or α alone, regardless of misspecification of covariance model, variance model, or variance and covariance models, respectively. Such construction involves parametrizing (7.1) in order to generate mutual orthogonality among the resulting score-estimating equations and parameter estimates. The resulting estimating functions will then satisfy a joint optimality criterion (e.g., Godambe and Thompson 1989).

Suppose that the covariance is a nuisance component and may be misspecified. Consider a class of models with density

$$f(y_k) = \Delta_k^{-1} \exp[y_k^T \theta_k + z_k^T \rho_k + c_k(\pi, y_k)], \tag{7.10}$$

where θ_k, ρ_k are implicitly specified by (μ_k, σ_k) in the first two components of (7.2) and $c_k(\pi, y_k)$ is a specified function of y_k indexed by a 'dependence' parameter vector π. By applying the chain rule, one is able to obtain score equations for (α, δ) and π

$$\frac{1}{\sqrt{K}} \sum_k \bar{D}_k^t \bar{V}_k^{-1} \bar{f}_k = 0,$$

$$\frac{1}{\sqrt{K}} \sum_k \{\partial c_k(\pi, y_k) - E c_k(\pi, Y_k) - \bar{B}_k \bar{f}_k\} = 0,$$

where

$$\bar{D}_k = \begin{pmatrix} \dfrac{\partial \mu_k}{\partial \alpha} & 0 \\ \dfrac{\partial \sigma_k}{\partial \alpha} & \dfrac{\partial \sigma_k}{\partial \delta} \end{pmatrix}, \qquad \hat{f}_k = \begin{pmatrix} y_k - \mu_k \\ \tilde{z}_k - \sigma_k \end{pmatrix},$$

$\bar{V}_k$ is a variance matrix for $(Y_k^T, \tilde{Z}_k^T)$, $\partial c_k(\pi, Y_k) = \partial c_k(\pi, Y_k)/\partial \pi$ with observed vector $\partial c_k(\pi, y_k)$ and $\bar{B}_k = \text{Cov}[\partial c_k(\pi, Y_k), \hat{f}_k]\bar{V}_k^{-1}$. Assuming correct specification of mean and variance, the first estimating equation yields consistent estimation of (α, δ), regardless of whether dependence structure is correctly specified. The information matrix for the $(\hat{\alpha}, \hat{\delta}, \hat{\pi})$ is block diagonal with blocks

$$\frac{1}{K} \sum_k \bar{D}_k^T \bar{V}_k^{-1} \bar{D}_k \qquad \text{and} \qquad \frac{1}{K} \sum_k \text{var}[\partial c_k(\pi, Y_k)] - \bar{B}_k \bar{V}_k \bar{B}_k^T.$$

Thus the $(\hat{\alpha}, \hat{\delta})$ and $\hat{\pi}$ are asymptotically independent and have an asymptotic normal distribution. Similarly, to obtain consistent estimation of (α, β) with possible misspecification of the variance component, one can generate score-estimating equations by replacing $z_k^T \rho_k$ with $w_k^T \omega_k$ in (7.10) (details omitted).

To obtain consistent estimation of α alone, consider a class of models (Zhao et al. 1991)

$$f(y_k) = \Delta_k^{-1} \exp[y_k^T \theta_k + c_k(\pi, y_k)], \tag{7.11}$$

where θ_k is implicitly specified by the μ_k in the first component of (7.2). Again by applying the chain rule, one is able to obtain score equations for α, π. The score equation for α is identical to the estimating equation proposed by Liang and Zeger (1986) and that for π is given by

$$\frac{1}{K} \sum_k \{\partial c_k(\pi, y_k) - Ec_k(\pi, Y_k) - \text{Cov}[\partial c_k(\pi, Y_k), Y_k] \sum (y_k)^{-1}(y_k - \mu_k)\} = 0.$$

Thus the first estimating equation yields consistent estimators of α regardless of the variance matrix $\sum(y_k)$ which depends on $c_k(\pi, y_k)$. The information matrix is block diagonal with blocks

$$\frac{1}{K} \sum_k \left(\frac{\partial \mu_k}{\partial \alpha}\right)^T \sum(y_k)^{-1} \left(\frac{\partial \mu_k}{\partial \alpha}\right)$$

and

$$\frac{1}{K} \sum_k \text{Var}[\partial c_k(\pi, Y_k)] - \text{Cov}[\partial c_k(\pi, Y_k), Y_k] \sum(y_k)^{-1} \text{Cov}[Y_k, \partial c_k(\pi, Y_k)].$$

Thus the estimators $\hat{\alpha}$ and $\hat{\pi}$ are asymptotically independent and have asymptotic normal distribution. Note, however, that these applications of models (7.10) and (7.11) with certain 'canonical' parameters modelled directly are fundamentally different from the application of (7.1) described previously, in which marginal mean, variance, and covariance parameters are modelled, while the induced models for canonical parameters vary with k in a complex fashion.

We end this section by noting that the computation of pseudo-maximum likelihood estimates based on GQM (7.1) is cumbersome in general due to the integral expressions for the log-likelihood function, score equations and information matrix if the Y_k included continuous random variables. Specifically to compute the likelihood function, one needs to obtain canonical parameters by solving equation (7.2), which involve integrations with respect to the whole sample space of Y_k in each 'block', $k = 1, 2, \ldots, K$. To obtain fourth moments in V_k, one has to compute the integrals in (7.8). As noted above, estimating function (7.2) involving working variance matrices V_k can avoid these difficult computations. It is also of interest to consider a submodel of GQM in which the computation of integrals can be avoided in the calculation of pseudo-maximum likelihood estimates.

7.3   The generalized normal model

The generalized normal model (GNM) is a subclass model in GQM with $c(y) \equiv 0$. The GNM continues to admit both continuous and discrete random variables but continuous random variables in Y are restricted to have domain $(-\infty, +\infty)$. The distribution function of GNM with respect to a Lebesgue or counting measure $v(y)$ is given by

$$f(y) = \Delta^{-1} \exp(y^T\theta + z^T\rho + w^T\omega), \qquad (7.12)$$

where θ, ρ, ω are implicitly specified by (7.2) with $c(y) \equiv 0$. Clearly, the GNM is a multivariate normal distribution if all random variables are continuous. If all random variables are discrete, the GNM is a long-linear model in terms of canonical parameters. Moreover, the GNM admits mixtures of continuous and discrete random variables. Let $Y^T = (Y_c^T, Y_d^T)$ where Y_c, Y_d are continuous and discrete random vector with means μ_c, μ_d, variance matrices Σ_c, Σ_d and covariance Σ_{cd}. It can be shown that the marginal distribution of Y_d is still a GNM with canonical parameters vectors $\theta^*, \rho^*, \omega^*$ which are functions of μ_d, Σ_d, and conditional distributions of Y_c given $Y_d = y_d$ which is multivariate normal with mean $\mu_c + \Sigma_{cd}\Sigma_d^{-1}(y_d - \mu_d)$ and variance $\Sigma_c - \Sigma_{cd}\Sigma_d^{-1}\Sigma_{cd}^T$. Note that Lauritzen and Wermuth (1989) utilized conditional Gaussian distributions for continuous variates in proposing a joint distribution and discrete random variables.

Let us consider K independent random observations $Y_1, Y_2, \ldots, Y_K$ where each Y_k is assumed to be from a GNM (7.12) with specified $(\mu_k, \sigma_k, \zeta_k)$. The score equation and information matrix under GNM are again given by (7.9). Provided that the $\tilde{D}_k, \tilde{f}_k, \tilde{V}_k$ are available, a Newton–Raphson algorithm may be used to obtain the maximum likelihood estimators. Starting from an initial value $(\alpha_0, \delta_0, \beta_0)$, one updates to $(\alpha_1, \delta_1, \beta_1)$ via

$$\begin{pmatrix} \alpha_1 \\ \delta_1 \\ \beta_1 \end{pmatrix} = \begin{pmatrix} \alpha_0 \\ \delta_0 \\ \beta_0 \end{pmatrix} + I_0^{-1} u_0,$$

where I_0 and u_0 are values of I and u at the initial value.

The $\tilde{f}_k$ are readily available from specified $(\mu_k, \sigma_k, \zeta_k)$, such as those in (7.3), for $k = 1, \ldots, K$. The $\tilde{D}_k$ are derivative matrices of $(\mu_k, \sigma_k, \zeta_k)$ with respect to (α, δ, β). A convenient, but computationally intensive, option is to calculate those derivative matrices numerically. A more efficient, but sometimes algebraically tedious, option is to obtain explicit expressions for the derivative matrices.

However, the $\tilde{V}_k$, involving third and fourth moments, are not directly available. We consider the computation of $\tilde{V}_k$ in three cases. First, when all random variables are continuous, GNM is simple a distribution function of multivariate normal distribution with the third and fourth centralized moment $\zeta_{kijl}^{ccc} = 0$ and $\zeta_{kijlm}^{cccc} = \zeta_{kij}\zeta_{klm} + \zeta_{kil}\zeta_{kjm} + \zeta_{kim}\zeta_{kjl}$, respectively, where the superscripts 'ccc' and 'cccc' indicate that the random variable i, j, l and i, j, l, m are continuous, respectively. Also, let superscripts, for example, 'ddcc' indicate that random variable i, j and l, m are continuous and discrete, respectively. Secondly, when all random variables are discrete, it appears necessary to compute the canonical parameter estimates $((\hat{\theta}_k, \hat{\rho}_k, \hat{\omega}_k)$ by solving equations (7.2) and thus to obtain density function estimator

$$\hat{f}(y_k) = \Delta_k^{-1} \exp(y^{\mathrm{T}} \hat{\theta}_k + z^{\mathrm{T}} \hat{\rho}_k + w^{\mathrm{T}} \hat{\omega}_k).$$

Then the $\tilde{V}_k$ is given by (7.8) in which the integration is replaced by the summation over the entire sample space of Y_k. Finally, suppose that Y_k^{T} contains both continuous and discrete random variables. Let us partition $Y_k^{\mathrm{T}} = (Y_{ck}^{\mathrm{T}}, Y_{dk}^{\mathrm{T}})$, where Y_{ck} and Y_{dk} are continuous and discrete random vectors without loss of generality. Let $\mu_{ck}, \mu_{dk}, \Sigma_{ck}, \Sigma_{dk}$ and Σ_{cdk} be means and variance matrices and covariances matrix of Y_{ck}, Y_{dk}. Let $B_k = \Sigma_{cdk}\Sigma_{dk}^{-1}$. The third and fourth moments of discrete random variables are obtained via computing canonical parameters. By utilizing conditional multivariate normal distribution, one is able to obtain the third and fourth moments without integration over continuous random variables (Zhao and Prentice 1991).

The third centralized moments are given by

$$\zeta_{kijl}^{ddd} = \sum_{y_d} (y_{di} - \mu_{dki})(y_{dj} - \mu_{dkj})(y_{dl} - \mu_{dkl})\hat{f}(y_d),$$

$$\zeta_{kijl}^{ddc} = \sum_{s_1} b_{kls_1} \zeta_{kijs_1}^{ddd},$$

$$\zeta_{kijl}^{dcc} = \sum_{s_1,s_2} b_{kjs_1} b_{kls_2} \zeta_{kis_1s_2}^{ddd},$$

$$\zeta_{kijl}^{ccc} = \sum_{s_1,s_2,s_3} b_{kis_1} b_{kjs_2} b_{kls_3} \zeta_{ks_1s_2s_3}^{ddd},$$

where $\hat{f}(y_d)$ is a distribution function of Y_d evaluated at $\hat{\theta}_{kd}^*, \hat{\rho}_{kd}^*, \hat{\omega}_{kdd}^*$ and b_{kij} is the ijth element in B_k.

The fourth-order centralized moment is given by

$$\zeta_{kijlm}^{dddd} = \sum_{y_d} (y_{di} - \mu_{dki})(y_{dj} - \mu_{dkj})(y_{dl} - \mu_{dkl})(y_{dm} - \mu_{dkm})\hat{f}(y_d),$$

$$\zeta_{kijlm}^{dddc} = \sum_{s_1} b_{kms_1} \zeta_{kijls_1}^{dddd},$$

$$\zeta_{kijlm}^{ddcc} = \sigma_{dkij}\sigma_{cklm} + \sum_{s_1,s_2} b_{kls_1} b_{kms_2} \zeta_{kijs_1s_2}^{dddd},$$

$$\zeta_{kijlm}^{dccc} = \sigma_{ckjl}^* \sum_{s_1} b_{kms_1} \sigma_{dkis_1} + \sigma_{ckjm}^* \sum_{s_1} b_{kls_1} \sigma_{dkis_1} + \sigma_{cklm}^* \sum_{s_1} b_{kjs_1} \sigma_{dkis_1}$$
$$+ \sum_{s_1,s_2,s_3} b_{kjs_1} b_{kls_2} b_{kms_3} \zeta_{kis_1s_2s_3}^{dddd},$$

$$\zeta_{kijlm}^{cccc} = \sigma_{kij}^*\sigma_{klm}^* + \sigma_{kil}^*\sigma_{kjm}^* + \sigma_{kim}^*\sigma_{kjl}^* + \sigma_{kij}^* \mathbf{b}_{kl}^T \Sigma_{kdd} \mathbf{b}_{km} + \sigma_{kil}^* \mathbf{b}_{kj}^T \Sigma_{kdd} \mathbf{b}_{km}$$
$$+ \sigma_{kim}^* \mathbf{b}_{kj}^T \Sigma_{kdd} \mathbf{b}_{kl} + \sigma_{kjl}^* \mathbf{b}_{ki}^T \Sigma_{kdd} \mathbf{b}_{km} + \sigma_{kjm}^* \mathbf{b}_{ji}^T \Sigma_{kdd} \mathbf{b}_{kl}$$
$$+ \sigma_{klm}^* \mathbf{b}_{ki}^T \Sigma_{kdd} \mathbf{b}_{kj} + \sum_{s_1,s_2,s_3,s_4} b_{kis_1} b_{kjs_2} b_{kls_3} b_{kms_4} \zeta_{ks_1s_2s_3s_4}^{dddd},$$

where σ_{dkij} is the ijth element in Σ_{dk}, σ_{ckij}^* is the ijth element in $\Sigma_{ck} - B_k \Sigma_{dk} B_k$ and $\mathbf{b}_{ki}$ is the ith row of B_k.

7.4 Examples

Three examples from longitudinal studies, studies with multiple responses and genetic studies on familial pedigree are briefly discussed below:

Example 1a Consider a longitudinal study on K independent subjects. Let $Y_k^T = (Y_{k1}, Y_{k2}, \ldots, Y_{kn_k})$ be a sequence of observations recorded at time

$t_{k1}, t_{k2}, \ldots, t_{kn_k}$ and let $x_{ki}, \dot{x}_{ki}, \ddot{x}_{ki}$, as described above, be corresponding regression vectors at t_{ki}. Interest will often focus on the regression effect of $x_{ki}, \dot{x}_{ki}, \ddot{x}_{ki}$ on the mean μ_{ki}, variance σ_{ki} and covariance ζ_{kij}. Two distinct approaches based on an 'observation-driven model' and a 'parameter-driven model' have been developed for longitudinal data analysis. The approach based on a parameter-driven model is illustrated in Example 1b.

In the observation-driven model approach, interest focuses, for example, on the regression effects of $x_{ki}, \dot{x}_{ki}, \ddot{x}_{ki}$ on the mean $\mu_{ki} = \mu(x_{ki}^T \alpha)$, variance $\sigma_{ki} = \phi(\dot{x}_{ki}^T \delta) v(\mu_{ki})$ and covariance

$$\zeta_{kij} = \left[\sum_{l=1}^{j-i+1} q(\ddot{x}_{kl}^T \beta) \right] \sigma_{ki},$$

where $q(\ddot{x}_{kl}^T \beta)$ is a derived autoregression coefficient. For example, under an observation-driven model for continuous random variables, the dependence between two adjacent observations is induced by an autoregression parameter q_{ki} via, for example, a first-order autoregression

$$Y_{ki+1} - \mu_{ki+1} = q_{ki}(Y_{ki} - \mu_{ki}) + E_{ki},$$

where E_{ki} is an independent random error. Based on this autoregression, the induced covariance is derived. With interest in simultaneous estimation of mean, variance, and autoregression parameters, one may assume that Y_k is from a GNM and carry out a maximum likelihood estimation of parameters of interest.

Example 1b Let us consider a parameter-driven model for longitudinal data. For simplicity of discussion, let us focus on a model, formulated by Laird and Ware (1982), for continuous random variables. Suppose that Y_k is from a multivariate normal distribution with mean $x_k \alpha + \dot{x}_k \theta$, where $x_k = (x_{k1}, \ldots, x_{kn_k})$ and $\dot{x}_k = (\dot{x}_{k1}, \ldots, \dot{x}_{kn_k})$, and a common variance δ. The parameter θ is a latent random vector and has a multivariate normal distribution with mean zero and variance

$$\mathcal{B} = \begin{pmatrix} \beta_{11} & \beta_{12} & \cdots & \beta_{1q} \\ \beta_{12} & \beta_{22} & \cdots & \beta_{2q} \\ \vdots & \vdots & \ddots & \vdots \\ \beta_{1q} & \beta_{2q} & \cdots & \beta_{qq} \end{pmatrix}.$$

Thus the Y_k, following these two assumptions, has a multivariate normal distribution with mean vector $\mu_k = x_k \alpha$ and variance matrix $\delta + \dot{x}_k^T \mathcal{B} \dot{x}_k$. This parameter-driven model, sometimes referred to as a random-effect model, yields a joint distribution of Y_k from GNM with mean $\mu_{ki} = x_{ki}^T \alpha$, variance $\sigma_{ki} = \delta + \dot{x}_{ki}^T \mathcal{B} \dot{x}_{ki}$ and covariance $\zeta_{kij} = \dot{x}_{ki}^T \mathcal{B} \dot{x}_{kj}$. Interest will focus on joint estimation of the mean, variance and covariance parameters simultaneously.

Example 2 Consider now a study with n endpoints from each of K independent subjects. Let $Y_k^T = (Y_{k1}, \ldots, Y_{kn})$ denote the response vector from subject k, $k = 1, \ldots, K$. Often interest will focus on the effect of a covariate vector x_k on the mean response $\mu_{ki} = \mu_i(x_k^T \alpha)$. In some other instances, the variability of the responses will also be of interest. In studies of quality control, for example, the effect of covariate vector $\dot{x}_k$ on the variance scalar ϕ_k in $\sigma_{ki} = \phi_i(\dot{x}_k^T \delta)v(\mu_{ki})$ may be of particular interest. Moreover, interest can even reside in the extent of dependence among responses, as may occur, for example, in studies assessing the risk and benefits of a toxic drug used for therapeutic purposes. In such circumstances, the effect of a covariate vector $\ddot{x}_k$ on the correlation coefficient γ_{kij} in covariance $\zeta_{kij} = \gamma(\beta_{ij} + \ddot{x}_k^T \beta)\sqrt{(\sigma_{ki}\sigma_{kj})}$ will be of interest. For such problems, one may assume that Y_k arises from a GNM with specified $(\mu_k, \sigma_k, \zeta_k)$, obtain maximum likelihood estimator $(\hat{\alpha}, \hat{\delta}, \hat{\beta})$ and carry out inferences on parameters in the response means, variances, and covariances.

Example 3 Consider K independent families and suppose that a family consists of two parents and a variable number of siblings. Let Y_{k1}, Y_{k2} and Y_{ki}, $i = 3, \ldots, n_k$ be phenotypes of two parents and $n_k - 2$ siblings. The dependence among family members is induced primarily by genetic variation and thus may be of fundamental interest. Now assume that the phenotypes $Y_{k1}, Y_{k2}, Y_{k3}, \ldots, Y_{kn_k}$ jointly arise from a GNM with homogeneous means $\mu_p, \mu_p, \mu_0, \ldots, \mu_0$ and variances $\sigma_p, \sigma_p, \sigma_0, \ldots, \sigma_0$ and with correlation coefficient of phenotypes between two parents (γ_{pp}), a parent and a sibling (γ_{po}) and two siblings (γ_{00}) (Wette *et al.* 1988). Thus the covariance of phenotypes is given by

$$\zeta_{kij} = \begin{cases} \sigma_p \gamma_{pp} & i \neq j = 1, 2; \\ (\sigma_p \sigma_0)^{1/2} \gamma_{po} & i = 1, 2; j = 3, \ldots, n_k; \\ \sigma_0 \gamma_{00} & i \neq j = 3, \ldots, n_k. \end{cases}$$

Based on GNM, the maximum likelihood estimators are obtained and estimated parameters, especially correlation coefficients ($\gamma_{pp}, \gamma_{0p}, \gamma_{00}$), can be examined.

7.5 Conclusion

The generalized quadratic model provides a parametric framework to study generalized estimating equations for means, variances, and covariances just as generalized linear models do for mean parameters. This framework allows a broad class of estimating functions to be generated for parameters in the means, variances, and covariances of a multivariate response that may contain discrete and continuous components. The special case termed a generalized normal model also provides a parametric framework to study estimation functions based on a multivariate normal distribution without assuming normality (e.g. Crowder 1985).

In conclusion, the quadratic exponential family yields simple score-estimating equations with respect to the parameters in means, variances, and covariances. The estimated parameters are consistent and have normal distribution under mild regularity assumptions and are efficient if the assumed model holds. The integral form of third and fourth moments, however, prohibits the direct application of these score equations unless all response variables are discrete. A special case, the GNM, is useful in accommodating both continuous and discrete random variables with tolerable computational burden. More generally, application involving working variance matrices leads to estimation procedures that are computationally and conceptually simple. In this paper the specification of the GQM and the derivation of score equations and information matrices assumed all response variables to be non-binary. Model (7.1) needs to be restricted by requiring the entries in ρ to be zero for binary response variables, giving rise to some straightforward modification of the estimating equations and information matrix (Zhao and Prentice 1990).

7.6   Acknowledgement

This research is supported by grant GM-24472 CA-53996 from the National Institutes of Health.

References

Crowder, M. (1985). Gaussian estimation for correlated binomial data. *J. Roy. Stat. Soc.* B, **47**, 229–37.

Godambe, V. P. and Thompson, M. E. (1989). An extension of quasi-likelihood estimation (with discussion). *J. Stat. Plan. Inf.*, **22**, 137–72.

Gourieroux, C., Monfort, A., and Trognon, A. (1984). Pseudo maximum likelihood methods: theory. *Econometrica*, **52**, 681–700.

Laird, N. M. and Ware, J. H. (1982). Random-effects models for longitudinal data. *Biometrics*, **32**, 963–74.

Lauritzen, S. L. and Wermuth, N. (1989). Graphical models for association between variables, some of which are qualitative and some quantitative. *Ann. Stat.*, **17**, 31–57.

Liang, K. Y. and Zeger, S. L. (1986). Longitudinal data analysis using generalized linear models. *Biometrika*, **73**, 13–22.

Prentice, R. L. (1988). Correlated binary regression with covariates specific to each binary observation. *Biometrics*, **44**, 1033–48.

Prentice, R. L. and Zhao, L. P. (1991). Estimating equations for parameters in means and covariances of multivariate discrete and continuous responses. *Biometrics*. (In press.)

Wette, R., McGue, M. K., Rao, D. C., and Cloninger, C. R. (1988). On the properties of maximum likelihood estimators of familial correlations under variable sibship sizes. *Biometrics*, **44**, 717–25.

White, H. (1982). Maximum likelihood estimation of misspecified models. *Econometrica*, **50**, 1–25.

Zeger, S. L. and Liang, K. Y. (1986). Longitudinal data analysis for discrete and continuous outcomes. *Biometrics*, **42**, 121–30.

Zhao, L. P. and Prentice, R. L. (1990). Correlated binary regression using a quadratic exponential model. *Biometrika*, **77**, 642–8.

Zhao, L. P. and Prentice, R. L. (1991). Multivariate modelling of mean, variance and covariance using a generalization of the normal distribution. Submitted for publication.

Zhao, L. P., Prentice, R. L., and Self, S. (1991). Multivariate mean parameter estimation using a partly exponential model. Submitted for publication.

PART 3

Stochastic processes

8
Generalized score tests for composite hypotheses
I. V. Basawa

ABSTRACT

Three large sample tests, namely, Rao's likelihood score test, Neyman's $C(\alpha)$ test and a generalized score test are reviewed in a unified general setting involving possibly dependent, and not necessarily identically distributed observations. An application to a first-order autoregressive-moving average model is discussed. The test statistics considered in this paper are based on certain estimating functions. The Rao statistic is based on the likelihood estimating function and the Neyman statistic requires partly a general score function and partly the likelihood estimating function. The general score test is based on an arbitrary estimating function. It is shown in the application that the choice of an optimal estimating function according to the Godambe criterion also leads to asymptotical optimal tests for composite hypotheses.

8.1 Introduction

Most estimating problems in statistical research can be reduced to a study of suitable estimating equations. Consider an estimating equation $R_n(\theta) = 0$, where $R_n(\theta)$ is a function of the observations $X(n) = (X_1, X_2, \ldots, X_n)$, and an unknown parameter θ. Moment estimates, least squares, maximum likelihood, partial likelihood, quasi-likelihood, least absolute deviation and other minimum-distance methods including the minimum chi-squared method, all lead to certain estimating equations involving a suitable score function (or estimating function). The standard approach used in statistical estimation theory includes the following steps:

(a) Justify the method used to obtain the estimating equation on some intuitive ground such as optimizing a certain desirable objective function. Maximizing the likelihood and minimizing the error sum of squares are examples of such an optimization criterion.
(b) Study the sampling properties of the estimator $\hat{\theta}_n$ obtained as a solution of the equation $R_n(\theta) = 0$.

(c) Compare possible estimating functions, say $R_{n1}(\theta)$ and $R_{n2}(\theta)$ (corresponding to two different methods of estimation in (a)) on the basis of the properties of the estimators $\hat{\theta}_{1n}$ and $\hat{\theta}_{2n}$ obtained by solving $R_{n1}(\theta) = 0$, and $R_{n2}(\theta) = 0$ respectively.

If for instance, $\hat{\theta}_{1n}$ has a smaller mean-square error than $\hat{\theta}_{2n}$, one may conclude that $R_{n1}(\theta)$ is better than $R_{n2}(\theta)$ as an estimating function. Since it is not always possible to compare the mean-square errors (or other desirable characteristics of the sampling distributions) for finite samples, one generally adopts an asymptotic approach. If the limit distributions of $\hat{\theta}_{in}$, $i = 1, 2$, are available, one can compare for instance, the variances of the limit distributions, and choose an estimator with a smaller variance. One can then conclude that $R_{1n}(\theta)$ is asymptotically better than $R_{2n}(\theta)$ if the asymptotic variance of $\hat{\theta}_{1n}$ is smaller than that of $\hat{\theta}_{2n}$.

Godambe (1960, 1985) argues that it is possible to compare two (or more) estimating functions *directly* rather than using the above indirect approach via the properties (asymptotic or otherwise) of the estimators they yield. Godambe defined an intrinsic information type criterion which can be used to select an optimum estimating function from among the possible functions in a given class. This criterion does not rely on asymptotics, and hence can also be used in finite samples. The Godambe approach has an advantage over the traditional approach in that it seeks to select $R_n(\theta)$ directly based on a simple criterion. The sampling properties of the estimators obtained from the selected estimating function become a secondary matter in this approach. However, in most applications of the Godambe criterion it turns out that the estimators so obtained do indeed exhibit good sampling properties asymptotically.

It is well known that the estimating functions can also be used as a basis for constructing test statistics. Consequently, an optimum Godambe score function can yield good estimators, as well as good test statistics. In this paper, we consider three test statistics for testing composite hypotheses. These statistics are based on the likelihood score function (Rao statistic), a score function which is partly a likelihood score and partly a general function (Neyman's $C(\alpha)$-statistic), and finally a general score function. See also Basawa (1985), and Basawa and Koul (1979, 1988) for some related work. Here we use the traditional approach of comparing test statistics via their asymptotic powers. This in turn reduces to a comparison of the non-centrality parameters of the limiting non-null limit distributions of the statistics. However, we shall illustrate with a specific application to a time-series model, and show that our asymptotic approach eventually involves a computation of an efficiency factor which is related to the Godambe criterion, thus establishing the practical significance of the Godambe criterion, at least in the particular appication considered here.

The paper is organized as follows. The basic assumptions are discussed in Section 8.2. The Rao score statistic and Neyman's $C(\alpha)$ statistic are presented in Sections 8.3 and 8.4 respectively. A general score statistic is introduced in Section 8.5. Finally, an application to a time-series model, ARMA (1,1), is discussed in Section 8.6.

8.2 The basic framework

Let $X = \{X_1, X_2, \ldots\}$ denote a sequence of possibly dependent random variables defined on a probability space $(R^\infty, B^\infty, P_\theta)$, θ being a k-dimensional unknown parameter taking values in $\Omega \subset R^k$. Denote the sample vector of n observations $(X_1, X_2, \ldots, X_n)$ by $X(n)$. If B^n denotes the σ-field generated by $X(n)$, and P_θ^n the restriction of P_θ to (R^n, B^n), we note that $X(n)$ is defined on the probability space (R^n, B^n, P_θ^n).

Let $\theta^T = (\alpha^T, \beta^T)$, where α, and β are of order $s \times 1$ and $(k - s) \times 1$ respectively. Consider the problem of testing the composite hypothesis $H: \alpha = \alpha_0$ against $K: \alpha \neq \alpha_0$, where α_0 is a specified vector. Note that β is an unknown nuisance parameter in the problem.

For any fixed θ, let $R_n(\theta)$ be a $k \times 1$ vector of measurable functions defined on (R^n, B^n). Thus, the components of $R_n(\theta)$ are some functions of $X(n)$ and θ. The vector $R_n(\theta)$ is chosen appropriately in a given estimation problem using specific criterion such as maximum likelihood, least squares, conditional least squares, least absolute deviation, minimum distance (with a specified distance function), and a quasi-likelihood, etc. Let $\hat{\theta}_n$ denote an appropriate solution of the estimating equation $R_n(\theta) = 0$. We shall refer to $R_n(\theta)$ as a *generalized score function*.

For any fixed $\theta_0 \in \Omega$, define a neighbourhood of θ_0 as follows:

$$N_n(\theta_0) = \{\theta \in \Omega, |\delta_n(\theta - \theta_0)| \leqslant h\}, \tag{8.1}$$

where δ_n is a sequence of $k \times k$ non-random symmetric positive definite matrices with the norms $\|\delta_n\|$ tending to infinity as $n \to \infty$, and h is a positive real number. Here, for any $k \times 1$ vector a, $|a|$ denotes $(\sum a_i^2)^{1/2}$, and for any $k \times k$ matrix A, $\|A\|$ denotes $\{\text{trace } A^T A\}^{1/2}$, where A^T is the transpose of A.

We shall assume that the score function $R_n(\theta)$ satisfies the following conditions:

(A.1) (*Local linearity*). There exists a $k \times k$ random matrix $W_n(\theta_0)$ such that

$$[\delta_n^{-1}\{R_n(\theta) - R_n(\theta_0)\} - \delta_n^{-1}W_n(\theta_0)(\theta - \theta_0)] \xrightarrow{P} 0,$$

for all $\theta \in N_n(\theta_0)$.

(A.2) (*Stability*). There exists a possibly random non-singular $k \times k$ matrix $W(\theta_0)$ such that

$$\delta_n^{-1}W_n(\theta_0)\delta_n^{-1} \xrightarrow{P} W(\theta_0).$$

(A.3) *(Asymptotic (mixed) normality)*. There exists a possibly random non-singular $k \times k$ matrix $G(\theta_0)$ such that

$$\delta_n^{-1} R_n(\theta_0) \xrightarrow{\text{d}} G^{1/2}(\theta_0)Z,$$

where Z is a $k \times 1$ vector of independent $N(0, 1)$ random variables which are independent of $G(\theta_0)$.

The limits, $\xrightarrow{\text{P}}$, and $\xrightarrow{\text{d}}$ denote the probability and the distribution convergence respectively.

Remarks   If $p_n(\cdot; \theta)$ denotes the density corresponding to P_θ^n, and assuming that $p_n(\cdot; \theta)$ is differentiable with respect to θ at least twice, the likelihood score function and the sample Fisher information matrix are respectively given by

$$S_n(\theta) = \{\ln p_n(x(n); \theta)\}' \qquad \text{and} \qquad W_n(\theta) = S_n'(\theta), \tag{8.2}$$

where the primes denote the derivative with respect to θ. Under further regularity conditions on $p_n(\cdot; \theta)$, one can check that the (likelihood) function $S_n(\theta)$ satisfies (A.1)–(A.3) with $W(\theta_0) = G(\theta_0) = F(\theta_0)$, where $F(\theta_0)$ is the limiting Fisher information. For ergodic type models $F(\theta_0)$ is non-random, whereas it is random for *non-ergodic* models, see Basawa and Koul (1979), Basawa and Prakasa Rao (1980), and Basawa and Scott (1983).

We shall use the notation $S_n(\theta)$ specifically for the likelihood score function, whereas $R_n(\theta)$ denotes an arbitrary score function. Our aim in this paper is to construct test statistics for testing H using the score function as a basis. From now on we shall assume for simplicity that δ_n is a diagonal matrix, as in Basawa (1985).

8.3   Rao's score statistic

Consider the problem of testing $H: \alpha = \alpha_0$, treating β as a nuisance parameter, where $\theta^T = (\alpha^T, \beta^T)$, as defined previously. The Rao statistic was discussed previously under a slightly different set-up by Basawa and Koul (1979), specifically for the non-ergodic situation. We now reformulate the Rao statistic under the general framework given in Section 8.2.

Let $S_n(\theta)$ be the likelihood score function defined in (2.2). Let us suppose that the density $p_n(\cdot; \theta)$ is sufficiently regular so that $S_n(\theta)$ satisfies conditions (A.1) to (A.3) of Section 8.2, for an appropriate choice of δ_n. Note that $W(\theta_0) = G(\theta_0)$ throughout this section. Let $\Omega = \Omega_1 \times \Omega_2$, and suppose $\alpha \in \Omega_1$, and $\beta \in \Omega_2$. Denote

$$\Omega_H = \{\theta = (\alpha_0, \beta)^T, \beta \in \Omega_2\}.$$

We shall denote an element of Ω_H by θ_H, i.e. $\theta_H = (\alpha_0, \beta)^T, \beta \in \Omega_2$. Let t_n^0 denote a suitable preliminary estimate of $\theta_H \in \Omega_H$ such that $\delta_n(t_n^0 - \theta_H)$ is bounded in probability. Define the one-step maximum likelihood estimator, restricted under H, by

$$\tilde{\theta}_n^0 = t_n^0 - W_n^{-1}(t_n^0)S_n(t_n^0). \tag{8.3}$$

Suppose $S_n(\theta)$ is partitioned as $S_n^T(\theta) = (U_n^T(\theta), V_n^T(\theta))$ where $U_n(\theta)$ is of order $s \times 1$, representing the derivatives with respect to α, and $V_n(\theta)$ is of order $(k - s) \times 1$ consisting of the derivatives with respect to β. Let $\hat{\theta}_n^0 = (\alpha_0, \hat{\beta}_0)^T$, where $\hat{\beta}_0$ is a consistent solution of $V_n(\theta_H) = 0$. Under appropriate regularity conditions it can be shown that $\hat{\theta}_n^0$ is asymptotically equivalent to $\tilde{\theta}_n^0$ in the sense that $\delta_n(\hat{\theta}_n^0 - \tilde{\theta}_n^0)$ converges to zero in probability.

We can now define the Rao score statistic as

$$T_{1n} = S_n^T(\tilde{\theta}_n^0)W_n^{-1}(\tilde{\theta}_n^0)S_n(\tilde{\theta}_n^0), \tag{8.4}$$

where $\tilde{\theta}_n^0$ is given by (8.3). As noted earlier, the estimator $\tilde{\theta}_n^0$ can be replaced by $\hat{\theta}_n^0$ without affecting the limit distribution of T_{1n} under H.

We have

$$T_{1n} = \begin{pmatrix} U_n(\tilde{\theta}_n^0) \\ V_n(\tilde{\theta}_n^0) \end{pmatrix}^T \begin{pmatrix} W_n^{11}(\tilde{\theta}_n^0) & W_n^{12}(\tilde{\theta}_n^0) \\ W_n^{21}(\tilde{\theta}_n^0) & W_n^{22}(\tilde{\theta}_n^0) \end{pmatrix} \begin{pmatrix} U_n(\tilde{\theta}_n^0) \\ V_n(\tilde{\theta}_n^0) \end{pmatrix},$$

where $W_n^{ij}(\cdot)$ are the appropriate submatrices of $W_n^{-1}(\cdot)$. Since $V_n(\tilde{\theta}_n^0) = o_p(1)$, it follows that

$$T_{1n} = U_n^T(\tilde{\theta}_n^0)W_n^{11}(\tilde{\theta}_n^0)U_n(\tilde{\theta}_n^0) + o_p(1). \tag{8.5}$$

It can further be verified that

$$U_n(\tilde{\theta}_n^0) = U_n(\theta_H) - W_{n12}(\theta_H)W_{n22}^{-1}(\theta_H)V_n(\theta_H) + o_p(1), \tag{8.6}$$

where $W_{n_{ij}}(\cdot)$ are the appropriate submatrices of $W_n(\cdot)$. The result in (8.6) shows that $U_n(\tilde{\theta}_n^0)$ is approximately equal to the regression of $U_n(\theta_H)$ on $V_n(\theta_H)$. Furthermore, we have

$$W^{11}(\theta) = (W_{11}(\theta) - W_{12}(\theta)W_{22}^{-1}(\theta)W_{21}(\theta))^{-1}. \tag{8.7}$$

Replacing θ_0 by θ_H in (A.1)–(A.3), it is easily verified via (8.5)–(8.7), assuming $W_n(\theta)$ to be continuous in θ, that under H,

$$T_{1n} \xrightarrow{d} \chi^2(s) \quad \text{as } n \to \infty. \tag{8.8}$$

Under some further regularity conditions one can derive the non-null limit distribution of T_{1n} under the sequence of alternative hypotheses

$$K_n: \theta = \theta_n^0, \quad \text{where } \theta_n^{0T} = (\alpha_n^{0T}, \beta^T), \tag{8.9}$$

where $\alpha_n^0 = \alpha_0 + \delta_{n1}^{-1}\eta$, δ_{n1} is of order $s \times s$ and η is an $s \times 1$ vector of real numbers. One can show that under K_n, assuming $W(\theta)$ to be non-random,

$$T_{1n} \xrightarrow{d} \text{non-central } \chi^2(s, \lambda^2), \tag{8.10}$$

where

$$\lambda^2 = \eta^{\mathrm{T}}(W_{11}(\theta_H) - W_{12}(\theta_H)W_{22}^{-1}(\theta_H)W_{21}(\theta_H))\eta. \qquad (8.11)$$

Note that if $W(\theta)$ is random, the limit distribution in (8.10) can be considered as a mixture of non-central chi-squared distributions with λ acting as a mixing random variable. It may be noted that Basawa and Koul (1979) considered the statistic $T_{1n}^* = S_n^{\mathrm{T}}(\tilde{\theta}_n^0)S_n(\tilde{\theta}_0)$ which has a more complicated limit distribution both under H and K_n. The rescaled score statistic T_{1n} defined in (8.4) has a simpler limit distribution. In particular, under H, T_{1n} continues to have a limiting chi-squared distribution even when $W(\theta)$ is random.

8.4   Neyman's $C(\alpha)$ statistic

Neyman (1959, 1979) introduced an interesting class of statistics which he called $C(\alpha)$ tests in honour of Cramer. The main idea of $C(\alpha)$ tests can be described as follows. We use the general set-up of Section 8.2 even though Neyman originally presented his tests for the classical situation of independent observations. Let

$$R_n(\theta) = \begin{pmatrix} R_{n1}(\theta) \\ R_{n2}(\theta) \end{pmatrix} \qquad \text{and} \qquad S_n(\theta) = \begin{pmatrix} S_{n1}(\theta) \\ S_{n2}(\theta) \end{pmatrix}$$

be respectively an arbitrary score function, and the likelihood score function, respectively, where $R_{n1}(\theta)$ and $S_{n1}(\theta)$ are of order $s \times 1$, and $R_{n2}(\theta)$, $S_{n2}(\theta)$ are $(k - s) \times 1$ vectors. Consider the 'hybrid' score vector, viz.

$$R_n^*(\theta) = \begin{pmatrix} R_{n1}(\theta) \\ S_{n2}(\theta) \end{pmatrix}. \qquad (8.12)$$

Let us now suppose that $R_n^*(\theta)$ satisfies the conditions (A.1)–(A.3) of Section 8.2 with $W_n(\cdot)$, $W(\cdot)$, and $G(\cdot)$ replaced by $W_n^*(\cdot)$, $W^*(\cdot)$, and $G^*(\cdot)$ respectively. Note that in general, $W^* \neq G^*$. Consider the 'regression' of $R_{n1}(\theta_H)$ on $S_{n2}(\theta_H)$ given by

$$Z_n(\theta_H) = R_{n1}(\theta_H) - W_{n12}^*(\theta_H)W_{n22}^{*-1}(\theta_H)S_{n2}(\theta_H), \qquad (8.13)$$

where W_{n12}^*, and W_{n22}^* are the appropriate partitioned matrices of W_n^*. Since the expression $Z_n(\theta_H)$ in (8.13) depends on an unknown nuisance parameter β, it cannot be used directly as a test statistic for testing $H: \alpha = \alpha_0$. As in Section 8.3, we now replace θ_H in $Z_n(\theta_H)$ by the restricted maximum likelihood estimator $\tilde{\theta}_n^0$ (or $\tilde{\theta}_n^0$), and obtain the Neyman $C(\alpha)$, statistic $Z_n(\tilde{\theta}_n^0)$. Note that, under H,

$$Z_n(\tilde{\theta}_n^0) = R_{n1}(\tilde{\theta}_n^0) + o_p(1) = Z_n(\theta_H) + o_p(1). \qquad (8.14)$$

It is easily verified that, under H,

$$\delta_{1n}^{-1} Z_n(\tilde{\theta}_n^0) \xrightarrow{d} C^{1/2}(\theta_H) Z, \tag{8.15}$$

where Z is a $s \times 1$ vector of independent $N(0, 1)$ random variables, and

$$C(\theta) = \{G_{11}^*(\theta) - W_{12}^*(\theta) W_{22}^{*-1}(\theta) G_{21}^*(\theta)\}$$
$$- \{G_{12}^*(\theta) - W_{12}^*(\theta) W_{22}^{*-1}(\theta) G_{22}^*(\theta)\}(W_{22}^{*-1}(\theta))^{\mathrm{T}} W_{12}^{*\mathrm{T}}(\theta). \tag{8.16}$$

The expression in (8.16) is simply the limiting covariance matrix of $\delta_{1n}^{-1} Z_n(\theta_H)$. Consider now the test statistic

$$T_{2n} = Z_n^{\mathrm{T}}(\tilde{\theta}_n^0) \delta_{1n}^{-1} C^{-1}(\tilde{\theta}_n^0) \delta_{1n}^{-1} Z_n(\tilde{\theta}_0). \tag{8.17}$$

From (8.15), it is clear that, under H,

$$T_{2n} \xrightarrow{d} \chi^2(s).$$

Under the sequence of alternatives K_n as defined in Section 8.3, we have

$$T_{2n} \xrightarrow{d} \text{non-central } \chi^2(s, \lambda^2),$$

with

$$\lambda^2 = \eta^{\mathrm{T}} A(\theta_H) \eta, \tag{8.18}$$

where $A(\theta) = B^{-1}(\theta)$ with $B(\theta)$ being the matrix consisting of the first s rows and s columns of $\{W^{*\mathrm{T}}(\theta) G^{*-1}(\theta) W^*(\theta)\}^{-1}$.

Neyman (1959, 1979) considered the problem of deriving an optimal $C(\alpha)$ test in the classical situation of independent observations. In the present framework, this problem is equivalent to choosing the score function $R_{n1}(\theta)$ in (4.1) so as to maximize the non-centrality parameter λ^2 in (4.7). Using essentially similar arguments as in Neyman (1959), one can show, not surprisingly, that an optimum choice for $R_{n1}(\theta)$ is $S_{n1}(\theta)$. Consequently, an optimum $C(\alpha)$ test is simply the Rao score statistic T_{1n} discussed in Section 8.3.

8.5   Generalized score statistic

Both the Rao and the Neyman statistics require the knowledge of the likelihood score function $S_n(\theta)$. While the Rao statistic is directly based on $S_n(\theta)$, the Neyman statistic is more flexible, in that the score function $R_{n1}(\theta)$ in (8.12) is allowed to be arbitrary, and the knowledge of $S_n(\theta)$ is indirectly involved in computing $C(\theta_H)$ in (8.16). In some situations, $S_n(\theta)$ is either not available or it is too complex to be of much use. In other circumstances, one may prefer to use a robust method of estimation, in which case the robust score function $R_n(\theta)$ is generally different from the likelihood score function $S_n(\theta)$. In this section, we consider a statistic which is based entirely on $R_n(\theta)$. See also Basawa (1985).

To do this, we simply replace $S_{n2}(\theta)$ in (8.12) by $R_{n2}(\theta)$. Then W_n^*, W^*, and G^* in the previous section are replaced by W_n, W, and G as specified in (A.1)–(A.3). We now consider the regression of $R_{n1}(\theta_H)$ on $R_{n2}(\theta_H)$.

$$\xi_n(\theta_H) = R_{n1}(\theta_H) - W_{n12}(\theta_H)W_{n22}^{-1}(\theta_H)R_{n2}(\theta_H). \qquad (8.19)$$

Since θ_H involves the unknown nuisance parameter β we need to replace it by an appropriate estimate as in Sections 8.3 and 8.4. In both the Rao and Neyman statistics we used the restricted maximum likelihood estimator of β restricted under $H: \alpha = \alpha_0$. However, in the present section, such an estimate is not available since the entire likelihood function is unavailable. Instead, we consider a generalized M-estimator suggested in Basawa (1985), viz.

$$\tilde{\beta}_n = \beta_n + W_{n22}^{-1}R_{n2}(\tilde{\theta}_H), \qquad (8.20)$$

where $R_n^T(\theta) = (R_{n1}^T(\theta), R_{n2}^T(\theta))$ with $R_{n1}(\theta)$ being of order $(s \times 1)$, etc, $\tilde{\theta}_H^T = (\alpha_0^T, \beta_n^T)$, and β_n is a preliminary estimate of β such that $\delta_{n1}(\theta)(\tilde{\beta}_n - \beta)$ is bounded in probability. Under appropriate regularity conditions on $R_n(\theta)$, one can obtain an estimate β_n^* of β by solving $R_{n2}(\theta_H) = 0$ for β, and then verify that $\tilde{\beta}_n^*$ and $\tilde{\beta}_n$ are asymptotically equivalent. Denote by $\tilde{\theta}_n^{*T} = (\alpha_0^T, \tilde{\beta}_n^*)$. We then consider $\xi_n(\tilde{\theta}_n^*)$ as a test statistic. Define

$$T_{3n} = \xi_n^T(\tilde{\theta}_n^*)\delta_{1n}^{-1}C^{-1}(\tilde{\theta}_n^*)\delta_{1n}^{-1}\xi_n(\tilde{\theta}_n^*), \qquad (8.21)$$

where the matrix C is as defined in (8.16) except that G_{ij}^*, W_{ij}^* in (8.16) are now replaced by G_{ij} and W_{ij} respectively.

The limiting null distribution of T_{31} can be seen to be identical with that of T_{2n}. The non-null distribution of T_{31} is non-central chi-squared with the non-centrality parameter defined as in (8.18), but now W^* and G^* are replaced by W and G respectively. The optimum choice of $R_n(\theta)$, in the sense of maximizing the non-centrality parameter, turns out to be the likelihood score functions $S_n(\theta)$, as one would expect.

8.6   An appplication to a time-series model

Consider an autoregressive-moving average model of order (1, 1), i.e. ARMA (1, 1), $\{X = X_1, X_2, \ldots\}$. The sequence of random variables $\{X_n\}$ satisfies the difference equation,

$$X_n - \alpha X_{n-1} = \varepsilon_n - \beta \varepsilon_{n-1}, \quad \text{with } X_0 = 0, \quad |\alpha| < 1, \quad |\beta| < 1, \ (8.22)$$

where $\{\varepsilon_n\}$ are independent $N(0, 1)$ unobservable random errors. Given a sample of observations $X(n) = (X_1, X_2, \ldots, X_n)$, suppose we are interested in testing the null hypothesis $H: \alpha = \alpha_0$ against a sequence of alternatives $K_n: \alpha = \alpha_0 + h/\sqrt{n}$, treating the moving-average parameter β as an unknown nuisance parameter. Here $\theta^T = (\alpha, \beta)$. First, consider the likelihood score vector $S_n(\theta)$ with $S_n^T(\theta) = (U_n(\theta), V_n(\theta))$ given by

$$U_n(\theta) = \sum_{t=1}^{n} \varepsilon_t(\theta) \frac{\partial \varepsilon_t(\theta)}{\partial \alpha} \quad \text{and} \quad V_n(\theta) = \sum_{t=1}^{n} \varepsilon_t(\theta) \frac{\partial \varepsilon_t(\theta)}{\partial \beta}, \quad (8.23)$$

where $\varepsilon_t(\theta) = X_t - \alpha X_{t-1} + \beta \varepsilon_{t-1}(\theta)$ are computed recursively with $\varepsilon_0(\theta) = 0$. It can easily be verified that the conditions (A.1)–(A.3) of Section 8.2 are satisfied with $\delta_n = n^{1/2}$, and

$$W(\theta) = G(\theta) = \begin{pmatrix} (1 - \alpha^2)^{-1} & -(1 - \alpha\beta)^{-1} \\ -(1 - \alpha\beta)^{-1} & (1 - \beta^2)^{-1} \end{pmatrix} = F(\theta), \quad (8.24)$$

where $F(\theta)$ denotes the Fisher information matrix. See, for instance, Box and Jenkins (1976).

The Rao statistic for testing $H: \alpha = \alpha_0$ is given by

$$T_{1n} = n^{-1} U_n^2(\tilde{\theta}_n^0) \left\{ \frac{(1 - \alpha_0^2)(1 - \alpha_0 \tilde{\beta}_n)^2}{(\alpha_0 - \tilde{\beta}_n)^2} \right\}, \quad (8.25)$$

where $\tilde{\theta}_n^{0T} = (\alpha_0, \tilde{\beta}_n)$, and $\tilde{\beta}_n$ is a solution of the equation $V_n(\tilde{\theta}_n^0) = 0$.

Consider now a generalized score function $R_n^T(\theta) = (U_n^*, V_n^*(\theta))$, where

$$U_n^* = \sum_{t=1}^{n} \left[g\{\varepsilon_t(\theta)\} \frac{\partial g\{\varepsilon_t(\theta)\}}{\partial \alpha} \right], \quad V_n^*(\theta) = \sum_{t=1}^{n} \left[g\{\varepsilon_t(\theta)\} \frac{\partial g\{\varepsilon_t(\theta)\}}{\partial \beta} \right]. \quad (8.26)$$

In (8.26), the function $g(\cdot)$ is differentiable, $0 < E\{g^2(\varepsilon_1)\} < \infty$, and $E\{g'(\varepsilon_1)\} \neq 0$, where $g'(u)$ denotes the derivation of g with respect to u. The general score function defined in (8.26) was used by Basawa et al. (1985) to derive robust test statistics. Here we shall use the score function $R_n(\theta)$ to illustrate the test statistic in Section 8.5.

It can be verified that

$$W(\theta) = E\{g'(\varepsilon_1)\} E\{\varepsilon_1 g(\varepsilon_1)\} F(\theta) \quad \text{and} \quad G(\theta) = E^2\{g^2(\varepsilon_1)\} F(\theta)), \quad (8.27)$$

where $F(\theta)$ is the Fisher information matrix defined in (8.24). It is clear that, in general, $W(\theta) \neq G(\theta)$. In the special case, when $g(u) \equiv u$, the generalized score function reduces to the likelihood score function, and we recover the identity $W(\theta) = G(\theta) = F(\theta)$.

Returning to the generalized score function in (8.26) we may note that the general score statistic in (8.21) reduces to

$$T_{3n} = n^{-1} \{U_n^*(\tilde{\theta}_n^*)\}^2 \left\{ \frac{(1 - \alpha_0^2)(1 - \alpha_0 \tilde{\beta}_n^*)^2}{(\alpha_0 - \tilde{\beta}_n^*)^2} \right\} [E_{\tilde{\theta}_n^*}^2 \{g^2(\varepsilon_1)\}]^{-1}, \quad (8.28)$$

where $\tilde{\theta}_n^{*T} = (\alpha_0^T, \tilde{\beta}_n^{*T})$, and $\tilde{\beta}_n^*$ is obtained as a solution of $V_n^*(\theta_H) = 0$.

The non-centrality parameters λ_1 and λ_2, corresponding to T_{1n} and T_{2n} respectively, are given by

$$\lambda_1 = \left\{ \frac{(\alpha_0 - \beta)^2}{(1 - \alpha_0^2)(1 - \alpha_0\beta)^2} \right\} \quad \text{and} \quad \lambda_2 = \lambda_1 \{Eg'(\varepsilon_1)\}^2 \{Eg^2(\varepsilon_1)\}^{-1} \rho^2_{\varepsilon_1, g(\varepsilon_1)},$$

(8.29)

where $\rho_{\varepsilon_1, g(\varepsilon_1)}$ denotes the correlation coefficient between ε_1 and $g(\varepsilon_1)$. The asymptotic relative efficiency of T_{2n} with respect to T_{1n} is given by

$$\lambda_2/\lambda_1 = \{Eg'(\varepsilon_1)\}^2 \{Eg^2(\varepsilon_1)\}^{-1} \rho^2_{\varepsilon_1, g(\varepsilon_1)},$$

(8.30)

which is maximized for the choice $g(u) \equiv u$.

Note that the relative efficiency in (8.30) is proportional to the Godambe criterion, viz. $\{Eg'(\varepsilon_1)\}^2 \{Eg^2(\varepsilon_1)\}^{-1}$.

In order to construct a Neyman $C(\alpha)$ statistic, consider $R_n^{*T}(\theta) = (U_n^*(\theta), V_n(\theta))$, where $U_n^*(\theta)$ and $V_n(\theta)$ are defined in (8.26) and (8.23) respectively. The $C(\alpha)$ statistic T_{2n} given in (8.17) is then given by the statistic defined in (8.28) with $\tilde{\theta}_n^*$ replaced by $\tilde{\theta}_n^0$, defined after (8.25). It is seen that for our particular choices of $R_n(\theta)$ and $R_n^*(\theta)$ above, the two statistics T_{3n} and T_{2n} will have asymptotically the same limit distributions, both under H, and under K_n.

Finally, note that the assumption $|\alpha| < 1$ used above ensures that G and W in (A.1)–(A.3) are non-random. For $|\alpha| > 1$, one needs to use a different normalizing sequence δ_n. One then obtains non-degenerate random variables G and W leading to the non-ergodic situation. The Rao statistic for a non-ergodic autoregressive process was discussed by Basawa and Koul (1979). Basawa et al. (1984) discuss Rao statistic for several time-series models again assuming an ergoci set-up.

8.7   Acknowledgements

This work was partially supported by a grant from the Office of Naval Research.

References

Basawa, I. V. (1985). Neyman–LeCam tests based on estimating functions. In *Proceedings of the Berkeley Conference in honor of Jerzy Neyman and Jack Kiefer*, **2**, 811–26. Wadsworth, Monterey.

Basawa, I. V., Billard, L., and Srinivasan, R. (1984). Large sample tests for time series models. *Biometrika*, **71**, 203–6.

Basawa, I. V., Huggins, R. M., and Standte, R. G. (1985). Robust tests for time series with an application to first-order autoregressive processes. *Biometrika*, **72**, 559–71.

Basawa, I. V. and Koul, H. L. (1979). Asymptotic tests of composite hypotheses for non-ergodic type stochastic processes. *Stoch. Processes Appl.*, **9**, 291–305.

Basawa, I. V. and Koul, H. L. (1988). Large-sample statistics based on quadratic dispersion. *Int. Stat. Rev.*, **56**, 199–219.

Basawa, I. V. and Prakasa Rao, B. L. S. (1980). Asymptotic inference for stochastic processes. *Stoch. Processes Appl.*, **10**, 221–54.

Basawa, I. V. and Scott, D. J. (1983). *Asymptotic optimal inference for non-ergodic models.* Lecture Notes in Statistics, Vol. 17. Springer-Verlag, New York.

Box, G. E. P. and Jenkins, G. M. (1976). *Time series analysis.* Wiley, New York.

Godambe, V. P. (1960). An optimum property of regular maximum likelihood equation. *Ann. Math. Stat.*, **31**, 1208–11.

Godambe, V. P. (1985). The foundation of finite sample estimation in stochastic processes. *Biometrika*, **72**, 419–28.

Neyman, J. (1959). Optimal asymptotic tests of composite hypotheses. In U. Grenander (ed.), *Probability and statistics*, H. Cramer Volume, pp. 213–34. Almquist and Wiksell, Uppsala.

Neyman, J. (1979). $C(\alpha)$ tests and their use. *Sankhya* A, **41**, 1–21.

9
Quasi-likelihood, stochastic processes, and optimal estimating functions

A. F. Desmond

ABSTRACT

Recent work on inference for stochastic processes via estimating functions is reviewed. The connections between the notions of quasi-likelihood (Wedderburn 1974) and quasi-scores (Godambe 1985) are outlined in a general semi-parametric framework. The emphasis is on obtaining optimal estimating functions within the martingale structure of Godambe and Heyde (1987). By focusing on properties such as linearity and unbiasedness, applied to estimating functions, rather than estimators, finite sample optimality properties similar in spirit to classical ideas such as the Gauss–Markov theorem are obtained. There are close connections also with the work of McCullagh (1983), who emphasizes local unbiasedness and local linearity and the asymptotic properties of maximum quasi-likelihood estimates.

9.1 Introduction

Point estimation, as traditionally presented in standard textbooks, e.g. Rao (1973), focuses on properties of estimators which are statistics i.e. functions of the observations alone. The class of estimators is then frequently restricted to a subclass of estimators, such as those which are unbiased, linear, equivariant, etc., and an optimal estimator among this subclass is sought. Optimality is frequently defined in terms of minimizing some objective function such as mean squared error, variance or some other loss function. Some examples of this approach are the classical theory of minimum variance unbiased estimation, minimum risk equivariant estimation, and so on. While this approach leads to mathematically attractive and elegant solutions, the emphasis on unbiasedness may have unfortunate consequences (Cox and Hinkley 1974, p. 253). Also, comparing the sampling distributions of estimators via their variance is not always sensible (Barnard and Sprott 1979).

An alternative approach is to focus on functions $g(x, \theta)$ of the data x and the unknown parameters θ, and to study estimators as solutions of

$$g(x, \theta) = 0. \tag{9.1}$$

(9.1) is referred to as an estimating equation, while the function g itself is termed an estimating function. In the case where θ is a vector of dimension k, say, g itself is also a vector function of dimension k. Many approaches to estimation eventually lead to solving systems of equations such as (9.1), e.g. maximum likelihood, minimum chi-squared, least squares, etc. Much of the literature on statistical inference is devoted to studying asymptotic properties of estimates generated in this way, properties such as consistency, asymptotic efficiency and asymptotic normality. In the theory of estimating functions it is the estimating function itself which is regarded as primary, rather than the estimator derived from it. Thus, one focuses on optimality criteria applied directly to the estimating function. Similarly, in combining data from different experiments, one investigates optimal combinations of estimating functions rather than estimators (Heyde 1987).

Godambe (1960) derived a certain optimality property of the maximum likelihood equation using the estimation equation approach. Briefly, he showed for the one-parameter case that among all unbiased estimating functions, i.e. those with $E_\theta g = 0$, the one with the minimal value of the optimality index

$$J(g) = \frac{E_\theta\{g^2\}}{\left[E_\theta\left(\dfrac{\partial g}{\partial \theta}\right)\right]^2} \tag{9.2}$$

is $g = \partial \log f / \partial \theta$, where $f(x, \theta)$ is a probability density completely specified except for an unknown parameter θ.

Kale (1962) and Bhapkar (1972) extend this result to the multiparameter case. For an introduction to the literature on estimating functions up to 1986 see Desmond (1989).

In this article we will discuss recent approaches to inference via estimating functions in situations where the underlying distribution is *not* completely specified. For example, in the next section we deal with quasi-likelihood models, where a mean-variance relationship is specified in parametric terms but the underlying distribution itself is unknown. Thus, the true score function is unavailable, while the quasi-score function provides an optimal surrogate according to criterion (9.2) (Godambe 1985).

In Section 9.3 the usefulness of the estimating function approach in inference for stochastic processes is illustrated using recent examples. It is perhaps worth noting in this regard, that Durbin (1960), in an important early paper, found that focusing on linear estimating functions for certain time-series models provided more satisfactory methods for dealing with the problems involved in estimating parameters of autoregressive models.

9.2   Quasi-likelihood and estimating functions

As we have seen in the previous section, Godambe's (1960) result (together with its multiparameter extensions) shows that the score function is the optimal estimating function in parametric situations, i.e. situations where the form of the underlying distribution is known up to an unknown (possibly vector) parameter. Denote the class of underlying distributions by $\mathscr{F}_0 = \{f_\theta: \theta \in \Omega)$, where Ω is a Euclidean space. Suppose, now, we broaden the class of underlying distributions to include, for example, semi-parametric models. One way of formalizing this is to assume that the underlying model is selected from a class which is a union (possibly uncountable) of parametric models. We write $\mathscr{F} = \bigcup_{N \in \mathscr{N}} \mathscr{F}_{0,N}$, say, where the space $\mathscr{N}$ is frequently an abstract space as in semi-parametric models. If the 'true' value N_0 of the nuisance function were known, then the score function, $U_{(\theta, N_0)} = \partial \log f(\theta, N_0)/\partial \theta$ would provide optimal estimation as before. However, in this more general situation, the true score function is *not* known.

This broader framework includes many models occurring in practice. Examples are:

(i) Nuisance parameter models, where $\mathscr{N}$ is also a Euclidean space (e.g. Neyman and Scott (1948)).

(ii) Semi-parametric models, such as Cox's regression model (Cox 1972), where the space $\mathscr{N}$ is related to the function space of baseline hazard functions and θ is the vector of regression coefficients.

(iii) Location models, i.e. $\mathscr{F} = \{f(x - \theta): f \text{ unknown}\}$.

(iv) Scale models, $\mathscr{F} = \{f(x/\theta): f \text{ unknown}\}$.

(v) Linear regression models with unspecified error distributions.

(vi) Generalized linear models with unspecified error distribution (for example, quasi-likelihood models).

Since the true underlying score function $U_{(\theta, N_0)}$ is unknown in these types of situations, intuitively one ought to look for an estimating equation which is 'closest' to $U_{(\theta, N_0)}$ in some sense. Two possibilities are

(i) Find $g^* \in \mathscr{G} = $ class of unbiased estimating equations such that

$$\text{Correlation}_F\left(g^*, \frac{\partial \log f_{(\theta, N_0)}}{\partial \theta}\right) \geq \text{Correlation}_F\left(g, \frac{\partial \log f_{(\theta, N_0)}}{\partial \theta}\right)$$

for all $g \in \mathscr{G}, F \in \mathscr{F}$.

(ii) Find g^* such that

$$E_F\left(g^* - \frac{\partial \log f_{(\theta, N_0)}}{\partial \theta}\right)^2 \leq E_F\left(g - \frac{\partial \log f_{(\theta, N_0)}}{\partial \theta}\right)^2$$

for all $g \in \mathscr{G}, F \in \mathscr{F}$.

Then (i) and (ii) are reasonable optimality criteria in *finite* samples. Obvious questions are (a) how are these criteria related? and (b) how do they relate to the criterion (9.2)? Fortunately, Godambe and Thompson (1987) have shown that all three criteria are equivalent and that, while for the broad class $\mathscr{G}$ of all unbiased estimating functions it is *not* possible in general to find a uniquely optimal estimating function, by restricting attention to an appropriate subclass $\mathscr{G}_1$ of $\mathscr{G}$, a unique (up to a multiplicative factor) optimum can be achieved. The uniquely optimal g^* is referred to as a quasi-score function or pseudo-score function and was introduced originally in the context of finite sample estimation for stochastic processes (Godambe 1985); (see the next section). The quasi-score function g^* has the further important (albeit asymptotic) optimality property that confidence intervals based on g^* are asymptotically shortest within $\mathscr{G}_1$. This recalls the well-known property (Wilks 1938) of the score function in fully specified parametric models. The integral of the quasi-score function with respect to θ will be referred to as a quasi-likelihood function.

The term quasi-likelihod was originally introduced by Wedderburn (1974) in the context of generalized linear models. However, historically, the theory has roots in two key ideas of Gauss:

(A) the Gauss–Markov theorem on least-squares estimation;
(B) the fact that maximum likelihood and least squares coincide uniquely for the normal distribution within the location family of distributions.

Recall that the original Gauss–Markov theorem assumes a constant variance, but no further distributional assumptions are necessary. Wedderburn assumed that the variance is a specified function of the mean together with a possibly unknown scalar multiplicative parameter. His main motivation in introducing quasi-likelihood was to relax one of the basic assumptions of the theory of generalized linear models, namely, that the error distribution be of a given exponential family form. However, looked at from a different point of view, one can regard it as an extension of the above ideas of Gauss. To see this, consider the class of distributions $\mathscr{F}$ which consist of product measures on $\mathbb{R}_n$ and satisfy

$$E_F(Y_i) = \alpha_i\{\theta(F)\} \quad \text{and} \quad \mathrm{Var}_F(Y_i) = k(F)\sigma_i^2\{\theta(F)\},$$

$$i = 1, \ldots, n, \quad F \in \mathscr{F}. \quad (9.3)$$

Note that we have made explicit the dependence on $F \in \mathscr{F}$ of both the expectation E and the parameter θ. The functions α_i and σ_i^2 are specified differentiable functions of θ, while k is assumed functionally independent of θ. This is rather more general than the usual framework of generalized linear models in that

(i) no link function specifing α_i in terms of a linear function of θ is assumed.

(ii) σ_i^2 is assumed to depend on θ in quite a general way not specifically through α_i.

In the simplest situation, where θ is a *real* parameter, it has been shown (see, e.g. Chapter 1) that the optimal estimating function in the class of linear Gauss-consistent estimating functions is given by quasi-likelihood. By a linear equation, here we mean an equation of the form

$$\sum_{i=1}^{n} \omega_i(\theta)(Y_i - \alpha_i(\theta)) = 0$$

where the $\omega_i(\theta)$ are appropriate weights. Such estimating equations are Gauss-consistent in the sense that, if the Y_i are observed without error, then the solution of this equation equals the true value θ (cf. Sprott 1983). Thus Wedderburn's quasi-likelihood can also be viewed as the optimal Gauss-consistent linear estimating equation. Godambe and Thompson (1987) show that this extends to the vector-parameter situation.

Other important contributions to the theory of quasi-likelihood are those of McCullagh (1983), Firth (1987) and Crowder (1987). All of these papers focus on asymptotic properties of quasi-likelihood as opposed to the finite sample criterion of Godambe. McCullagh shows that analogues of the usual results for ordinary likelihoods apply in the independent case, e.g. the maximum quasi-likelihood estimate (MQLE) is consistent, asymptotically normal, etc. MQLEs are also shown to satisfy an asymptotic optimality property similar in spirit to that in the Gauss–Markov theorem. McCullagh claims greater generality for this result, in that global linearity is replaced by local linearity, and unbiasedness is replaced by local unbiasedness. However, this greater generality is offset by the asymptotic nature of the optimality results, as opposed to Gauss's original finite-sample optimality (Heyde and Seneta 1977). By focusing on assumptions of linearity and unbiasedness applied directly to the estimating functions, rather than the estimates themselves, the spirit of the original finite sample justification by Gauss is more closely preserved. In this connection it is worth re-emphasizing that Gauss did not invoke either unbiasedness or asymptotics.

Firth (1987) focuses on efficiency of quasi-likelihood estimation relative to some particular exponential family distributions. Efficiency is defined in terms of relative asymptotic variances. Crowder (1987) similarly emphasizes the asymptotic variance of estimates. While the optimality criterion in these papers is mathematically similar to that of Godambe and Thompson (1987, 1989), the philosophical framework of the two approaches is quite distinct. With some presumption, it might be claimed that the treatment of Godambe and Thompson is closer to the spirit of Gauss, whereas that of McCullagh, Firth, and Crowder is logically closer to that of Laplace (cf. Heyde and Seneta 1977).

9.3 Optimal estimation for stochastic processes via estimating equations

One area of statistics which is undergoing vigorous study at present is that of inference and estimation for stochastic processes (see, for example, Basawa and Prakasa Rao 1980). Much of this work is likelihood-based, although there are situations in which it is difficult to write down the likelihood function. For examples, see Cox and Hinkley (1974, p. 16) or Jacod (1987). Also, many of these studies involve asymptotic considerations and a great deal of literature now exists on asymptotic normality and consistency of estimators for special processes, extending earlier work for independent processes. A unified treatment of much of this via martingale limit theorem is now possible (e.g. Hall and Heyde 1980, especially Chapter 6).

However, there are two alternative directions in inference for stochastic processes worthy of consideration. First, there is the question of finite-sample optimality properties as opposed to asymptotics. Secondly, the question of methods which do not involve the determination of the likelihood, i.e. second-order methods such as least squares and quasi-likelihood and more general semi-parametric models.

A first step towards establishing some finite-sample results for stochastic processes is initiated in Godambe (1985). (It is worth remarking, perhaps, that, in a recent extensive review of inference for stochastic processes (Moore 1987), this is the only reference which deals explicitly with finite-sample optimality.) In this paper an analogue of the Gauss–Markov theorem appropriate for discrete-time stochastic processes is proved under mild conditions. The approach is via estimating equations as follows:

Let $Y_1, \ldots, Y_n$ be a sample in discrete time from a stochastic process whose probability measure depends on some unknown parameter θ. Let $\mathscr{F}$ be the underlying class of distributions on $\mathbb{R}_n$ and $\theta = \theta(F)$, $F \in \mathscr{F}$ a real parameter. We will suppress the dependence of θ on F for notational convenience in what follows. Denote by $\mathscr{F}_i$ the σ-algebra generated by $Y_1, \ldots, Y_i$, $1 \leqslant i \leqslant n$ and $E_i(\cdot) = E(\cdot / \mathscr{F}_i)$ the conditional expectation operator given this σ-algebra. Further, suppose there exist (measurable) functions $h_i = h_i(Y_1, \ldots, Y_i; \theta)$ satisfying

$$E_{i-1}\{h_i(Y_1, \ldots, Y_i; \theta)\} = 0, \qquad i = 1, \ldots, n, F \in \mathscr{F}, \qquad (9.4)$$

that is the h_i, $1 \leqslant i \leqslant n$, are martingale differences. Note that (9.4) implies

$$E(h_i h_j) = 0, \qquad i \neq j, \qquad (9.5)$$

where E is unconditional expectation, so that the h_i are orthogonal.

Such functions are easily found. For example, in the AR(1) process

$$Y_i = \theta Y_{i-1} + \varepsilon_i$$

where the ε_i are independent zero-mean random variables with constant variances, a natural choice is

$$h_i = Y_i - \theta Y_{i-1}, \qquad i = 1, \ldots, n.$$

More generally,

$$h_i = Y_i - E_{i-1}(Y_i), \qquad i = 1, \ldots, n, \qquad (9.6)$$

i.e. the residual between Y_i and its 'best' predictor $E_{i-1}(Y_i)$ based on $Y_1 \ldots Y_{i-1}$, satisfies (9.4) and (9.5).

Now consider the class $\mathscr{G}_1$ of 'linear' martingale estimating equations

$$\mathscr{G}_1 = \left\{ g \colon g = \sum_{i=1}^{n} a_{i-1} h_i \right\}, \qquad (9.7)$$

where the coefficients a_{i-1} are functions of $Y_1, \ldots, Y_{i-1}$ and θ. Note that in (9.7) the h_i are specified and the class $\mathscr{G}_1$ is generated by varying the a_{i-1}. Also, because of (9.4), $\mathscr{G}_1$ is a subset of the class $\mathscr{G}$ of all unbiased estimating equations, i.e.

$$E(g) = 0, \qquad \text{all } g \in \mathscr{G}_1.$$

Then it can be shown that

$$g^* = \sum_{i=1}^{n} a_{i-1}^* h_i$$

is optimal, according to criterion (9.2), *within the class* $\mathscr{G}_1$, for the choice

$$a_{i-1}^* = \frac{-E_{i-1}(\partial h_i / \partial \theta)}{E_{i-1}(h_i^2)}, \qquad (9.8)$$

where the appropriate derivatives and expectations are assumed to exist.

For the AR(1) process with $h_i = Y_i - \theta Y_{i-1}$ and $\mathrm{Var}(h_i) = \sigma^2$, constant, this yields

$$a_{i-1}^* = \frac{-Y_{i-1}}{\sigma^2},$$

whence

$$g^* = \sum_{i=1}^{n} Y_{i-1}(Y_i - \theta Y_{i-1}).$$

More generally, let $h_i = Y_i - E_{i-1}(Y_i)$ as in (9.6) and

$$\mathscr{F} = \{F \colon E_{i-1}(h_i) = 0 \quad \text{and} \quad E_{i-1}(h_i^2) = \sigma^2(F), \quad i = 1, \ldots, n\},$$

where $\sigma^2(F)$ is independent of θ. Then the optimum estimating equation according to (9.2) is

$$g^* = \sum_{i=1}^{n} \{Y_i - E_{i-1}(Y_i)\} \left\{ \frac{\partial E_{i-1}(Y_i)}{\partial \theta} \right\} = 0. \qquad (9.9)$$

This is the same as the equation obtained by minimizing $\sum \{Y_i - E_{i-1}(Y_i)\}^2$, i.e. the sum of squares of the errors of best prediction. This procedure is referred to as conditional least squares by Klimko and Nelson (1978). If we are prepared to assume normally distributed errors (9.9) is also equivalent to maximum likelihood estimation. Thus for constant variance we have a situation similar to the independent case. However, when we allow σ^2 to depend on θ, similar results to those for the independent case may be derived (see Godambe 1985, Section 3.2, for details). A naïve (albeit initially plausible) application of weighted least squares here would minimize the weighted (conditional) sum of squares

$$S = \sum_{i=1}^{n} \frac{\{Y_i - E_{i-1}(Y_i)\}^2}{\sigma_i^2(\theta)},$$

where $\sigma_i^2(\theta) = \mathrm{Var}_{i-1}(Y_i) = E_{i-1}(h_i^2)$.

This leads to the estimating equation

$$\sum_{i=1}^{n} \left\{ \frac{Y_i - E_{i-1}(Y_i)}{\sigma_i^2(\theta)} \right\} \frac{\partial E_{i-1}(Y_i)}{\partial \theta} + \frac{1}{2} \sum_{i=1}^{n} \left\{ \frac{Y_i - E_{i-1}(Y_i)}{\sigma_i^2(\theta)} \right\}^2 \frac{\partial \log \sigma_i^2(\theta)}{\partial \theta} = 0.$$

$$(9.10)$$

However, it is not difficult to show that this does not in general provide consistent estimation of θ, as pointed out for the independent case by Crowder (1986) and for aggregate Markov chains by McLeish (1984, Section 2). Nevertheless, both McLeish (1984) and Kalbfleisch and Lawless (1984) (see also Kalbfleisch et al. 1983), indicate how, by ignoring the dependence of $\sigma_i^2(\theta)$ on θ, an appropriately modified version of weighted (conditional) least squares produces consistent estimation of θ. In our present context this approach leads to the estimating equation

$$g^* = \sum_{i=1}^{n} \left\{ \frac{Y_i - E_{i-1}(Y_i)}{\sigma_i^2(\theta)} \right\} \frac{\partial E_{i-1}(Y_i)}{\partial \theta} = 0, \qquad (9.11)$$

which from (9.8) is also the optimal estimating equation within $\mathscr{G}_1$. Furthermore, (9.11) is probabilistically closer to the likelihood equation, under the Gaussian assumption, than equation (9.10)

Now since g^*, given by (9.11), is optimal according to criterion (9.2) within $\mathscr{G}_1$, it follows that (i) and (ii) of the previous section are also satisfied. Thus the optimal g^* is referred to as the quasi-score function. Note that, as in Wedderburn's treatment, only second-moment assumptions are necessary.

Thavaneswaran and Thompson (1986) generalize this result to continuous-time semi-martingales, i.e. they consider a process of the form

$$X_t = V_{t,\theta} + H_{t,\theta},$$

where V_t is predictable and locally of bounded variation, and H_t is a local martingale (for definitions, see Shiryaev 1981). This fairly general model includes many models important in applications, e.g. Gill's model (Andersen *et al.* 1982), Aalen's model (Aalen 1978), diffusion processes, etc.

Restricting attention to the class of linear estimating functions

$$\mathscr{L} = \left\{ g_{t,\theta} = \int_0^t a_{s,\theta} \, \mathrm{d}H_{s,\theta} \right\}$$

for some predictable process $\{a_t\}$ and assuming standard conditions on $(X_t, \mathscr{F}_t; t \geq 0)$ they find the optimal estimating function in the class $\mathscr{L}$ is given by

$$g_{t,\theta}^* = \int_0^t a_{s,\theta}^* \, \mathrm{d}H_{s,\theta}$$

where

$$a_{s,\theta}^* = \frac{E(\mathrm{d}\dot{H}_{s,\theta})}{E(\mathrm{d}\langle H_t \rangle^2)},$$

the dot denoting differentiation with respect to θ, and $\langle H \rangle^2$ being the quadratic characteristic of H_t, assumed to exist. For V_t linear in θ this produces similar results to least squares and maximum likelihood.

A general framework for studying optimal estimation in stochastic processes via quasi-likelihood is given by Godambe and Heyde (1987). The framework focuses on martingale estimating functions, both in discrete and continuous time, and assumes a p-dimensional parameter θ. The class of underlying probability measures for the r-dimensional process $\{X_t, 0 \leq t \leq T\}$ is assumed to be a *union* $\mathscr{P} = \{P_\theta\}$ of parametric families as in the previous section. Attention is focused on the class $\mathscr{G}$ of zero-mean, square-integrable estimating functions $G_T = G_T\{X_t, 0 \leq t \leq T\}$, i.e. for which $EG_T(\theta) = 0$ for each $P_\theta \in \mathscr{P}$, and on the class $\mathscr{M} \subseteq \mathscr{G}$ of martingale estimating functions. All estimating functions are of dimension p.

For each $P_\theta \in \mathscr{P}$ we assume the existence of a p-dimensional score function $U_T(\theta)$. Since we have a union of parametric families it is not known *which* score function applies. However, using the motivation suggested in the previous section, it is reasonable to look for an estimating function G_T which has minimum dispersion distance from U_T or maximum correlation with U_T. Thus we need p-dimensional analogues of (i) and (i)) of the previous section.

An analogue of (i) is to choose the estimating function G_T^* which maximizes the vector correlation between G_T and U_T, that is

$$\rho^2(G_T, U_T) = \frac{|E(G_T U_T')|^2}{|E(G_T G_T')| \, |E(U_T U_T')|} \tag{9.12}$$

as defined, for example, by Hotelling (1936). An analogue of (ii) is to choose G_T^* to maximize

$$E(U_T - G_T)(U_T - G_T)' \tag{9.13}$$

in the partial order of non-negative-definite matrices. A further criterion analogous to (9.2) is to maximize

$$E(\dot{G}_T)'(EG_T G'_T)^{-1}(E\dot{G}_T) \tag{9.14}$$

in the partial order of non-negative definite matrices. (Note that for these criteria to be meaningful it is necessary to restrict attention to the subclass $\mathscr{G}_1$ of $\mathscr{G}$ for which the appropriate derivatives and inverses exist.) Again it is found that these three criteria are equivalent in $\mathscr{G}_1$, which extends to the multiparameter case the results of Godambe and Thompson (1987).

In the semiparametric situation under consideration, it is again the case that a uniquely optimal estimating function may not exist for the full class $\mathscr{G}_1$. However, it is possible to obtain an optimum by restricting attention to a meaningful subclass of $\mathscr{G}_1$. This leads to the following:

Definition G_T^* is said to be O_F-optimal within $\mathscr{G}_2$ (that is finite-sample optimal) if

$$E(G_T^*)'(EG_T^* G_T^{*\prime})^{-1}(EG_T^*) - (E\dot{G}_T)'(EG_T G'_T)^{-1}(E\dot{G}_T)$$

is non-negative-definite for all $G_T \in \mathscr{G}_2 \subset \mathscr{G}_1$, $\theta \in \Omega$ and $P_\theta \in \mathscr{P}$.

For the reasons cited previously, the O_F-optimal estimating function G_T^* is referred to as the quasi-score function.

To illustrate these ideas consider the semimartingale process $\{X_T, T \geqslant 0\}$ treated by Hutton and Nelson (1986)

$$X_T = \int_0^T f_t(\theta) \, d\lambda_t + m_T(\theta) \tag{9.15}$$

where $\{f_t(\theta)\}$, $\{\lambda_t\}$ are increasing predictable processes and $m_T(\theta)$ is a local, square-integrable martingale with characteristic $\langle m(\theta) \rangle_T$ given by

$$\langle m(\theta) \rangle_T = \int_0^T a_t(\theta) \, d\lambda_t,$$

where $\{a_t(\theta)\}$ is a predictable process. Assume also that $f_t(\theta)$ is almost surely differentiable with respect to θ. $\{X_t\}$ is r-dimensional and θ is p-dimensional.

In order to obtain an optimal estimating function it is necessary to restrict consideration to a meaningful subclass $\mathscr{G}_2$. Consider the class of estimating functions defined by

$$\mathscr{G}_2 = \left\{ G: G_T(\theta) = \int_0^T \alpha_s(\theta) \, dm_s(\theta) \right\}$$

where $\{\alpha_s(\theta)\}$ is a $p \times r$ predictable process whose elements are almost surely continuously differentiable with respect to θ. This is the continuous-time multiparameter analogue of the class $\mathscr{L}$ treated by Godambe (1985) and is motivated by similar considerations (cf. Section 4 of Godambe 1985).

Within the class $\mathscr{G}_2$, the optimum estimating function, according to criterion (9.2), can be shown to be

$$Q_T(\theta) = \int_0^T \dot{f}_s'(\theta) a_s^+(\theta) \, dm_s(\theta), \tag{9.16}$$

where $a_s^+(\theta)$ is the Moore–Penrose pseudo inverse of $a_s(\theta)$. For the particular process given by (9.15) the quasi-score function $Q_T(\theta)$ may also be written as

$$Q_T(\theta) = \int_0^T (d\bar{m}_s(\theta))'(d\langle m(\theta)\rangle_s)^+ \, dm_s(\theta), \tag{9.17}$$

where $d\bar{m}_t(\theta) = E(d\dot{m}_t(\theta)|\mathscr{F}_{t-})$, and $\{\mathscr{F}_t, t \geqslant 0\}$ is a standard filtration. This immediately suggests an extension to more general settings not necessarily described by the model (9.15).

9.4 Quadratic estimating functions

The previous section has focused on optimal estimating equations within subclasses $\mathscr{G}_1$ of linear estimating functions. For example, in the discussion of the AR(1) process following (9.7), the h_i have been chosen as linear functions of the data. (Note, however, that the final optimal estimating function is quadratic in the observations.) The restriction to linearity, while mathematically convenient, is not an essential ingredient of the theory. It does, however, have statistical implications. For example, Wedderburn's quasi-likelihood equations for the independent case which are optimal according to criterion (9.2) within the linear class (an analogue of the Gauss–Markov theorem) will typically be suboptimal within a larger quadratic class. Crowder (1987), Firth (1987) and Godambe and Thompson (1989) have studied such quadratic estimating equations and derived the optimal weights in the independent case. Heyde (1987) has studied these in the more general context of dependent observations (Heyde treats the combination of arbitrary (not necessarily quadratic) estimating functions). In the special case of AR(1) he combines the optimal linear equation given by our equation (9.9) with the optimal quadratic equation based on the choice

$$h_i = (Y_i - \theta Y_{i-1})^2 - \sigma^2, \qquad i = 1, \ldots, n.$$

Unfortunately, optimal quadratic estimating equations require knowledge of higher-order properties of the data such as skewness γ and kurtosis κ; see, for example, Crowder's equation (4.1) for the independent case or Heyde's (11) for the AR(1) case. Thus, increased efficiency will typically be at the expense of reduced robustness. Since it is rare in practice that third or fourth moment properties will be available with any precision, it is not clear under what circumstances use of quadratic estimating equations, as opposed to the simpler linear quasi-likelihood estimation, is warranted. Indeed, even for linear optimal estimating equations the sensitivity of estimation to misspecification of the variance function is an important consideration. Crowder (1987) describes a situation, his Example 2, where misspecification of the variance function produces very poor results asymptotically. Sensitivity of the optimal quadratic estimating equations to misspecification of γ and κ is even less certain and needs to be studied more, Crowder (1987) gives an example where guessed values of γ and κ give good results. Dean and Lawless (1989) perform efficiency calculations which suggest that quadratic estimating functions can perform well within the context of the negative-binomial mixed Poisson model. The efficiency calculations of Firth (1987) indicate that Gaussian estimation, which leads to a particular quadratic estimating equation, performs well for error distributions within the log-gamma family, provided the skewness is not too large (see Firth, 1987, Table 1).

Finally, it should be emphasized that the martingale theory of the previous section is quite general, including Wedderburn's quasi-likelihood, quadratic estimating equations, and their extensions to stochastic processes by appropriate choices of the basic martingales $m_T(\theta)$ in (9.15). Furthermore, as indicated by Kulkarni and Heyde (1987) it is possibly to further robustify the estimating functions by judicious choice of $m_T(\theta)$ and their weights $\alpha_T(\theta)$.

9.5 Acknowledgements

This research was supported by a grant from the Natural Sciences and Engineering Research Council of Canada through Grant No. A85584. I would also like to thank a referee for thoughtful and constructive comments and Professor Godambe for stimulating my interest in this topic.

References

Aalen, O. (1978). Nonparametric inference for a family of counting processes. *Ann. Stat.*, **6**, 701–26.

Andersen, P. K., Borgan, Ø., Gill, R. D., and Keiding, N. (1982). Linear nonparametric tests for comparison of counting process, with applications to censored survival data (with discussion). *Int. Stat. Rev.*, **50**, 219–58. Correction, **52**, 225.

Barnard, G. A. and Sprott, D. A. (1979). Theory of estimation. Technical Report, STAT-79-05, Department of Statistics, University of Waterloo.

Basawa, I. V. and Prakasa Rao, B. L. S. (1980). *Statistical inference for stochastic processes*. Academic Press, London, New York.

Bhapkar, V. P. (1972). On a measure of efficiency of an estimating equation. *Sankhya*, *A***34**, 467–72.

Cox, D. R. (1972). Regression models and life-tables (with discussion). *J. Roy. Stat. Soc.* B, **34**, 187–202.

Cox, D. R. and Hinkley, D. V. (1974). *Theoretical statistics*. Chapman and Hall, London.

Crowder, M. (1986). On consistency and inconsistency of estimating equations. *Econometric Theory*, **2**, 305–30.

Crowder, M. (1987). On linear and quadratic estimating functions. *Biometrika*, **74** (3), 591–7.

Dean, C. and Lawless, J. F. (1989). Contribution to discussion of paper by Godambe, V. P. and Thompson, M. E., *J. Stat. Plan. Inf.*, **22**, 155–8.

Desmond, A. F. (1989). The theory of estimating equations. In S. Kotz, N. L. Johnson, and C. B. Read (ed.), *Encyclopedia of Statistical Sciences*. Supplement Volume. Wiley, New York.

Durbin, J. (1960). Estimation of parameters in time-series regression models. *J. Roy. Stat. Soc.* B, **22**, 139–53.

Firth, D. (1987). On the efficiency of quasi-likelihood estimation. *Biometrika*, **74** (2), 233–45.

Godambe, V. P. (1960). An optimum property of regular maximum likelihood estimation. *Ann. Math. Stat.*, **31**, 1208–12.

Godambe, V. P. (1985). The foundations of finite sample estimation in stochastic processes. *Biometrika*, **72**, 419–28.

Godambe, V. P. and Heyde, C. C. (1987). Quasi-likelihood and optimal estimation. *Int. Stat. Rev.*, **55** (3), 231–44.

Godambe, V. P. and Thompson, M. E. (1987). Logic of least squares revisited. Technical Report, STAT-87-06, Department of Statistics, University of Waterloo.

Godambe, V. P. and Thompson, M. E. (1989). An extension of quasi-likelihood estimation (with discussion). *J. Stat. Plan. Inf.*, **22**, 137–72.

Hall, P. G. and Heyde, C. C. (1980). *Martingale limit theory and its application*. Academic Press, New York.

Heyde, C. C. (1987). On combining quasi-likelihood estimating functions. *Stoch. Processes. Appl.*, **25**, 281–7.

Heyde, C. C. and Seneta, E. (1977). *I. J. Bienaymé: statistical theory anticipated*. Springer, New York.

Hotelling, H. (1936). Relations between two sets of variables. *Biometrika*, **28**, 321–77.

Hutton, J. E. and Nelson, P. I. (1986). Quasi-likelihood estimation for semi-martingales. *Stoch. Processes Appl.*, **22**, 245–57.

Jacod, J. (1987). Partial likelihood process and asymptotic normality. *Stoch. Processes Appl.*, **26**, 47–71.

Kalbfleisch, J. D. and Lawless, J. F. (1984). Least-squares estimation of transition probabilities from aggregate data. *Canad. J. Stat.*, **12**, 169–82.

Kalbfleisch, J. D., Lawless, J. F., and Vollmer, W. M. (1983). Estimation in Markov models from aggregate data. *Biometrics*, **39**, 907–19.

Kale, B. K. (1962). An extension of the Cramer–Rao inequality for statistical estimation functions. *Skand. Actuar.*, **45**, 80–90.

Klimko, L. A. and Nelson, P. I. (1978). On conditional least squares estimation for stochastic processes. *Ann. Stat.*, **6**, 629–42.

Kulkarni, P. M. and Heyde, C. C. (1987). Optimal robust estimation for discrete time stochastic processes. *Stoch. Processes Appl.*, **26**, 267–76.

McCullagh, P. (1983). Quasi-likelihood functions. *Ann. Stat.*, **11**, 59–67.

McLeish, D. L. (1984). Estimation for aggregate models: The aggregate Markov chain. *Canad. J. Stat.*, **12**, 265–82.

Moore, M. (1987). Inférence statistique dans les processes stochastiques: Aperçu historique. *Canad. J. Stat.*, **15** (3), 185–207.

Neyman, J. and Scott, E. L. (1948). Consistent estimates based on partially consistent observations. *Econometrica*, **16**, 1–32.

Rao, C. R. (1973). Linear statistical inference and its applications (2nd edn). Wiley, New York.

Shiryaev, A. N. (1981). Martingales: recent developments, results and aplications. *Int. Stat. Rev.*, **49**, 199–233.

Sprott, D. A. (1983). Gauss, Carl Friedrich, In S. Kotz and N. L. Johnson (ed.), *Encyclopedia of Statistical Sciences*, Vol. 3. Wiley, New York, pp. 305–8.

Thavaneswaran, A. and Thompson, M. E. (1986). Optimal estimation for semi-martingales. *J. Appl. Prob.*, **23**, 409–17.

Wedderburn, R. W. M. (1974). Quasi-likelihood functions, generalized linear models and the Gauss–Newton method. *Biometrika*, **61**, 439–47.

Wilks, S. S. (1938). Shortest average confidence intervals from large samples. *Ann. Math. Stat.*, **9**, 166–75.

10
On optimal estimating functions for partially specified counting process models
P. E. Greenwood and W. Wefelmeyer

ABSTRACT

Consider a semimartingale with predictable characteristics depending on a one-dimensional parameter ϑ. Suppose that the associated martingale problem does not have a unique solution. We are interested in efficient estimation of ϑ. Thavaneswaran and Thompson (1986) and Heyde (1987) give versions of Godambe's optimality concept based on estimating functions in this setting. For a counting process model, we compare their concepts with asymptotic efficiency based on the partially specified likelihood process. We give conditions under which the notions are asymptotically compatible.

10.1 Introduction

An optimality criterion for estimating functions was introduced by Godambe (1985, 1987) for discrete-time stochastic processes, and generalized to continuous-time semimartingales by Thavaneswaran and Thompson (1986). Another optimality criterion for continuous-time semimartingales was introduced by Heyde (1987). These criteria were studied further by Godambe and Heyde (1987) and Heyde (1988). For a simple counting process model, we spell out regularity conditions under which these optimality criteria are asymptotically compatible with asymptotic efficiency in the sense of Hájek's convolution theorem.

Assume that we observe a counting process X and a 'covariate' process Y on $[0, t]$, and that the Doob–Meyer decomposition of X with respect to the filtration generated by X and Y is of the form

$$dX(s) = dM_\tau(s) + a_\tau(s)\, ds.$$

The underlying distribution is, in general, not uniquely determined by the intensity process a_τ. Such a model is called *partially specified* (see Greenwood 1988). We assume for convenience that the parameter τ is in a one-dimensional parameter space Θ.

Godambe and Heyde (1987) consider estimating functions of the form

$$G_t(\tau) = \int_0^t f_\tau \, dM_\tau,$$

where f_τ is a predictable process. As an *asymptotic optimality criterion* for estimating functions they suggest minimizing

$$\bar{G}_t(\vartheta)^{-2}\langle G(\vartheta)\rangle_t, \tag{10.1}$$

where $\bar{G}$ is the compensator of the derivative G' of G. As a *non-asymptotic optimality criterion* they suggest minimizing

$$(E_\vartheta G_t'(\vartheta))^{-2} E_\vartheta G_t(\vartheta)^2. \tag{10.2}$$

They show that optimality in both senses is achieved if f_τ is chosen to be the logarithmic derivative $a_\tau^{-1} a_\tau'$ of the intensity process.

In section 10.2 we give conditions under which a solution $\hat{\vartheta}_t$ of the estimating equation

$$G_t(\tau) = 0$$

is asymptotically normal as $t \to \infty$:

$$c_t(\hat{\vartheta}_t - \vartheta) \sim N(0, (f_\vartheta, a_\vartheta^{-1} a_\vartheta')^{-2}\|f_\vartheta\|^2).$$

Here $\|v\|$ is a certain seminorm determined by the limit in probability of Cauchy averages over paths of the process v, and (v, w) is the corresponding inner product. We note that the expressions (10.1) and (10.2) tend to the asymptotic variance of the estimator. Hence one can say that the optimality criteria for estimating functions are compatible with asymptotic efficiency in the sense of minimal asymptotic variance among solutions of estimating equations.

In Section 10.3 we show that the asymptotic variance bound for solutions of estimating equations is, in fact, an asymptotic variance bound for a much larger class of estimators. To describe this class, fix ϑ and consider the *partially specified log-likelihood process*

$$L_{t\vartheta\tau} = \int_0^t \log(a_\vartheta^{-1} a_\tau) \, dX - \int_0^t (a_\tau - a_\vartheta) \, ds.$$

For a counting process setting, the partially specified likelihood process was introduced by Gill (1985) and Slud (1986, Chapter 6), generalizing the discrete-time version proposed by Cox (1975). A corresponding concept for general semimartingales was studied by Jacod (1987, 1990a). Inference based on the partially specified likelihood process for semimartingales was considered by Sørensen (1988b, 1990) and Jacod (1990a,b).

Following Greenwood and Wefelmeyer (1990a), we introduce a family of distributions $\bar{P}_{t\vartheta\tau}$, $\tau \in \Theta$, by

$$d\bar{P}_{t\vartheta\tau} = \exp(L_{t\vartheta\tau}) \, dP_\vartheta.$$

We prove local asymptotic normality of this family and obtain a convolution theorem which shows that the asymptotic variance bound in Section 10.2 holds for all *regular* estimators. We then show that the solutions of estimating equations are, in fact, regular.

10.2 Solutions of estimating equations

Let $(\Omega, \mathscr{F})$ be a measurable space. Let X be a counting process and Y an arbitrary d-dimensional process on the time interval $[0, \infty)$. Let $\mathbb{F} = (\mathscr{F}_t)_{t \geqslant 0}$ denote the filtration generated by X and Y. Let Θ be a one-dimensional parameter space. For each $\tau \in \Theta$ let a_τ be a predictable process. Assume that there exists a distribution P on $\mathscr{F}$ such that X has a Doob–Meyer decomposition of the form

$$X(t) = M_\tau(t) + \int_0^t a_\tau(s) \, ds, \tag{10.3}$$

where M_τ is a martingale and a_τ is the *intensity process*.

The underlying distribution P on $\mathscr{F}$ is, in general, not uniquely determined by specifying the intensity process a_τ. In other words: the martingale problem need not have a unique solution. Let $\mathscr{P}_\tau$ denote the family of all distributions under which the intensity process of X is a_τ. For convenience we assume that the intensities are positive.

We observe a path of X and Y over the time interval $[0, t]$ and want to estimate τ. The model may be interpreted as a regression model, with covariate process Y. Some versions with infinite-dimensional parameters are treated in Greenwood and Wefelmeyer (1990a,b, 1991).

This model fits into the general setting of Godambe and Heyde (1987, Section 2). In Section 5 they introduce estimating functions as follows. For each $\tau \in \Theta$ let f_τ be a predictable process. Fix $t > 0$. Under an appropriate integrability condition on f_τ we have

$$\int \int_0^t f_\tau \, dM_\tau \, dP = 0, \qquad P \in \mathscr{P}_\tau, \quad \tau \in \Theta.$$

This means that $\tau \to \int_0^t f_\tau \, dM_\tau$ is an *estimating function*. The corresponding *estimation equation* is

$$\int_0^t f_\tau \, dM_\tau = 0. \tag{10.4}$$

Of course, in order to use a solution of the estimating equation as an estimator, we have to write it as a deterministic function of the paths of X and Y on $[0, t]$. This can be achieved by writing $f_\tau(t)$ and $a_\tau(t)$ as deterministic functions of t and the paths of X and Y on $[0, t)$.

In the classical case of i.i.d. observations $x_1, \ldots, x_n$, estimating equations are of the form

$$0 = \sum_{i=1}^{n} f_\tau(x_i) = n \int f_\tau(x) \, dF_n(x) = n \int f_\tau(x) \, d(F_n - F_\tau)(x),$$

where F_n and F_τ denote the empirical and the true distribution function, respectively. Here $M_\tau = n(F_n - F_\tau)$. Occasionally, the name estimating *function* is used for $\tau \to f_\tau$, not for $\tau \to \int f_\tau(x) \, dM_\tau(x)$.

We want to compare the optimality criteria of Thavaneswaran and Thompson (1986) and Heyde (1987) for estimating functions with the asymptotic behaviour when $t \to \infty$ of solutions of corresponding estimating equations. For this, we will derive a stochastic approximation for such solutions. It will be convenient to make the following definitions.

Fix $\vartheta \in \Theta$ and $P \in \mathscr{P}_\vartheta$. Let $b_t \to \infty$ for $t \to \infty$. Let $V(b_t)$ be the set of real-valued predictable processes v such that there exists a positive number $\|v\|$ with

$$b_t^{-2} \int_0^t v^2 a_\vartheta \, ds = \|v\|^2 + o_P(1). \tag{10.5}$$

This defines a seminorm on $V(b_t)$. Note that $\|v\|$ is far from being a norm since it is only determined by the tail behaviour of v.

We call two processes $v \in V(b_t)$ and $w \in V(c_t)$ *compatible* if there exists a real number (v, w) with

$$b_t^{-1} c_t^{-1} \int_0^t v w a_\vartheta \, ds = (v, w) + o_P(1). \tag{10.6}$$

Continuity and differentiability with respect to the seminorm $\| \ \|$ can be introduced as follows. A process-valued function $\tau \to f_\tau$ is called b_t-continuous at ϑ if f_ϑ is in $V(b_t)$ and, for $\varepsilon_t \to 0$,

$$b_t^{-2} \sup_{|\tau - \vartheta| \leq \varepsilon_t} \int_0^t (f_\tau - f_\vartheta)^2 a_\vartheta \, ds = o_P(1). \tag{10.7}$$

A process-valued function $\tau \to f_\tau$ is called b_t-*differentiable at ϑ with derivative d_ϑ* if d_ϑ is in $V(b_t)$ and, for $\varepsilon_t \to 0$,

$$b_t^{-2} \sup_{|\tau - \vartheta| \leq \varepsilon_t} \int_0^t (f_\tau - f_\vartheta - (\tau - \vartheta) d_\vartheta)^2 a_\vartheta \, ds = o_P(\varepsilon_t^2). \tag{10.8}$$

Note that d_ϑ is determined only up to $\| \ \|$-equivalence.

An estimator $\hat{\vartheta}_t$ is called c_t-*consistent* at ϑ if

$$P \circ c_t(\hat{\vartheta}_t - \vartheta), \qquad t \geqslant 0, \text{ is tight.} \tag{10.9}$$

We can now state a result on stochastic approximations and the asymptotic distribution of solutions of estimating equations.

Proposition 10.1 *Assume that* $\tau \to a_\vartheta^{-1} a_\tau$ *is* c_t-*differentiable at* ϑ *with derivative* k_ϑ *and that* $\tau \to f_\tau$ *is* b_t-*continuous at* ϑ. *Assume that* f_ϑ *and* k_ϑ *are compatible, with* $(f_\vartheta, k_\vartheta) \neq 0$. *Then any* c_t-*consistent solution* $\hat{\vartheta}_t$ *of the estimating equation* (2.2) *admits a stochastic approximation of the form*

$$c_t(\hat{\vartheta}_t - \vartheta) = c_t \left(\int_0^t f_\vartheta k_\vartheta a_\vartheta \, ds \right)^{-1} \int_0^t f_\vartheta \, dM_\vartheta + o_P(1)$$

$$= b_t^{-1}(f_\vartheta, k_\vartheta)^{-1} \int_0^t f_\vartheta \, dM_\vartheta + o_P(1). \tag{10.10}$$

Assume, in addition, the following Lindeberg condition:

$$b_t^{-2} \int_0^t f_\vartheta^2 1_{\{|f_\vartheta| > \varepsilon b_t\}} a_\vartheta \, ds = o_P(1), \qquad \varepsilon > 0. \tag{10.11}$$

Then $\hat{\vartheta}_t$ *is asymptotically normal:*

$$P \circ c_t(\hat{\vartheta}_t - \vartheta) \Rightarrow N(0, (f_\vartheta, k_\vartheta)^{-2} \|f_\vartheta\|^2). \tag{10.12}$$

The proof is based on a Taylor expansion; see Section 10.4. Note that the convergence rate of $\hat{\vartheta}_t$ is always c_t, regardless of the rate b_t associated with the estimating function. The rate of a plausible estimating function is in no way related to the rate of k_ϑ. For example, multiplying an estimating function by some deterministic function of time (and, possibly, the parameter) does not change the estimator. This is why we allow estimating functions with rates b_t different from c_t. Note also that the stochastic approximation is valid only for c_t-consistent $\hat{\vartheta}_t$. Strong (global) regularity conditions are needed even to ensure the existence of consistent solutions (without a rate). For the optimal estimating function, such conditions are given by Hutton and Nelson (1986, p. 251, Theorem 3.1) for general semimartingales, and by Sørensen (1988a) in the case of diffusions with jumps.

By the Schwarz inequality,

$$(f_\vartheta, k_\vartheta)^{-2} \|f_\vartheta\|^2 \geqslant \|k_\vartheta\|^{-2} = (k_\vartheta, k_\vartheta)^{-2} \|k_\vartheta\|^2. \tag{10.13}$$

If $\tau \to k_\tau$ is b_t-continuous at ϑ, we can apply Proposition 10.1 for $f_\tau = k_\tau$. We see that the function $\tau \to k_\tau$ leads to an estimator which is asymptotically efficient among solutions of estimating equations.

Because of (10.13), we can consider minimizing

$$(f_\vartheta, k_\vartheta)^{-2} \| f_\vartheta \|^2$$

or maximizing the 'correlation'

$$\| f_\vartheta \|^{-1} \| k_\vartheta \|^{-1} (f_\vartheta, k_\vartheta)$$

as an *asymptotic efficiency criterion*. These expressions have appeared in other, but related contexts, e.g. Barnard (1973), McLeish (1984).

Let us now compare this with the optimality criteria of Thavaneswaran and Thompson (1986, Section 2) and Heyde (1987, Section 2). They assume that $\tau \to a_\tau(s)$ is P-a.e. differentiable with derivative $a'_\tau(s)$, say. Following the notation of Godambe and Heyde (1987), we write

$$G_t(\tau) = \int_0^t f_\tau \, \mathrm{d}M_\tau$$

and denote by $\bar{G}(\vartheta)$ the compensator of the derivative $G'(\vartheta) = \partial_\vartheta G(\vartheta)$ of $G(\vartheta)$. Under their conditions,

$$\langle G(\vartheta) \rangle_t = \int_0^t f_\vartheta^2 a_\vartheta \, \mathrm{d}s,$$

$$G'_t(\vartheta) = \int_0^t f'_\vartheta \, \mathrm{d}M_\vartheta - \int_0^t f_\vartheta a'_\vartheta \, \mathrm{d}s,$$

$$\bar{G}_t(\vartheta) = - \int_0^t f_\vartheta a'_\vartheta \, \mathrm{d}s.$$

The asymptotic optimality criterion of Heyde (1987, p. 284) is based on the *martingale information* defined as

$$\langle G(\vartheta) \rangle_t^{-1} \bar{G}_t(\vartheta)^2.$$

Assume, in addition to pointwise differentiability of $\tau \to a_\tau$, that $a_\vartheta^{-1} a'_\vartheta$ is a c_t-derivative of $\tau \to a_\vartheta^{-1} a_\tau$ at ϑ. Then we obtain for the inverse of the martingale information,

$$c_t^2 \bar{G}_t(\vartheta)^{-2} \langle G(\vartheta) \rangle_t = (f_\vartheta, a_\vartheta^{-1} a'_\vartheta)^{-2} \| f_\vartheta \|^2 + o_P(1). \qquad (10.14)$$

The estimating function based on $\tau \to a_\tau^{-1} a'_\tau$ is

$$G_t^*(\tau) = \int_0^t a_\tau^{-1} a'_\tau \, \mathrm{d}M_\tau. \qquad (10.15)$$

Godambe and Heyde (1987, Section 5) show that

$$\langle G(\vartheta) \rangle_t^{-1} \bar{G}_t(\vartheta)^2 \geqslant \langle G^*(\vartheta) \rangle_t^{-1} \bar{G}_t^*(\vartheta)^2. \qquad (10.16)$$

This is their *asymptotic optimality criterion*. Recall that the c_t-derivative k_ϑ

of $\tau \to a_\vartheta^{-1} a_\tau$ at ϑ, introduced in Proposition 10.1, is uniquely determined up to $\|\ \|$-equivalence. In particular, $\|k_\vartheta - a_\vartheta^{-1} a_\vartheta'\| = 0$, and (10.13) can be written as

$$(f_\vartheta, a_\vartheta^{-1} a_\vartheta')^{-2}\|f_\vartheta\|^2 \geq \|a_\vartheta^{-1} a_\vartheta'\|^{-2} = (a_\vartheta^{-1} a_\vartheta', a_\vartheta^{-1} a_\vartheta')^{-2}\|a_\vartheta^{-1} a_\vartheta'\|^2. \quad (10.17)$$

From (10.14) we see that the asymptotic optimality criterion (10.16) is asymptotically compatible with the 'asymptotic efficiency criterion' (10.17). A proof of (10.16) for general semimartingales is given in Jacod (1990b, Theorem 4.19).

The non-asymptotic optimality criterion of Thavaneswaran and Thompson (1986, p. 411) is based on

$$\left(\int G_t'(\vartheta)\, dP\right)^{-2} \int G_t(\vartheta)^2\, dP.$$

Note that this expression depends on the choice of P in $\mathscr{P}_\vartheta$. Under our assumptions and appropriate integrability conditions,

$$c_t^2 \left(\int G_t'(\vartheta)\, dP\right)^{-2} \int G_t(\vartheta)^2\, dP = (f_\vartheta, a_\vartheta^{-1} a_\vartheta')^{-2}\|f_\vartheta\|^2 + o_P(1). \quad (10.18)$$

Thavaneswaran and Thompson (1986, Section 2) show that

$$\left(\int G_t'(\vartheta)\, dP\right)^{-2} \int G_t(\vartheta)^2\, dP \geq \left(\int G_t^{*\prime}(\vartheta)\, dP\right)^{-2} \int G_t^*(\vartheta)\, dP. \quad (10.19)$$

This is Godambe and Heyde's *non-asymptotic criterion*. From (10.18) we see that it is asymptotically compatible with the 'asymptotic efficiency criterion' (10.17).

Note that the optimality criteria involve a derivative of the function f_τ. This is not the case for our 'asymptotic efficiency criterion' (10.17). In fact, to prove asymptotic normality of solutions of estimating equations (Proposition 10.1), differentiability of f_τ was not involved.

Since $\tau \to a_\tau^{-1} a_\tau'$ is optimal in the sense of the criteria (10.16) and (10.19),

$$G_t^*(\tau) = \int_0^t a_\tau^{-1} a_\tau'\, dM_\tau$$

is called a *quasi-score function*. A solution of the corresponding estimating equation is called a *quasi-maximum likelihood estimator*.

10.3 Asymptotic efficiency

In the preceding section we have seen that the quasi-maximum likelihood estimator is asymptotically efficient among solutions of estimating equations.

In this section we will show that it is asymptotically efficient in a much larger class of estimators. An asymptotic efficiency concept can be introduced as follows.

Fix $\vartheta \in \Theta$. For $\tau \in \Theta$ the *partially specified log-likelihood* between τ and ϑ is defined as

$$L_{t\vartheta\tau} = \int_0^t \log(a_\vartheta^{-1} a_\tau) \, dX - \int_0^t (a_\tau - a_\vartheta) \, ds. \qquad (10.20)$$

If the martingale problem (see Jacod and Shiryaev 1987, p. 132) has a unique solution, relation (10.20) is a representation of the log-likelihood process of X. As noted at the beginning of Section 10.2, this will, in general, not be the case here.

Fix $P \in \mathscr{P}_\vartheta$. Let $\tau \to a_\tau(s)$ be P-a.e. differentiable with derivative $a_\tau'(s)$, say. The *partial score function* is obtained by differentiating the partially specified log-likelihood $\tau \to L_{t\vartheta\tau}$. We have

$$\partial_{\tau=\vartheta} L_{t\vartheta\tau} = \int_0^t a_\vartheta^{-1} a_\vartheta' \, dM_\vartheta = G_t^*(\vartheta).$$

We see that the partial score function is the same as the quasi-score function. Hence the quasi-maximum likelihood estimator can be interpreted as a *partial maximum likelihood estimator*.

Following Greenwood and Wefelmeyer (1990a), we introduce a model whose likelihood is the partially specified likelihood $\exp(L_{t\vartheta\tau})$. Define $\bar{P}_{t\vartheta\tau}$ by

$$d\bar{P}_{t\vartheta\tau} = \exp(L_{t\vartheta\tau}) \, dP. \qquad (10.21)$$

Note that $\bar{P}_{t\vartheta\tau}$ depends on the choice of P in $\mathscr{P}_\vartheta$, but $L_{t\vartheta\tau}$ does not.

To prove local asymptotic normality, we introduce a local parametrization

$$r \to \vartheta_{tr} = \vartheta + c_t^{-1} r, \qquad r \in \mathbb{R},$$

and write

$$L_{tr} = L_{t\vartheta\vartheta_{tr}}, \qquad \bar{P}_{tr} = \bar{P}_{t\vartheta\vartheta_{tr}}.$$

Proposition 10.2 *Assume that* $\tau \to a_\vartheta^{-1} a_\tau$ *is* c_t-*differentiable at* ϑ *with derivative* k_ϑ *fulfilling the Lindeberg condition*

$$c_t^{-2} \int_0^t k_\vartheta^2 1_{\{|k_\vartheta| > \varepsilon c_t\}} a_\vartheta \, ds = o_p(1), \qquad \varepsilon > 0.$$

Then

$$L_{tr} = rc_t^{-1} \int_0^t k_\vartheta \, dM_\vartheta - \tfrac{1}{2} r^2 \|k_\vartheta\|^2 + o_P(1) \qquad (10.22)$$

with

$$P \circ c_t^{-1} \int_0^t k_\vartheta \, dM_\vartheta \Rightarrow N(0, \|k_\vartheta\|^2). \tag{10.23}$$

The proof is indicated in Section 10.4. Local asymptotic normality for another parametric counting process model and related asymptotic efficiency results have already been obtained by Kutoyants (1984, p. 122, Section 4.3). The assumption of c_t-differentiability on $\tau \to a_\vartheta^{-1} a_\tau$, see (10.8), can be weakened to a Hellinger-type differentiability by replacing $a_\vartheta^{-1} a_\tau$ by the square root $a_\vartheta^{-1/2} a_\tau^{1/2}$ and τ by $\vartheta + c_t^{-1} r$. The condition then reads

$$\int_0^t (a_\vartheta^{-1/2} a_{\vartheta_{tr}}^{1/2} - 1 - \tfrac{1}{2} c_t^{-1} r k_\vartheta)^2 a_\vartheta \, ds = o_P(1).$$

A convolution theorem can now be stated as follows. Call an estimator $\hat{\vartheta}_t$ *regular* at ϑ with *limit R* if

$$\bar{P}_{tr} \circ c_t(\hat{\vartheta}_t - \vartheta_{tr}) \Rightarrow R, \qquad r \in \mathbb{R}. \tag{10.24}$$

Theorem 10.1   *Assume (10.22)–(10.24). Then there exists a distribution S such that*

$$P \circ \left(\|k_\vartheta\|^{-2} c_t^{-1} \int_0^t k_\vartheta \, dM_\vartheta, \, c_t(\hat{\vartheta}_t - \vartheta) - \|k_\vartheta\|^{-2} c_t^{-1} \int_0^t k_\vartheta \, dM_\vartheta \right)$$
$$\Rightarrow N(0, \|k_\vartheta\|^{-2}) \times S.$$

A proof of the theorem can be obtained, e.g. by adapting Bickel's proof of Hájek's convolution theorem, as in Droste and Wefelmeyer (1984, p. 135, Theorem 2.3). The theorem implies the assertion of the convolution theorem in its usual form:

$$P \circ c_t(\hat{\vartheta}_t - \vartheta) \Rightarrow N(0, \|k_\vartheta\|^{-2}) * S. \tag{10.25}$$

This justifies calling an estimator $\hat{\vartheta}_t$ *asymptotically efficient* at ϑ if

$$P \circ c_t(\hat{\vartheta}_t - \vartheta) \Rightarrow N(0, \|k_\vartheta\|^{-2}).$$

According to Proposition 10.1, the asymptotic distribution of the quasi- (or partial) maximum likelihood estimator is $N(0, \|k_\vartheta\|^{-2})$. By the convolution theorem, this estimator is asymptotically efficient among all regular estimators. We will show that the class of regular estimators contains the solutions of estimating equations.

Let us call an estimator $\hat{\vartheta}_t$, *asymptotically linear* at ϑ with *influence function* $v \in V(b_t)$ if

$$c_t(\hat{\vartheta}_t - \vartheta) = b_t^{-1} \int_0^t v \, dM_\vartheta + o_P(1).$$

We see from the convolution theorem that an estimator which is asymptotically efficient at ϑ is asymptotically linear at ϑ with influence function $\|k_\vartheta\|^{-2} k_\vartheta \in V(c_t)$.

The regular estimators among the asymptotically linear ones can be characterized as follows. Let $\hat{\vartheta}_t$ be asymptotically linear at ϑ with influence function $v \in V(b_t)$. Assume that v and k_ϑ are compatible. Note first that $b_t^{-1} \int_0^t v \, dM_\vartheta$ and $rc_t^{-1} \int_0^t k_\vartheta \, dM_\vartheta$ are jointly asymptotically normal with variances $\|v\|^2$ and $r^2 \|k_\vartheta\|^2$ and covariance $r(v, k_\vartheta)$. By LeCam's third lemma (e.g. LeCam 1986, p. 468, Corollary),

$$\bar{P}_{tr} \circ c_t(\hat{\vartheta}_t - \vartheta) \Rightarrow N(r(v, k_\vartheta), \|v\|^2).$$

Hence

$$\bar{P}_{tr} \circ c_t(\hat{\vartheta}_t - \vartheta_{tr}) \Rightarrow N(r(v, k_\vartheta) - r, \|v\|^2).$$

The estimator $\hat{\vartheta}_t$ is regular at ϑ if and only if the bias $r(v, k_\vartheta) - r$ vanishes, i.e. if and only if its influence function v fulfills

$$(v, k_\vartheta) = 1.$$

This characterization is well known for i.i.d. observations; see e.g. Rieder (1980, p. 108) and Bickel (1981, p. 17).

According to Proposition 10.1, an estimator based on a function $\tau \to f_\tau$ is asymptotically linear at ϑ with influence function

$$v = (f_\vartheta, k_\vartheta)^{-1} f_\vartheta \in V(b_t).$$

For this influence function we have

$$(v, k_\vartheta) = (f_\vartheta, k_\vartheta)^{-1} (f_\vartheta, k_\vartheta) = 1.$$

Hence solutions of estimating equations are regular.

Care is required in extending the statements made here to more general semimartingale models. The Doob–Meyer decomposition $X = M_\vartheta + A_\vartheta$ used here and in Hutton and Nelson (1986), Thavaneswaran and Thompson (1986), Godambe and Heyde (1987, Section 5) and Heyde (1988, Section 4) leads to an optimal estimating function which is *not*, in general, asymptotically efficient. The reason is that the (partial) score function is a sum of stochastic integrals with respect to the martingales coming from the canonical representation of the observed martingale; these martingales are obtained by further decomposing the martingale M_ϑ in the Doob–Meyer decomposition. This was pointed out by Sørensen (1990, 1991) in connection with diffusions with jumps and general jump processes. The partial score function of a general partially specified semimartingale model is obtained in Jacod (1990b).

10.4 Proofs

Proof of Proposition 10.1 Define r_t by

$$a_{\hat{\vartheta}_t} = a_\vartheta (1 + (\hat{\vartheta}_t - \vartheta)k_\vartheta + r_t).$$

The estimating equation (10.4) can be written as

$$0 = \int_0^t f_{\hat{\vartheta}_t} \, dM_{\hat{\vartheta}_t} = \int_0^t f_\vartheta \, dM_\vartheta - (\hat{\vartheta}_t - \vartheta) \int_0^t f_\vartheta k_\vartheta a_\vartheta \, ds + R_t \quad (10.26)$$

with

$$R_t = \int_0^t (f_{\hat{\vartheta}_t} - f_\vartheta) \, dM_\vartheta - \int_0^t (f_\vartheta + f_{\hat{\vartheta}_t} - f_\vartheta) r_t a_\vartheta \, ds$$

$$- (\hat{\vartheta}_t - \vartheta) \int_0^t (f_{\hat{\vartheta}_t} - f_\vartheta) k_\vartheta a_\vartheta \, ds.$$

Since $\tau \to a_\vartheta^{-1} a_\tau$ is c_t-differentiable and $\hat{\vartheta}_t$ is c_t-consistent, we have

$$\int_0^t r_t^2 a_\vartheta \, ds = o_P(1).$$

Since $\tau \to f_\tau$ is b_t-continuous,

$$b_t^{-2} \int_0^t (f_{\hat{\vartheta}_t} - f_\vartheta)^2 a_\vartheta \, ds = o_P(1).$$

It is easy to check that

$$b_t^{-1} R_t = o_P(1).$$

On the other hand

$$b_t^{-2} \left\langle \int_0^\cdot f_\vartheta \, dM_\vartheta \right\rangle_t = b_t^{-2} \int_0^t f_\vartheta^2 a_\vartheta \, ds = \|f_\vartheta\|^2 + o_P(1).$$

Hence R_t is negligible in (10.26). The processes f_ϑ and k_ϑ are compatible, i.e.

$$b_t^{-1} c_t^{-1} \int_0^t f_\vartheta k_\vartheta a_\vartheta \, ds = (f_\vartheta, k_\vartheta) + o_P(1).$$

The stochastic approximation (10.10) now follows from (10.26). Because of the Lindeberg condition (10.11), asymptotic normality (10.12) follows from the central limit theorem; see e.g. Jacod and Shiryaev (1987, p. 405, Theorem 5.4).

Proof of Proposition 10.2   Define

$$q(x) = \log(1 + x) - x + \tfrac{1}{2}x^2.$$

Write the partially specified log-likelihood process as

$$L_{tr} = \int_0^t \log(a_\vartheta^{-1} a_{\vartheta_{tr}})\, dX - \int_0^t (a_{\vartheta_{tr}} - a_\vartheta)\, ds$$

$$= \int_0^t (a_\vartheta^{-1} a_{\vartheta_{tr}} - 1)\, dM_\vartheta - \tfrac{1}{2}\int_0^t (a_\vartheta^{-1} a_{\vartheta_{tr}} - 1)^2 a_\vartheta\, ds + R_{tr}$$

with

$$R_{tr} = \int_0^t q(a_\vartheta^{-1} a_{\vartheta_{tr}} - 1)\, dX - \tfrac{1}{2}\int_0^t (a_\vartheta^{-1} a_{\vartheta_{tr}} - 1)^2\, dM_\vartheta.$$

For the remainder we have $R_{tr} = o_P(1)$ by arguments exactly analogous to those in Greenwood and Wefelmeyer (1990a, Section 2). Next we use the differentiability assumption (10.8) on $\tau \to a_\vartheta^{-1} a_\tau$ for $\tau = \vartheta_{tr} = \vartheta + c_t^{-1}r$ to replace $a_\vartheta^{-1} a_{\vartheta_{tr}} - 1$ by $c_t^{-1} r k_\vartheta$. We obtain

$$L_{tr} = rc_t^{-1}\int_0^t k_\vartheta\, dM_\vartheta - \tfrac{1}{2}r^2 c_t^{-2}\int_0^t k_\vartheta^2 a_\vartheta\, ds + o_P(1). \qquad (10.27)$$

The process k_ϑ is in $V(c_t)$, i.e.

$$c_t^{-2}\int_0^t k_\vartheta^2 a_\vartheta\, ds = \|k_\vartheta\|^2 + o_P(1).$$

The stochastic approximation (10.22) now follows from (10.27). Asymptotic normality (10.23) is obtained as in the proof of Proposition 10.1.

10.5   Acknowledgement

The authors are grateful to Jean Jacod, Richard Gill and Michael Sørensen for access to unpublished work, and to the referee for several suggestions and references.

The work was supported by NSERC, Canada.

References

Barnard, G. A. (1973). Maximum likelihood and nuisance parameters. *Sankhyā Ser. A*, **35**, 133–8.

Bickel, P. J. (1981). Quelques aspects de la statistique robuste. In P. L. Hennequin (ed.), *Ecole d'Eté de Probabilités de Saint-Flour*, IX-1979, pp. 1–72, Lecture Notes in Mathematics **876**, Springer-Verlag, Berlin.

Cox, D. R. (1975). Partial likelihood. *Biometrika*, **62**, 69–76.

Droste, W. and Wefelmeyer, W. (1984). On Hájek's convolution theorem. *Statistics and Decisions*, **2**, 131–44.

Gill, R. D. (1985). Notes on product integration, likelihood and partial likelihood for counting processes, non-informative and independent censoring. Unpublished manuscript.

Godambe, V. P. (1985). The foundations of finite sample estimation in stochastic processes. *Biometrika*, **72**, 419–28.

Godambe, V. P. (1987). The foundations of finite sample estimation in stochastic processes—II. In Yu. Prohorov, V. V. Sazonov (ed.), *Proceedings of the 1st World Congress of the Bernoulli Society*, Vol. 2, *Mathematical Statistics, Theory and Applications*, pp. 49–54, VNU Science Press, Utrecht.

Godambe, V. P. and Heyde, C. C. (1987). Quasi-likelihood and optimal estimation. *Int. Statist. Rev.*, **55**, 231–44.

Greenwood, P. E. (1988). Partially specified semimartingale experiments. In N. U. Prabhu (ed.), *Statistical Inference from Stochastic Processes*, pp. 1–17; *Contemporary Math.*, **80**, Amer. Math. Soc.

Greenwood, P. E. and Wefelmeyer, W. (1990a). Efficiency of estimators for partially specified filtered models. *Stoch. Proc. Appl.*, **32**, 353–70.

Greenwood, P. E. and Wefelmeyer, W. (1990b). Efficient estimating equations for nonparametric filtered models. In I. V. Basawa, N. U. Prabhu (ed.), *Statistical Inference in Stochastic Processes*, pp. 107–41, Marcel Dekker, New York.

Greenwood, P. E. and Wefelmeyer, W. (1991). Efficient estimation in a nonlinear counting process regression model. To appear in *Canad. J. Stat.*

Heyde, C. C. (1987). On combining quasi-likeihood estimation for semimartingales. *Stoch. Proc. Appl.* **25**, 281–7.

Heyde, C. C. (1988). Fixed sample and asymptotic optimality for classes of estimating functions. In N. U. Prabhu (ed.), *Statistical Inference from Stochastic Processes*, pp. 241–7, *Contemporary Math.*, **80**, Amer. Math. Soc.

Hutton, J. E. and Nelson, P. I. (1986). Quasi-likelihood estimation for semimartingales. *Stoch. Proc. Appl.*, **22**, 245–57.

Jacod, J. (1987). Partial likelihood process and asymptotic normality. *Stoch. Proc. Appl.*, **26**, 47–71.

Jacod, J. (1990a). Sur le processus de vraisemblance partielle. *Ann. Inst. H. Poincaré Probab. Statist.*, **26**, 299–329.

Jacod, J. (1990b). Regularity, partial regularity, partial information process for a filtered statistical model. *Probab. Theory Related Fields*, **86**, 305–35.

Jacod, J. and Shiryaev, A. N. (1987). *Limit theorems for stochastic processes*. Springer-Verlag, Berlin.

Kutoyants, Yu. A. (1984). *Parameter estimation for stochastic processes*. Heldermann, Berlin.

LeCam, L. (1986). *Asymptotic methods in statistical decision theory*. Springer-Verlag, New York.

McLeish, D. L. (1984). Estimation for aggregate models: The aggregate Markov chain. *Canad. J. Stat.*, **12**, 265–82.

Rieder, H. (1980). Estimates derived from robust tests. *Ann. Stat.*, **8**, 106–15.

Slud, E. (1986). *Martingale methods in statistics*. Book manuscript.

Sørensen, M. (1988a). A note on the existence of a consistent maximum likelihood estimator for diffusions with jumps. Research Reports No. 170, Dept. of Theoretical Statistics, Institute of Mathematics, University of Aarhus.

Sørensen, M. (1988b). Some asymptotic properties of quasi likelihood estimators for semimartingales. To appear in *Proceedings of the Fourth Prague Symposium on Asymptotic Statistics*.

Sørensen, M. (1990). On quasi likelihood for semimartingales. *Stoch. Proc. Appl.*, **35**, 331–46.

Sørensen, M. (1991). Likelihood methods for diffusions with jumps. In N. U. Prabhu, I. V. Basawa (ed.), *Statistical Inference in Stochastic Processes*, pp. 67–105, Marcel Dekker, New York.

Thavaneswaran, A. and Thompson, M. E. (1986). Optimal estimation for semimartingales. *J. Appl. Prob.*, **23**, 409–17.

11
Approximate confidence zones in an estimating function context

C. C. Heyde and Y.-X. Lin

ABSTRACT

This paper is concerned with competing approaches to the provision of asymptotic confidence zones for a vector-valued population characteristic of a general stochastic process. In particular, the case where the characteristic is a parameter is included. Both ergodic and non-ergodic situations are considered. The discussion is carried out in an estimating function framework and the use of a random norming to produce asymptotic normality may be essential. Some emphasis is given to consideration of different normings which may be used in the asymptotics.

11.1 Introduction

Our framework is that of a family of experiments defined in a statistical space $(\Omega, \mathscr{F}, \{P_\theta\})$, where $\theta \in \Theta$, an open subset of p-dimensional Euclidean space. The experiments are indexed by $t \in [0, \infty)$ and $\{\mathscr{F}_t\}$ denotes a standard filtration generated from these experiments. The framework covers discrete as well as continuous time and our object is to obtain asymptotic confidence statements about θ of maximum precision. The setting may be parametric or non-parametric; θ could be, for example, the mean of a stationary process.

A discussion concerning confidence zones of minimum size can be most naturally formulated in terms of properties of estimating functions; see Heyde (1989) and references therein. To this end, and with little loss in generality, we shall confine attention to the class $\mathscr{G}$ of zero mean, square integrable, $\mathscr{F}_t$-measurable, semimartingale estimating functions $G_t(\theta)$ which are vectors of dimension p such that

$$\dot{G}_t(\theta) = (\partial G_{t,i}/\partial \theta_j),$$

$E\dot{G}_t(\theta)$, $EG_t(\theta)G_t'(\theta)$ and $[G(\theta)]_t$ are (a.s.) non-singular for each $t > 0$, the prime denoting transpose and $[G(\theta)]_t$ the quadratic variation of $G_t(\theta)$. Here

$$[G(\theta)]_t = [G^{cm}(\theta)]_t + \sum_{0 < s \leqslant t} (\Delta G_s)(\Delta G_s)'.$$

where G^{cm} is the unique continuous martingale part of G and

$$\Delta G_s = G_s(\theta) - G_{s-}(\theta);$$

see e.g. Rogers and Williams (1987, p. 391) for a discussion.

The estimating function $G_t(\theta)$ produces an estimator (or estimators) θ_t^* as the solution of the estimating equation $G_t(\theta_t^*) = 0$.

An effective choice of G_t is vital and in the case where the likelihood function exists, is known, and is tractable, the score function (the derivative of the log likelihood with respect to θ) provides the benchmark and should ordinarily be used for G_t. It is well known, for example, that maximum likelihood (ML) estimation is associated with minimum-size asymptotic confidence zones under suitable regularity conditions (e.g. Hall and Heyde 1980, Chapter 6). We shall, however, take a more general approach, via a discussion of quasi-likelihood, for which properties similar to that of the ML estimator hold within what may be a more restricted setting.

11.2 The formulation

Asymptotic confidence statements about the 'parameter' θ are based on consideration of the asymptotic properties of a suitably chosen estimating function $G_t(\theta)$ as $t \to \infty$. For this purpose we shall assume, as is typically the case in regular problems of genuine physical relevance, that $G_t(\theta)$ has a limit distribution under some appropriate normalization.

Indeed, for $G_t(\theta) \in \mathcal{G}$, the result

$$(EG_t G_t')^{-1/2} G_t \overset{d}{\to} X \tag{11.1}$$

for some proper law X, which is not in general normal, but does not depend on the choice of G_t, seems to be uniquitous.

As a basis for obtaining confidence zones for θ we suppose that data is available for $0 \leqslant t \leqslant T$ and, letting θ_T^* be a solution of $G_T(\theta) = 0$, we use Taylor expansion to obtain

$$0 = G_T(\theta_T^*) = G_T(\theta) + \dot{G}_T(\theta_{1,T})(\theta_T^* - \theta), \tag{11.2}$$

where $\|\theta - \theta_{1,T}\| \leqslant \|\theta - \theta_T^*\|$, the norm denoting sum of squares of elements. Then, assuming that

$$(EG_T(\theta)G_T'(\theta))^{-1/2} \dot{G}_T(\theta_{1,T})(E\dot{G}_T(\theta))^{-1} E(G_T(\theta)G_T'(\theta))^{1/2} \overset{P}{\to} YI_p$$

for some random variable Y (> 0 a.s.) we have, as $T \to \infty$,

$$(EG_T(\theta)G_T'(\theta))^{-1/2}(E\dot{G}_T(\theta))(\theta_T^* - \theta) \overset{d}{\to} Z, \tag{11.3}$$

say, not depending on the choice of G_t.

The size of confidence zones for θ is then governed by the scaling 'information'

$$\mathscr{E}(G_t(\theta)) = (E\dot{G}_t(\theta))'(EG_t(\theta)G_t'(\theta))^{-1}(E\dot{G}_t(\theta))$$

and we prefer estimating function $G_{1,t}$ to $G_{2,t}$ if $\mathscr{E}(G_{1,t}) \geq \mathscr{E}(G_{2,t})$ for each $t \geq t_0$ in the partial order of non-negative-definite matrices. The best such estimating function, within a suitably chosen class $\mathscr{H} \subseteq \mathscr{G}$, is called a quasi-score estimating function and the choice of these is explored in some detail in the papers of Godambe and Heyde (1987), Heyde (1987, 1988, 1989), Sørensen (1989). Quasi-score estimating functions lead to quasi-likelihood estimators and have certain fixed sample optimality properties as well as asymptotic ones. Also, the quasi-likelihood estimator quite often coincides with the maximum likelihood estimator.

Unfortunately, quasi-score estimating functions not infrequently suffer from the problem of nuisance parameters. However, this problem can often be avoided through the use of asymptotic quasi-score estimating functions (Heyde and Gay 1989). These enjoy the same advantages as ordinary quasi-score estimating functions in terms of producing asymptotic confidence zones of minimum size.

To formulate the principle of asymptotic quasi-score estimating functions we first need to introduce the idea of asymptotically non-negative definite matrices.

Let $\{A_n\}, \{B_n\}$ be sequences of symmetric positive-definite matrices and $\{D_n\}$ a sequence of matrices such that $A_n - B_n + D_n$ is non-negative definite for each n and $\|D_n\| \to 0$ as $n \to \infty$. We shall then say that $A_n - B_n$ is *asymptotically non-negative definite*.

The concept of asymptotic quasi-likelihood is formalized in the following definition.

Definition Suppose that $G_t^* \in \mathscr{H} \subseteq \mathscr{G}$. If there is a positive function $\{\alpha_t, t > 0\}$ such that

$$\alpha_t^{-1}\{\mathscr{E}(G_t^*) - \mathscr{E}(G_t)\}$$

is asymptotically nonnegative for all $G_t \in \mathscr{H}$ as $t \to \infty$ and

$$\lim_{t \to \infty} \alpha_t^{-1}\|\mathscr{E}(G_t^*)\| > 0$$

we shall say that G_t^* is an *asymptotic quasi-score estimating function* within $\mathscr{H}$. A solution θ_t^* of $G_t^*(\theta) = 0$ will be called an *asymptotic quasi-likelihood estimator* within $\mathscr{H}$.

For more details see Heyde and Gay (1989). However, the following simple example illustrates the relevance of the idea.

Consider a varying environment population model of the form

$$X_t = \theta_t X_{t-1} + \varepsilon_t, \qquad t = 0, 1, 2, \ldots$$

where

$$E(\varepsilon_t|\mathscr{F}_{t-1}) = 0, \qquad E(\varepsilon_t^2|\mathscr{F}_{t-1}) = \sigma^2 X_{t-1}$$

$$E(\theta_t|\mathscr{F}_{t-1}) = \theta, \qquad \mathrm{var}(\theta_t|\mathscr{F}_{t-1}) = \eta^2, \qquad E(\varepsilon_t(\theta_t - \theta)|\mathscr{F}_{t-1}) = 0,$$

$\sigma^2, \eta^2 > 0$ are nuisance parameters and $\theta > 1$ is to be estimated. Then, the quasi-score estimating function based on the martingale differences $X_t - \theta X_{t-1}$ is

$$\sum_{k=1}^{t} (\sigma^2 + \eta^2 X_{k-1})^{-1}(X_k - \theta X_{k-1})$$

which contains the nuisance parameters σ^2 and η^2, no estimator being available for η^2. However, there is a convenient asymptotic quasi-score estimating function given by

$$\sum_{k=1}^{t} X_{k-1}^{-1}(X_k - \theta X_{k-1})$$

which presents no such problems.

Criteria for the choice of asymptotic quasi-score estimating functions are given in Heyde and Gay (1989).

We shall henceforth suppose that a quasi-score or asymptotic quasi-score estimating function $Q_t(\theta)$ has been chosen for which (11.3) holds and our considerations regarding confidence zones will be based on Q_t.

11.3 Confidence zones

Confidence zones based on (11.3) are mostly difficult to formulate unless Z is normally distributed and it is desirable, wherever possible, to renormalize to obtain asymptotic normality.

For $Q_t(\theta) \in \mathscr{G}$, the result

$$[Q]_t^{-1/2} Q_t \overset{d}{\to} N_p(0, I_p), \tag{11.4}$$

N_p denoting the p-variate normal, seems to encapsulate the most general form of asymptotic normality result. Norming by a random process such as $[Q]_t$ is essential in what is termed the non-ergodic case (for which $(E[Q]_t)^{-1}[Q]_t \overset{P}{\not\to}$ constant as $t \to \infty$). A simple example of this is furnished by the pure birth process N_t with intensity θN_{t-}, where we take for Q_t the score function and

$$Q_t = \theta^{-1}(N_t - 1) - \int_0^t N_{s-}\, \mathrm{d}s,$$

$$[Q]_t = \theta^{-2}(N_t - 1),$$

while

$$[Q]_t^{-1/2} Q_t \overset{d}{\to} N(0, 1), \qquad (EQ_t^2)^{-1/2} Q_t \overset{d}{\to} W^{1/2} N(0, 1)$$

and

$$(E[Q]_t)^{-1}[Q]_t \overset{a.s.}{\to} W,$$

where W has a gamma distribution with form parameter N_0 and shape parameter N_0 and $W^{1/2}N(0, 1)$ is distributed as the product of independent $W^{1/2}$ and $N(0, 1)$ variables.

To obtain confidence zones for θ from (11.4) we use the Taylor expansion (11.2) and then under appropriate continuity conditions for $\dot{Q}_t$,

$$\dot{Q}_T(\theta_{1,T})(\dot{Q}_T(\theta))^{-1} \xrightarrow{p} I_p$$

and, when (11.4) holds

$$[Q(\theta)]_T^{-1/2}\dot{Q}_T(\theta)(\theta_T^* - \theta) \xrightarrow{d} N_p(0, I_p) \tag{11.5}$$

as $T \to \infty$.

For the construction of confidence intervals we actually need this convergence to be uniform in compact intervals of θ; we shall use $\xrightarrow{u.d.}$ to denote such convergence and henceforth suppose that (11.5) holds in this mode. We shall also write

$$\bar{\mathscr{E}}(Q_t(\theta)) = (-\dot{Q}_t(\theta))'[Q(\theta)]_t^{-1}(-\dot{Q}_t(\theta)). \tag{11.6}$$

If C is a column vector of dimension p, convergence in (11.5) is mixing in the sense of Rényi (see Hall and Heyde 1980, p. 64) and $\bar{\mathscr{E}}(Q_T(\theta))$ behaves asymptotically like a constant matrix then, replacing $\bar{\mathscr{E}}(Q_T(\theta))$ by the estimated $\bar{\mathscr{E}}(Q_T(\theta_T^*))$, we obtain

$$P(C'\theta \in C'\theta_T^* \pm z_{\alpha/2}(C'\bar{\mathscr{E}}(Q_T(\theta_T^*))^{-1}C) \approx 1 - \alpha$$

where $\Phi(z_\beta) = 1 - \beta$, Φ denoting the standard normal distribution function. In particular this provides asymptotic confidence results for the individual elements of θ and also any nuisance parameters can conveniently be deleted.

Other confidence statements of broader scope may also be derived. Noting that

$$(\theta_T^* - \theta)'\bar{\mathscr{E}}(Q_T(\theta_T^*))(\theta_T^* - \theta) \xrightarrow{u.d.} \chi_p^2$$

and that

$$(\theta_T^* - \theta)'\bar{\mathscr{E}}(Q_T(\theta_T^*))(\theta_T^* - \theta) = \max_c \frac{(C'(\theta_T^* - \theta))^2}{C'(\bar{\mathscr{E}}(Q_T(\theta_T^*))^{-1}C'},$$

we have that

$$P\left(\max_c \frac{|C'(\theta_T^* - \theta)|}{(C'(\bar{\mathscr{E}}(Q_T(\theta_T^*))^{-1}C)^{1/2}} \leq \chi_{p,\alpha}\right) \approx 1 - \alpha,$$

where $\chi_{p,\alpha}^2$ is the upper α point of the chi-squared distribution with p degrees of freedom, so that

$$P(C'\theta_0 \in C'\theta_T^* \pm \chi_{p,\alpha}(C'(\bar{\mathscr{E}}(Q_T(\theta_T^*))^{-1}C)^{1/2} \quad \text{for all C}) \approx 1 - \alpha.$$

Also, if we let c_α be the set of all possible θ satisfying the inequality

$$(\theta_T^* - \theta)'\bar{\mathscr{E}}(Q_T(\theta_T^*))(\theta_T^* - \theta) \leq \chi_{p,\alpha}^2,$$

then simultaneous confidence intervals for functions $g_i(\theta)$, $i = 1, 2, \ldots, p$ with asymptotic confidence possibly greater than $(1 - \alpha)$ are given by

$$\left\{ \min_{\theta \in c_\alpha} g_i(\theta), \max_{\theta \in c_\alpha} g_i(\theta) \right\}, \qquad i = 1, 2, \ldots, p.$$

This allows us to deal conveniently with confidence intervals for non-linear functions of the components of θ.

Other methods are available in particular cases and, for example, it often happens, in cases where likelihood-based theory is available and tractable, that the ML ratio produces better-behaved confidence zones than the ML estimator (e.g. Cox and Hinkley (1974, p. 343)). In such a case, if $\theta = (\psi, \lambda)$ and $\Omega_0 = \{\psi_0, \lambda \in \Omega_\lambda\}$, then the set

$$\left\{ \Omega_0 \colon \sup_\Omega L_t(\theta) - \sup_{\Omega_0} L_T(\theta) \leqslant \tfrac{1}{2}\chi^2_{d,\alpha} \right\} \tag{11.7}$$

based on the likelihood function L, and with $d = \dim \psi$, may give an asymptotic confidence region for ψ of size α. It should be noted that (11.7) is invariant under transformation of the parameter of interest.

It is usually the case that the quasi-score or asymptotic quasi-score estimating function $Q_t(\theta)$ is a martingale and subject to suitable scaling,

$$-\dot{Q}_t(\theta) - [Q(\theta)]_t \tag{11.8}$$

$$-\dot{Q}_t(\theta) - \langle Q(\theta) \rangle_t \tag{11.9}$$

and

$$[Q(\theta)]_t - \langle Q(\theta) \rangle_t \tag{11.10}$$

are zero mean martingales, $\langle Q(\theta) \rangle_t$ being the quadratic characteristic of Q_t. The quantity $-\dot{Q}_t(\theta)$ is a generalized version of what is known as the observed information, while

$$EQ_t(\theta)Q_t'(\theta) = E[Q(\theta)]_t = E\langle Q(\theta) \rangle_t = -E\dot{Q}_t(\theta)$$

is a generalized Fisher information. The martingale relationships (11.8)–(11.10) and that fact that each of the quantities $-\dot{Q}_t(\theta)$, $[Q(\theta)]_t$ and $\langle Q(\theta) \rangle_t$ goes a.s. to infinity as t increases usually implies their asymptotic equivalence.

Which of the asymptotically equivalent forms should be used for asymptotic confidence zones based on (11.5) is unclear in general despite the fact that many special investigations have been conducted. Let $\mathscr{S}$ be the set of (possibly random) normalizing sequences $\{A_t\}$ which are positive definite and such that

$$A_t^{-1}(\theta)[Q(\theta)]_t^{-1/2}\dot{Q}_t(\theta) \xrightarrow{P} I_p \tag{11.11}$$

As $t \to \infty$. Then, each element of $\mathscr{S}$ determines an asymptotic confidence zone for θ, $\bar{\mathscr{E}}(Q_t(\theta))$ being replaced by $A_t'(\theta)A_t(\theta)$. Of course (11.11) ensures that zones will be similar for large T with high probability.

In the ergodic case when likelihoods rather than quasi-likelihoods are being considered, there is evidence to support the use of the observed information $-\ddot{Q}_t(\theta)$ rather than the expected information $EQ_t(\theta)Q_t'(\theta)$ (e.g. Efron and Hinkley 1978).

In the general case, Barndorff-Nielsen and Sørensen (1989) have used a number of examples to argue the advantages of (11.6) which, as we have seen, comes naturally from (11.4). On the other hand, Heyde (1987) proposed the form

$$(-\bar{Q}_t(\theta))'\langle Q(\theta)\rangle_t^{-1}(-\bar{Q}_t(\theta))$$

where $\bar{Q}_t(\theta)$ is a matrix of predictable processes such that $\dot{Q}_t(\theta) - \bar{Q}_t(\theta)$ is a martingale. This is a direct extension of the classical Fisher information and is closely related to $\mathscr{E}(Q_t(\theta))$.

The general situation is further complicated by the fact that $\bar{\mathscr{E}}(Q_t(\theta))$ ordinarily involves the unknown θ and consequently has to be replaced by $\bar{\mathscr{E}}(Q_t(\theta_t^*))$ in confidence statements. The variety of possibilities is such that each case should be examined on an individual basis.

Finally, some special comments on the non-ergodic case are appropriate. Suppose that

$$(E[Q]_t)^{-1}[Q]_t \xrightarrow{P} W(>0 \text{ a.s.}).$$

As always, a sequence of normalizing matrices $\{A_t\}$ is sought so that

$$A_T(\theta_T^* - \theta) \xrightarrow{d} N_p(0, I_p)$$

as $T \to \infty$. However, it has been argued that one should condition on the limit random variable W and then treat the unobserved value w of W as a nuisance parameter to be estimated. This approach has the attraction of reducing a non-ergodic model to an ergodic one but at the price of introducing asymptotics which, although plausible, may be very difficult to formalize in practice (e.g. Basawa and Brockwell (1984)). A related approach based on conditioning on an asymptotic ancillary statistic has been discussed by Sweeting (1986). This also poses considerable difficulties in formalization but may be useful in some cases. The problem of precise choice of normalization, of course, usually remains.

In conclusion, we also need to comment on the point estimate itself. If confidence statements are based on (11.5) and $\bar{\theta}_T$ is an estimator such that

$$[Q(\theta)]_T^{-1/2}\dot{Q}_T(\theta)(\bar{\theta}_T - \theta_T^*) \xrightarrow{P} 0$$

as $T \to \infty$, then

$$[Q(\theta)]_T^{-1/2}\dot{Q}_T(\theta)(\bar{\theta}_T - \theta) \xrightarrow{d} N_p(0, I_p)$$

clearly offers an alternative to (11.5). There may sometimes be particular reasons to favour a certain choice of estimator, such as in the example discussed towards the end of Section 2 above, where nuisance parameters can be avoided.

References

Barndorff-Nielsen, O. E. and Sørensen, M. (1989). Information quantities in non-classical settings. Research Report No. 192. Department of Theoretical Statistics, Aarhus University, Denmark.

Basawa, I. V. and Brockwell, P. J. (1984). Asymptotic conditional inference for regular non-ergodic models with an application to autoregressive processes. *Ann. Stat.*, **12**, 161–71.

Cox, D. R. and Hinkley, D. V. (1974). *Theoretical statistics*. Chapman and Hall, London.

Efron, B. and Hinkley, D. V. (1978). Assessing the accuracy of the maximum likelihood estimator: observed versus expected Fisher information. *Biometrika*, **65**, 457–87.

Godambe, V. P. and Heyde, C. C. (1987). Quasi-likelihood and optimal estimation. *Int. Stat. Rev.*, **55**, 231–44.

Hall, P. G. and Heyde, C. C. (1980). *Martingale limit theory and its application*. Academic Press, New York.

Heyde, C. C. (1987). On combining quasi-likelihood estimating function. *Stoch. Processes Appl.*, **25**, 267–76.

Heyde, C. C. (1988). Fixed sample and asymptotic optimality for classes of estimating functions. *Contemporary Math.*, **80**, 241–47.

Heyde, C. C. (1989). Quasi-likelihood and optimality for estimating functions: some current unifying themes. *Bull. Int. Stat. Inst.*, **53** (Book 1), 19–29.

Heyde, C. C. and Gay, R. (1989). On asymptotic quasi-likelihood estimation. *Stoch. Processes Appl.*, **31**, 223–36.

Rogers, L. C. G. and Williams, D. (1987). *Diffusions, Markov processes and martingales.* Vol. 2, *Itô Calculus.* Wiley, Chichester.

Sørensen, M. (1990). On quasi-likelihood for semimartingales. *Stoch. Processes Appl.*, **35**, 331–46.

Sweeting, T. J. (1986). Asymptotic conditional inference for the offspring mean of a supercritical Galton–Watson process. *Ann. Stat.*, **14**, 925–33.

12
Simplified and two-stage quasi-likelihood estimators

J. E. Hutton, O. T. Ogunyemi, and Paul I. Nelson

ABSTRACT

Properties of a quasi-likelihood estimator are typically obtained from an expansion of the quasi-likelihood in which the remainder term is asymptotically negligible. The remainder term, however, can be very difficult to control in some important applications, especially in models where both the mean and variance depend on the unknown parameter. In such cases we propose a modified quasi-likelihood based on the use of a simplified covariance matrix. Conditions for the consistency and asymptotic normality of the resulting estimators are developed. We illustrate this approach with a branching process with immigration model and present a simulation study of the small sample behaviour of the estimators.

12.1 Introduction

Godambe optimal estimating functions can be used in a wide variety of settings and require relatively few assumptions. Because of this generality, properties of maximum quasi-likelihood estimators (mqes) can be very difficult to obtain. Thus, for example, it may be very hard to even show that the quasi-score function has a root lying in the parameter space or that such a root has a limiting normal distribution. In addition, there may be instability in the numerical procedures used to find the mqe.

To deal with these problems, we propose some simplifications of the quasi-score function. First, the conditional covariance can be replaced with a simpler but asymptotically equivalent function. We call the resulting estimator a simplified maximum quasi-likelihood estimator (smqe). In a second optional stage, the smqe can be substituted for the unknown parameter in the conditional covariance and a root of the resulting modified quasi-score function used as an estimator. We develop asymptotic properties of these procedures and show that the two-stage procedure can lead to a fully efficient estimator. Our approach is illustrated for a branching process with immigration. We also present a simulation study comparing the simplified estimators for this model.

12.2   The model and notation

Let $\{y_k\}_{k=1}^\infty$ be a discrete-time stochastic process taking values in r-dimensional Euclidean space whose possible distributions $\{P_\theta\}$ are indexed by a p-dimensional parameter θ lying in an open subset $\Theta \subset R^p$. Our goal is to estimate θ from a single realization $\{y_k\}_{k=1}^n$. Let θ^* denote the 'true' θ and let all unsubscripted probabilities and expectations be taken with respect to θ^*. In Section 12.6 we describe how our results can be adapted to cover inference from continuous-time processes.

Let $\{F_k\}$ be a non-decreasing family of sigma fields such that y_k is measurable with respect to F_k, $k \geqslant 1$, F_0 = the trivial field and $y_o \equiv 0$. Assume that y_k is square integrable, $k \geqslant 1$. Form a regression type model by writing

$$y_k = E_\theta(y_k|F_{k-1}) + y_k - E_\theta(y_k|F_{k-1}) \equiv f_k(\theta) + \varepsilon_k(\theta). \qquad (12.1)$$

Note that the error terms $\{\varepsilon_k(\theta)\}$ are martingale differences with respect to $\{F_k\}$, so that

$$\left\{ m_n(\theta) \equiv \sum_{k=1}^n \varepsilon_k(\theta), F_n \right\}_{n \geqslant 1} \qquad (12.2)$$

is a P_θ martingale. The $r \times r$ quadratic characteristic $\langle m(\theta)\rangle_n$ and quadratic variation $[m(\theta)]_n$ of $\{m_n(\theta)\}$ are defined a.e. with respect to P_θ for $n \geqslant 1$ by

$$\langle m(\theta)\rangle_n = \sum_{k=1}^n E_\theta(\varepsilon_k(\theta)\varepsilon_k'(\theta)|F_{k-1}),$$

$$\equiv \sum_{k=1}^n a_k(\theta), \quad \text{say}, \qquad (12.3)$$

and

$$[m(\theta)]_n = \sum_{k=1}^n \varepsilon_k(\theta)\varepsilon_k'(\theta),$$

where we have set

$$a_k(\theta) = E_\theta(\varepsilon_k(\theta)\varepsilon_k'(\theta)|F_{k-1}), \qquad k \geqslant 1, \qquad (12.4)$$

the conditional covariance of $\varepsilon_k(\theta)$. Finally, the integrated process $\{x_n = \sum_{k=1}^n y_k\}$, for $n \geqslant 1$, can be expressed as

$$x_n = \sum_{k=1}^n f_k(\theta) + m_n(\theta). \qquad (12.5)$$

For a square matrix A, $A^\dagger$ denotes its Moore–Penrose pseudo-inverse, $|A|$ its determinant, $e_1(A)$ its smallest eigenvalues and $e_2(A)$ its largest eigenvalue. For an $r \times 1$ vector $A(\theta)$, $\dot{A}(\theta)$ denotes the $r \times p$ matrix of its derivatives with respect to $\theta^{p \times 1}$. For any matrix A, A' denotes the transpose of A and $\|A\|$ its Euclidean norm.

12.3 Simplified and two-stage estimators

For the model given in (12.5) and within the class of unbiased martingale estimating functions of the form

$$G_n(\theta) = \sum_{k=1}^{n} \alpha_k(\theta)\varepsilon_k(\theta),$$

where the $p \times r$ matrix $\alpha_k(\theta)$ is measurable with respect to F_{k-1}, $k \geqslant 1$, Hutton and Nelson (1986) showed that Godambe optimal estimating function (under some regularity conditions) is given by the quasi-score function:

$$Q_n(\theta) \equiv \sum_{k=1}^{n} \dot{f}'_k(\theta)a_k^{\dagger}(\theta)\varepsilon_k(\theta). \tag{12.6}$$

The maximum quasi-likelihood quasi-likelihood estimator $\hat{\theta}_n$ is defined as a root of (12.6).

In cases where the quasi-score function is difficult to work with, we propose a class of estimating functions of the form

$$\tilde{Q}_n(\theta, \{s_k(\theta)\}) \equiv \tilde{Q}_n(\theta) \equiv \sum_{k=1}^{n} f'_k(\theta)s_k^{\dagger}(\theta)\varepsilon_k(\theta) \tag{12.7}$$

where $\{s_k(\theta)\}$ is a sequence of $r \times r$ positive-semidefinite matrices with $s_k(\theta)$ measurable with respect to F_{k-1}, $k \geqslant 1$. The weight matrices $\{s_k(\theta)\}$ should be selected with the goal of yielding a fully efficient estimator whose properties are easier to obtain than those of the mqe. Accordingly, we call $\tilde{\theta}_n(\{s_k(\theta)\}) \equiv \tilde{\theta}_n$, $n \geqslant 1$, a simplified mqe (smqe) if

$$\tilde{Q}_n(\tilde{\theta}_n) = 0. \tag{12.8}$$

Note that the mqe is an smqe. We use $\{\hat{\theta}_n\}$ to denote the mqe and $\{\tilde{\theta}_n\}$ a general smqe.

An efficacious choice of $\{s_k(\theta)\}$ must be made on a case-by-case basis. In general, the greatest simplification occurs when $a_k(\theta)$ depends on θ but $s_k(\theta)$ is free of θ, $k \geqslant 1$, since this leads to both a numerically more tractable equation to solve and a much more manageable expansion of $\tilde{Q}_n(\theta)$ which is used to show the existence of the estimator and describe its asymptotic behaviour. See Examples 3.1 and 4.1. The conditional least-squares estimators of Klimko and Nelson (1978) are of this type with $s_k(\theta) = I$, the $r \times r$ identity matrix.

We now present an example which illustrates the potential advantages of modifying $\{a_k(\theta)\}_{k=1}^{n}$ to be free of θ.

Example 12.1 Suppose that for some positive constant θ, $E(y_k|F_{k-1}) = \theta$, $E((y_k - \theta)^2|F_{k-1})) = 1 + \theta u_{k-1}$, $k \geqslant 1$ where $\{u_k\}$ is a family of nonnegative random variables adapted to $\{F_k\}$. The quasi-score function for this model is

$$Q_n(\theta) = \sum_{k=1}^{n} \left(\frac{y_k - \theta}{1 + \theta u_{k-1}} \right). \tag{12.9}$$

As a simplification, take $s_k(\theta) = u_{k-1}$, $k \geqslant 1$, so that

$$\tilde{Q}_n(\theta, \{s_k(\theta)\}) = \sum_{k=1}^{n} (y_k - \theta) u_{k-1}^{\dagger}. \tag{12.10}$$

First, note that the smqe can be found explicitly as a root of (12.10), whereas the mqe must be obtained by iteratively finding a root of (12.9). Second, as will be seen in Section 12.4 (where this example will be discussed further) asymptotic properties of the smqe are easier to find than those of the mqe.

Another approach is to use what we call a two-stage smqe. Given observations $\{y_k\}_{k=1}^{n}$, in the first stage the matrices $\{s_k(\theta)\}$ are chosen to yield convenient, consistent smqes $\{\tilde{\theta}_m\}_{m=1}^{n}$, with $\tilde{\theta}_m$ computed from $\{y_k\}_{k=1}^{m}$, $m \geqslant 1$. In the second stage an smqe $\tilde{\theta}_n^{(2)}$ is defined by the equation:

$$\tilde{Q}_n(\tilde{\theta}_n^{(2)}, \{a_k(\tilde{\theta}_{k-1})\}) = 0. \tag{12.11}$$

Since $\tilde{\theta}_{k-1}$ is measurable with respect to F_{k-1}, $k \geqslant 1$, $\{\tilde{Q}_n(\theta^*), a_k(\tilde{\theta}_{k-1})\}$, $\{F_n\}_{n \geqslant 1}$ is a martingale (under some integrability conditions), a fact which is central to obtaining the existence and asymptotic properties of $\{\tilde{\theta}_n^{(2)}\}$. We also propose a related two-stage estimator defined as a solution to the system of equations:

$$\tilde{Q}_n(\theta, \{a_k(\tilde{\theta}_n)\}) = 0, \tag{12.12}$$

where only $\tilde{\theta}_n$, the first-stage smqe based on all the data, is used. Note that $\{\tilde{Q}_n(\theta, \{a_k(\tilde{\theta}_n)\}), F_n\}$ is not in general a martingale since $\tilde{\theta}_n$ need only be measurable with respect to F_n, $n \geqslant 1$. Hence, this estimator is not an smqe. Lacking the machinery of martingale theory, we cannot obtain the asymptotic properties of this type of estimator. However, our simulation study in Section 12.5 shows that it performs well for a branching process with immigration model.

An smqe $\{\tilde{\theta}_n\}$ may not be fully efficient compared to the mqe in the sense of having asymptotically a larger covariance matrix in the partial ordering of positive-definite matrices. This is true of the conditional least squares estimators (CLS) for a branching process with immigration. However, for this model the second stage smqe based on the CLS estimator is fully efficient. See Section 12.5. Bear in mind that in finite samples an asymptotically inefficient estimator can perform well.

12.4 Asymptotics

Given a simplified quasi-score function $\tilde{Q}_n(\theta, \{s_k(\theta)\}) \equiv \tilde{Q}_n(\theta)$, the existence and consistency of the smqe a.e. on an event E follows if:

for all sufficiently small $\delta > 0$, a.e. on E, ultimately in n,

$$\tilde{Q}_n(\theta) = 0 \qquad (12.13)$$

has a root in the sphere $\{\theta, \|\theta - \theta^*\| \leqslant \delta\}$.

Hutton and Nelson (1986) obtained (12.13) for the mqe by assuming that there exists a process $\{q_n(\theta)\}$ which, a.e. on E, ultimately in n attains a local maximum in the interior of $\{\theta, \|\theta - \theta^*\| \leqslant \delta\}$ for all sufficiently small δ and such that a.e. on E

$$\dot{q}_n(\theta) = Q_n(\theta). \qquad (12.14)$$

Since there is no such process $\{q_n(\theta)\}$ satisfying (12.14) for important applications like a branching process with immigration, we employ a more general approach here. Using the Brouwer fixed-point theorem, Aitchison and Silvey (1958) showed that if $Q_n(\theta)$ is continuous in θ a.e. on E, $n \geqslant 1$ and for all small $\delta > 0$, a.e. on E,

$$\limsup_{n \to \infty} \left(\sup_{\|\theta - \theta^*\| = \delta} (\theta - \theta^*)' \tilde{Q}_n(\theta) \right) < 0, \qquad (12.15)$$

then (12.13) holds and a consistent smqe $\{\tilde{\theta}_n\}$ exists a.e. on E. The measurability of $\tilde{\theta}_n$, $n \geqslant 1$, follows from the results of Jennrich (1969) applied to $\|\tilde{Q}_n(\theta)\|$.

Typically, the greatest difficulty in verifying (12.15) when p, the dimension of θ, is greater than one, lies in showing that $(\theta - \theta^*)' \tilde{Q}_n(\theta) < 0$, ultimately in n, *uniformly* in θ on $\{\theta: \|\theta - \theta^*\| = \delta\}$. We use an expansion of $\tilde{Q}_n(\theta)$ to give useful conditions under which (12.15) holds. Henceforth, assume that

$$E(\|\tilde{Q}_n(\theta^*)\|^2) < \infty, \qquad n \geqslant 1, \qquad (12.16)$$

so that $\{(\theta - \theta^*)' \tilde{Q}_n(\theta^*)\}$ is a square-integrable martingale with respect to $\{F_n, P_\theta^*\}$. We then have

$$\langle (\theta - \theta^*)' \tilde{Q}(\theta^*) \rangle_n = (\theta - \theta^*)' \langle \tilde{Q}(\theta^*) \rangle_n (\theta - \theta^*)$$

$$= (\theta - \theta^*)' \sum_{k=1}^{n} \dot{f}_k'(\theta^*) s_k^\dagger(\theta^*) a_k(\theta^*) s_k^\dagger(\theta^*) \dot{f}_k(\theta^*)(\theta - \theta^*). \qquad (12.17)$$

Let

$$\tilde{V}_n = \sum_{k=1}^{n} \dot{f}_k'(\theta^*) s_k^\dagger(\theta^*) \dot{f}_k(\theta^*) \qquad (12.18)$$

and note that $\dot{Q}_n(\theta^*) = -\tilde{V}_n + N_n$, where $\{N_n, F_n\}$ is a $p \times p$ zero-mean martingale. Using a first-order Taylor expansion of $(\theta - \theta^*)'\tilde{Q}_n(\theta)$, we obtain

$$(\theta - \theta^*)'\tilde{Q}_n(\theta) = (\theta - \theta^*)'\tilde{Q}_n(\theta^*) + (\theta - \theta^*)'(-\tilde{V}_n + N_n)(\theta - \theta^*) + \tilde{R}_n(\theta)$$

$$= (\theta - \theta^*)'\tilde{Q}_n(\theta^*) - (\theta - \theta^*)V_n(\theta - \theta^*) + R_n(\theta). \qquad (12.19)$$

where $\tilde{R}_n(\theta)$ and $R_n(\theta)$ are appropriate remainder terms.

Let $h(x)$ denote a positive, non-decreasing function defined on the non-negative reals such that

$$\int_0^\infty h^{-2}(x)\,\mathrm{d}x < \infty. \qquad (12.20)$$

Theorem 12.1 *Assume that (12.16) holds and that the elements of $\tilde{Q}_n(\theta)$ are a.e. differentiable with respect to θ, $n \geq 1$. Further, suppose that a.e. on an event E, as $n \to \infty$,*

$$\lim e_1(\langle \tilde{Q}(\theta^*) \rangle_n) = \infty, \qquad (12.21)$$

$$\lim \sup \left(h(e_2(\langle \tilde{Q}(\theta^*) \rangle_n))/e_1(\tilde{V}_n) \right) < \infty, \qquad (12.22)$$

where $h(\cdot)$ is a function described in (12.20),

$$\lim_{n \to \infty} \sup_{\|\theta - \theta^*\| = \delta} (R_n(\theta)/(\theta - \theta^*)'\tilde{V}_n(\theta - \theta^*)) \leq 0. \qquad (12.23)$$

Then, there is a family of estimators $\{\tilde{\theta}_n\}$ converging to θ^* a.e. on E and $\tilde{Q}_n(\tilde{\theta}_n) = 0$ a.e. on E ultimately in n.

Proof The proof follows from (12.15) using techniques similar to those employed by Hutton and Nelson (1986). Details are omitted.

Note that if the regression-type functions $\{f_k(\theta)\}$ defined in (12.1) are a.e. P_θ linear in θ and $\{s_k(\theta)\}$ is actually free of θ, then the remainder $R_n(\theta)$ in (12.19) is zero, $n \geq 1$, which facilitates the application of Theorem 12.1. Condition (12.22) may be interpreted as requiring that the approximating weights $\{s_k(\theta)\}$ asymptotically be not too far from the quasi-score weights $\{a_k(\theta)\}$. These observations will be illustrated in Example 12.2.

The asymptotic distribution of an smqe $\{\tilde{\theta}_n\}$ depends on the large-sample behaviour of the martingale $\{\tilde{Q}_n(\theta^*)\}$. Suppose that $\tilde{\theta}_n \to \theta^*$ and $\tilde{Q}_n(\tilde{\theta}_n) = 0$ a.e. on an event E, $n \geq 1$. Consider an expansion of the form

$$0 = \tilde{Q}_n(\tilde{\theta}_n)$$

$$= \tilde{Q}_n(\theta^*) - \tilde{V}_n(\tilde{\theta}_n - \theta^*) + r_n(\tilde{\theta}_n). \qquad (12.24)$$

If the remainder $r_n(\tilde{\theta}_n)$ is appropriately asymptotically negligible $Q_n(\theta^*)$ and $\tilde{V}_n(\tilde{\theta}_n - \theta^*)$ have the same asymptotic distribution.

Theorem 12.2 *Let $(\tilde{\theta}_n)$ be an smqe which converges a.e. to θ^* on an event E. Assume that there is a sequence of $p \times p$ matrices $\{H_n(\theta)\}$, possibly random, and an avent A such that for all $z \in R^p$,*

$$P\{H_n(\theta^*)\tilde{Q}_n(\theta^*) \leqslant z | A \cap E\} \to P(Z \leqslant z) \qquad (\text{stably}), \qquad (12.25)$$

where $Z \sim N_p(0, I)$, and for some $\delta > 0$,

$$P \lim_{n \to \infty} \sup_{\{\theta, \|\theta - \theta^*\| \leqslant \delta\}} \|(H_n(\theta^*)r_n(\theta))\| = 0. \qquad (12.26)$$

Then for any sequence of events E_n, E_n measurable with respect to F_n, $n \geqslant 1$, with $P(E_n \cap A) \to P(E \cap A) > 0$; for all $z \in R^p$, we have

$$P(H_n(\theta^*)\tilde{V}_n(\tilde{\theta}_n - \theta^*) \leqslant z | A \cap E_n) \to P(Z \leqslant z). \qquad (12.27)$$

Proof   The proof follows immediately from (12.24). For non-ergodic models the event A typically represents an event where for some positive norming function $\{k_n\}$ diverging to ∞, $k_n^{-1}[\tilde{Q}(\theta^*)]_n$ converges to a positive-definite random matrix. Depending on circumstances, a variety of central limit theorems can be used to obtain (12.25). See Hall and Heyde (1980) for some martingale central limit theorems.

If $\{\tilde{\theta}_n\}$ obtained from stage one converges to θ^* sufficiently rapidly, the mqe and the second-stage estimator $\{\tilde{\theta}_n^{(2)}\}$ will typically have the same limiting distribution. One way to use Theorem 12.2 to show this is to demonstrate that

$$\tilde{Q}_n(\theta, \{a_k(\tilde{\theta}_{k-1}) - a_k(\theta^*)\}) \to 0 \qquad (12.28)$$

at an appropriate rate so that

$$\{\tilde{Q}_n\{\theta, \{a_k(\tilde{\theta}_{k-1})\}\} \qquad \text{and} \qquad \{Q_n(\theta, \{a_k(\theta^*)\})\}$$

have the same limiting distribution.

Now, we show how Theorems 12.1 and 12.2 can be used to obtain the asymptotic properties of the smqe described in Example 12.1. Note that, unfortunately, nice asymptotic properties of the smqe do not necessarily carry over to the mqe even though the weights $\{s_k(\theta)\}$ are close to the quasi-score weights $\{a_k(\theta)\}$.

Example 12.2 (Example 12.1 continued)   From (12.17) and (12.18) we have that

$$\langle \tilde{Q}(\theta^*) \rangle_n = \sum_{k=1}^{n} (1 + \theta^* u_{k-1})(u_{k-1}^\dagger)^2,$$

and

$$\tilde{V}_n = \sum_{k=1}^{n} u_{k-1}^{\dagger}.$$

Then, a.e. on $E \equiv \{\sum_{k=1}^{\infty} u_{k-1}^{\dagger} = \infty, \liminf_{k \to \infty} (u_k) > 0\}$, as $n \to \infty$.

$$\langle \tilde{Q}(\theta^*) \rangle_n \to \infty \qquad \text{and} \qquad \sup_{n \to \infty} \langle \tilde{Q}(\theta^*) \rangle_n / \tilde{V}_n < \infty,$$

so that (12.21) and (12.22) hold with $h(x) = x$. Since $R_n(\theta) \equiv 0$, $n \geq 1$, (12.23) holds. Hence, from Theorem 12.4 it follows that the smqe is a.e. consistent on E. Some additional assumptions are required to guarantee that the smqe is asymptotically normally distributed. Suppose, for example, that there is a sequence of positive constants $\{k_n\}$ diverging to infinity, a positive constant $C(\theta)$ and a non-negative random variable $B(\theta)$ such that as $n \to \infty$,

$$P \lim(k_n^{-1}[\tilde{Q}(\theta^*)]_n) = B(\theta^*),$$

$$\lim(k_n^{-1}E([Q(\theta^*)]_n) = C(\theta^*),$$

$$P \lim\left(\max_{k \leq n} k_n^{-1/2}|y_k - \theta^*|u_{k-1}^{\dagger}\right) = 0.$$

Then, from Theorem 12.2 and Corollary 2.3 of Hutton and Nelson (1984), for all z we have

$$P([\tilde{Q}(\theta^*)]_n^{-1/2} \tilde{V}_n(\tilde{\theta}_n - \theta^*) \leq z | E \cap \{B(\theta^*) > 0\}) \to P(Z \leq z),$$

where $Z \sim N(0, 1)$. For a specific example, let $\{u_k^2, F_k\}$ be a standard Poisson process sampled at integer times and let $y_k = \theta + w_k(1 + \theta u_{k-1})^{1/2}$, $k \geq 1$, where $\{w_k\}$ are i.i.d., have finite fourth moment, with $E(w_k|F_{k-1}) = 0$, $E(w_k^2|F_{k-1}) = 1$, a.e., $k \geq 1$. The above then leads to $n^{1/4}(\tilde{\theta}_n - \theta^*) \to (\theta^*/2)^{1/2}Z$, in distribution.

There is a partial converse to Theorem 12.1 which gives necessary conditions for the existence of a consistent smqe.

Theorem 12.3 *Let the functions $\{f_k(\theta), k \geq 1\}$ defined in (12.1) be a.e. linear in θ and let the sequence $\{s_k(\theta)\}$ be a.e. free of θ. Assume the following:*

$$P(\tilde{Q}_n(\theta^*) = 0, n \geq 1) < 1, \tag{12.29}$$

$$\lambda(\theta^*) \equiv \sup_n (e_2(\langle \tilde{Q}(\theta^*) \rangle_n)) \in L^1(\Omega, F, P_{\theta^*}), \tag{12.30}$$

$$\eta \equiv \sup_n (e_2(\tilde{V}_n)) < \infty, \qquad \text{a.e. } P_{\theta^*}. \tag{12.31}$$

*Then, the smqe is **not** consistent a.e.*

Proof Suppose that $\{\tilde{\theta}_n\}$ is an a.e. consistent smqe and that (without loss of generality) $\tilde{Q}_n(\tilde{\theta}_n) = 0$ a.e., $n \geq 1$. Since the sequence $\{f_k(\theta)\}$ is linear in θ, and the sequence $\{s_k\}$ is free of θ, it follows from the definition of $\tilde{V}_n$ in (12.18) that

$$\dot{\tilde{Q}}_n(\theta^*) = -\tilde{V}_n.$$

Hence,

$$
\begin{aligned}
0^{p \times 1} &= \tilde{Q}_n(\tilde{\theta}_n) \\
&= \tilde{Q}_n(\theta^*) + \dot{\tilde{Q}}_n(\theta^*)(\tilde{\theta}_n - \theta^*) \\
&= \tilde{Q}_n(\theta^*) - \tilde{V}_n(\tilde{\theta}_n - \theta^*). \qquad (12.32)
\end{aligned}
$$

From (12.31) and the consistency of $\tilde{\theta}_n$, we have that

$$\| \tilde{V}_n(\tilde{\theta}_n - \theta^*) \| \leqslant \eta^2 \|\tilde{\theta}_n - \theta^*\|^2 \to 0.$$

Hence, from (12.32),

$$\tilde{Q}_n(\theta^*) \to 0 \quad \text{a.e.} \quad \text{as } n \to \infty. \qquad (12.33)$$

For any $p \times 1$ vector of constants d, let $u_n(d) = d'\tilde{Q}_n(\theta^*)$. Then $\{u_n(d), F_n\}$ is a real martingale which is L^2-bounded since

$$E(u_n^2(d)) \leqslant \|d\|^2 E(\lambda(\theta^*)).$$

Therefore, for all d, $\{u_n(d)\}$ is uniformly integrable. Thus, there exists an integrable random variable $u(d)$ such that

$$\lim_{n \to \infty} u_n(d) = u(d), \quad \text{and} \quad u_n(d) = E(u(d)|F_n) \quad \text{a.e.} \quad n \geqslant 1.$$

Since d is arbitrary, it follows from (12.33) that $\tilde{Q}_n(\theta^*) = 0$ a.e., $n \geqslant 1$. This completes the proof by contradicting (12.29).

Condition (12.21) can be used to check whether or not the statistical model provides enough information for the consistent estimation of θ. The following example is a simple illustration of a situation where the noise overwhelms the signal.

Example 12.3 Let $\{y_k\}$ be independent (θ, σ_k^2), σ_k^2 free of θ, $k \geqslant 1$, such that $\sum_{k=1}^{\infty} \sigma_k^{-2} < \infty$. Then, taking $s_k(\theta) = \sigma_k^2 \equiv a_k(\theta)$, i.e. using the mqe, $\langle Q(\theta) \rangle_n = V_n = \sum_{k=1}^{n} 1/\sigma_k^2$, which is bounded in n so that (12.30) and (12.31) hold. Theorem 12.3 yields that the mqe is not consistent. Further, suppose in fact that y_k has a normal distribution with the specified mean and variance, $k \geqslant 1$. Then as noted by Hutton and Nelson (1986), the existence of *any* consistent estimator requires that the Radon–Nikodym derivatives

$dP_\theta^{(n)}/dP_{\theta^*}^{(n)} \to 0$ a.e., P_θ^* as $n \to \infty$, where $P_\theta^{(n)}$ is the restriction of P_θ to F_n, $\theta \in \Theta$. Here,

$$\frac{dP_\theta^{(n)}}{dP_{\theta^*}^{(n)}} = \exp\left[-\tfrac{1}{2}\left((\theta - \theta^*)^2 \sum_{k=1}^n \frac{1}{\sigma_k^2} + 2(\theta - \theta^*) \sum_{k=1}^n \left(\frac{y_k - \theta^*}{\sigma_k^2} \right) \right) \right],$$

which does not converge to zero a.e. since

$$\left\{ \sum_{k=1}^n \left(\frac{y_k - \theta^*}{\sigma_k^2} \right), F_n \right\}$$

is an L^2-bounded martingale. Again, the problem in this example is not the concept of quasi-likelihood but the model itself.

12.5   Applications to a branching process with immigration

First, we show how Theorems 12.1 and 12.2 can be used to obtain a fully efficient two-stage smqe of the birth and immigration rates of a particular branching process model. As far as we know, the existence and consistency of the corresponding mqes have not yet been analytically verified. Maximum likelihood estimation is difficult here because of the complexity of the likelihood function and a study of the performance of the mle has not been carried out.

Second, we present a simulation study of a comparison among the mqe, a two-stage smqe, a two-stage estimator of the type defined in (12.12), and the conditional least-squares estimator.

Let $\{y_k\}$ be a discrete-time branching process with Bernoulli offspring distribution, p = probability of a birth, $q = 1 - p$ and independent Poisson immigration with mean λ at each generation, $0 < p < 1, 0 < \lambda$. This process fits into the framework of Section 12.2 with

$$\theta = (p, \lambda)'$$

$$f_k(\theta) = \lambda + py_{k-1}, \qquad k \geq 2,$$

$$a_k(\theta) = \lambda + pqy_{k-1}, \qquad k \geq 2,$$

$$\varepsilon_k(\theta) = y_k - \lambda - py_{k-1}, \qquad k \geq 2,$$

$$\tilde{Q}_n(\theta) = \left(\sum_{k=2}^n \varepsilon_k(\theta)y_{k-1}s_k^\dagger(\theta), \sum_{k=2}^n \varepsilon_k(\theta)s_k^\dagger(\theta) \right)'.$$

In the first stage, let $\tilde{\theta}_k = (\tilde{p}_k, \tilde{\lambda}_k)$ be the conditional least-squares estimators (i.e. an smqe with $s_k(\theta) \equiv I, k \geq 2$) based on $\{y_j\}_{j=1}^k, k \geq 2$. These estimators are consistent a.e. See Klimko and Nelson (1978). We now illustrate how Theorems 12.1 and 12.2 can be used to obtain the asymptotic properties of the second stage smqe $\{\tilde{\theta}_n^{(2)}\}$ defined as solutions to

$$\tilde{Q}_n(\theta, \{a_k(\tilde{\theta}_{k-1})\}) = 0.$$

For convenience, let

$$s_k \equiv a_k(\tilde{\theta}_{k-1}), \qquad k \geqslant 2.$$

We have

$$\tilde{V}_n = \sum_{k=2}^{n} \dot{f}_k'(\theta) s_k^{\dagger} \dot{f}_k(\theta) = \begin{pmatrix} \Sigma y_{k-1}^2/s_k & \Sigma y_{k-1}/s_k \\ \Sigma y_{k-1}/s_k & (n-1)/s_k \end{pmatrix}.$$

Theorem 1.1 of Billingsley (1961) implies that the ergodic theorem holds for $\{y_k\}$ no matter what the initial distribution. This observation justifies the convergence results that follows. Note the since s_k is free of θ and $f_k(\theta)$ is linear in θ, the remainder $R_n(\theta)$ in (12.19) is identically zero, $n \geqslant 2$. Conditions (12.21) and (12.22) of Theorem 12.1 hold with $h(x) = x$, since

$$\lim_{n \to \infty} \tilde{V}_n/n = \lim_{n \to \infty} \langle \tilde{Q}(\theta^*) \rangle_n/n = \begin{pmatrix} E(y_1^2/a_2(\theta^*)) & E(y_1/a_2(\theta^*)) \\ E(y_1/a_2(\theta^*)) & E(1/a_2(\theta^*)) \end{pmatrix}$$

$$\equiv \tilde{V} \quad \text{a.e.,} \tag{12.34}$$

which is a positive-definite matrix. The expectations in $\tilde{V}$ are with respect to the stationary distribution of $\{y_k\}$ determined by P_{θ^*}. To verify the limit in (12.34), we have, for example,

$$(\tilde{V}_n)_{11}/n = \frac{1}{n} \sum_{k=1}^{n-1} \frac{y_k^2}{s_{k+1}}$$

$$= \frac{1}{n} \sum_{k=1}^{n-1} \frac{y_k^2}{a_{k+1}(\theta^*)} \left(\frac{a_{k+1}(\theta^*)}{s_{k+1}} \right)$$

$$= \frac{1}{n} \sum_{k=1}^{n-1} \frac{y_k^2}{a_{k+1}(\theta^*)} (1 + \Delta_k).$$

Again by the ergodic theorem we have

$$\frac{1}{n} \sum_{k=1}^{n-1} \frac{y_k^2}{a_{k+1}(\theta^*)} \to E\left(\frac{y_1^2}{a_2(\theta^*)} \right) = (\tilde{V})_{11}.$$

Since $\tilde{\lambda}_k \to \lambda^*$, $\tilde{p}_k \to p^*$ a.e., it can easily be shown that $\Delta_k \to 0$ a.e. Hence

$$\frac{1}{n} \sum_{k=1}^{n-1} \frac{y_k^2}{a_{k+1}(\theta^*)} (1 + \Delta_k) \to (\tilde{V})_{11} \quad \text{a.e.}$$

Similarly, we have that $1/n\langle \tilde{Q}(\theta^*) \rangle_n \to \tilde{V}$ a.e. It can be shown using Theorem

12.2 and the martingale central limit theorem of Billingsley (1961), pp. 52–3, that in distribution

$$\sqrt{n}(\tilde{\theta}_n^{(2)} - \theta^*) \to N(0, \tilde{V}^{-1}). \tag{12.35}$$

The limiting covariance matrix given in (12.35) is the same one that would be obtained by the mqe itself if it satisfies the conditions of Theorem 12.2, a matter which we have been unable to decide. Hence, the mqe would be no better asymptotically than the smqe. Also, it follows from Godambe and Heyde (1987) that $C - \tilde{V}^{-1}$ is positive semidefinite, where C is the limiting covariance matrix of the normalized conditional least-squares estimators $\{\tilde{\theta}_n\}$.

We carried out a designed experiment to compare the mean-square errors of the mqe, the cls estimator, the smqe based on the cls as defined in (12.11) and the two-stage estimator (denoted mqe(2)) based on the cls as defined in (12.12). We generated 100 data sets from the branching process with immigration model described above using the IMSL routines GGBN and GGPOS for each of the following combinations of the parameters p, λ, n: $n = 10, 20, 40$; $p = 0.1, 0.3, 0.5, 0.7, 0.9$; $\lambda = 1, 5, 10, 20, 40$. We started the process both with $y_0 \equiv 1$ and with y_0 having the stationary distribution of the process. All four estimators were computed for each data set. Negative estimates of λ and estimates of p not in $[0, 1]$ were deleted. The remaining values were used to estimate the mean square errors denoted generically $\text{MSE}(\hat{p})$ and $\text{MSE}(\hat{\lambda})$, and their standard errors (rounded to three significant places). Representative results appear in Tables 12.1–6. Overall, we observed that the two-stage smqe defined in (12.11) tended to be different from the other three, particularly when estimating λ for large λ and small n. The other three estimators tended to be very close. Note that for small n, and large $\lambda/(1 - p)$, none of the estimators performed well.

To validate our interpretations of the estimated mean:square errors, for each n $p(\lambda)$ combination we carried out separate randomized block analyses of variance on the mses of the $\hat{p}$s ($\hat{\lambda}$s) using the values of $\lambda(p)$ as 'blocks' and the four estimators as treatments. With $y_0 \equiv 1$, there were 30 analyses of variance. Fourteen of these led to a statistically significant estimator effect at level 0.05. In 12 of these 14 cases the smqe was different from the other three, which were not statistically significantly different among themselves using Fisher's LSD, $\alpha = 0.05$. For y_0 taken from the stationary distribution of the process there were also 30 analyses of variance. Nine of these led to a statistically signficant estimator effect, $\alpha = 0.05$, and in all these cases a Fisher LSD analysis identified SMQE as being different from the other three, which were not different among themselves. Thus, the analyses of variance confirmed our impressions of the patterns in the tables of mean-square errors.

Table 12.1 y_0 from the stationary Poisson $(\lambda/(1 - p))$ distribution
MSE$(\hat{p})$ and their standard errors
$n = 10$

λ	p	CLS	MQE	MQE(2)	SMQE
1.0	0.3	0.035	0.040	0.038	0.033
		0.000	0.000	0.000	0.000
	0.5	0.063	0.062	0.063	0.060
		0.001	0.000	0.000	0.001
	0.7	0.117	0.114	0.115	0.105
		0.001	0.001	0.001	0.001
5.0	0.3	0.038	0.034	0.037	0.048
		0.000	0.000	0.000	0.000
	0.5	0.058	0.056	0.056	0.069
		0.000	0.000	0.000	0.001
	0.7	0.107	0.106	0.106	0.111
		0.000	0.000	0.000	0.001
10.0	0.3	0.035	0.034	0.034	0.049
		0.000	0.000	0.000	0.002
	0.5	0.054	0.054	0.054	0.053
		0.000	0.000	0.000	0.000
	0.7	0.090	0.090	0.090	0.084
		0.000	0.000	0.000	0.001

Table 12.2 y_0 from the stationary Poisson $(\lambda/(1-p))$ distribution
MSE($\hat{p}$) and their standard errors
$n = 20$

λ	p	CLS	MQE	MQE(2)	SMQE
1.0	0.3	0.026	0.025	0.025	0.024
		0.000	0.000	0.000	0.000
	0.5	0.041	0.039	0.039	0.040
		0.000	0.000	0.000	0.000
	0.7	0.060	0.060	0.060	0.062
		0.001	0.001	0.001	0.001
5.0	0.3	0.029	0.029	0.029	0.032
		0.000	0.000	0.000	0.000
	0.5	0.039	0.039	0.039	0.053
		0.000	0.000	0.000	0.001
	0.7	0.043	0.042	0.043	0.055
		0.001	0.001	0.001	0.001
10.0	0.3	0.026	0.026	0.026	0.043
		0.000	0.000	0.000	0.000
	0.5	0.039	0.039	0.039	0.057
		0.000	0.000	0.000	0.001
	0.7	0.052	0.052	0.052	0.068
		0.000	0.000	0.000	0.001

Table 12.3 y_0 from the stationary Poisson $(\lambda/(1 - p))$ distribution
MSE($\hat{p}$) and their standard errors
$n = 40$

λ	p	CLS	MQE	MQE(2)	SMQE
1.0	0.3	0.018	0.018	0.018	0.019
		0.000	0.000	0.000	0.000
	0.5	0.025	0.025	0.025	0.024
		0.000	0.000	0.000	0.000
	0.7	0.022	0.022	0.022	0.022
		0.000	0.000	0.000	0.000
5.0	0.3	0.017	0.018	0.018	0.019
		0.000	0.000	0.000	0.000
	0.5	0.021	0.021	0.021	0.026
		0.000	0.000	0.000	0.000
	0.7	0.021	0.021	0.021	0.026
		0.000	0.000	0.000	0.000
10.0	0.3	0.020	0.020	0.020	0.030
		0.000	0.000	0.000	0.000
	0.5	0.021	0.021	0.021	0.026
		0.000	0.000	0.000	0.000
	0.7	0.031	0.031	0.031	0.044
		0.000	0.000	0.000	0.000

Table 12.4 y_0 from the stationary Poisson $(\lambda/(1 - p))$ distribution
MSE$(\hat{\lambda})$ and their standard errors
$n = 10$

p	λ	CLS	MQE	MQE(2)	SMQE
0.3	1.0	0.199	0.210	0.205	0.265
		0.002	0.002	0.002	0.003
	5.0	3.450	3.288	3.383	3.416
		0.026	0.016	0.016	0.017
	10.0	9.064	8.839	8.938	13.092
		0.087	0.086	0.086	0.132
0.5	1.0	0.376	0.377	0.377	0.414
		0.004	0.004	0.004	0.004
	5.0	7.685	7.483	7.501	8.263
		0.057	0.055	0.055	0.083
	10.0	25.498	25.388	25.406	22.861
		0.145	0.098	0.099	0.223
0.7	1.0	2.353	2.297	2.307	2.116
		0.010	0.009	0.009	0.004
	5.0	34.901	34.716	34.740	36.912
		0.160	0.178	0.178	0.348
	10.0	107.512	107.496	107.558	101.103
		0.333	0.205	0.207	0.740

Table 12.5 y_0 from the stationary Poisson $(\lambda/(1 - p))$ distribution
MSE($\hat{\lambda}$) and their standard errors
$n = 20$

p	λ	CLS	MQE	MQE(2)	SMQE
0.3	1.0	0.137	0.139	0.138	0.138
		0.004	0.003	0.003	0.003
	5.0	1.577	1.615	1.615	1.904
		0.012	0.010	0.011	0.008
	10.0	6.431	6.523	6.518	8.119
		0.060	0.055	0.055	0.030
0.5	1.0	0.235	0.227	0.228	0.227
		0.001	0.000	0.000	0.000
	5.0	4.890	4.875	4.880	5.612
		0.033	0.040	0.040	0.068
	10.0	17.264	17.313	17.325	20.690
		0.011	0.038	0.037	0.189
0.7	1.0	0.922	0.918	0.923	0.923
		0.004	0.003	0.003	0.004
	5.0	12.486	12.255	12.362	14.069
		0.431	0.385	0.386	0.313
	10.0	62.880	62.836	62.874	75.284
		0.072	0.005	0.006	0.680

Table 12.6 y_0 from the stationary Poisson $(\lambda/(1 - p))$ distribution
$\mathrm{MSE}(\hat{\lambda})$ and their standard errors
$n = 40$

p	λ	CLS	MQE	MQE(2)	SMQE
0.3	1.0	0.069	0.074	0.074	0.075
		0.000	0.000	0.000	0.000
	5.0	1.006	1.048	1.047	1.028
		0.006	0.009	0.009	0.009
	10.0	4.614	4.666	4.664	5.447
		0.021	0.026	0.026	0.057
0.5	1.0	0.129	0.133	0.134	0.134
		0.001	0.001	0.001	0.001
	5.0	2.371	2.352	2.355	2.410
		0.021	0.021	0.021	0.024
	10.0	7.852	7.727	7.728	9.608
		0.062	0.071	0.071	0.085
0.7	1.0	0.276	0.272	0.273	0.284
		0.002	0.001	0.001	01001
	5.0	5.890	5.854	5.854	7.835
		0.003	0.004	0.004	0.032
	10.0	34.821	34.942	34.959	44.005
		0.280	0.290	0.290	0.441

Based on our study, we recommend the cls estimator for this branching process model with immigration model. It is much simpler to compute than the other estimators and performs as well. A similar conclusion was reached by McLeish (1984) for aggregate Markov chain models.

12.6 Continuous-time processes

Most of our results carry over directly to continuous-time processes $\{x_t\}$. The analogue to (12.5) is the semimartingale model defined by

$$x_T = \int_0^T f_t(\theta)\, d\lambda_t + m_T(\theta), \qquad T \geqslant 0, \qquad (12.36)$$

where $\{m_t(\theta), F_t\}$ is a cadlag locally square integrable martingale with respect to P_θ and $\{\lambda_t\}$ is a real, non-decreasing right continuous process with $\lambda_0 \equiv 0$. Note that (12.5) is a special case of (12.36) with λ_t equal to the greatest integer function. See Hutton and Nelson (1986) for the machinery needed to work with (12.36).

References

Aitchson, J. and Silvey, S. D. (1958). Maxilim likelihood estimation of parameters subject to restraints. *Ann. Math. Stat.*, **29**, 813–28.

Billingsley, P. (1961). *Statistical inference for Markov processes.* University Chicago Press, Chicago.

Godambe, V. P. and Heyde, C. C. (1987). Quasi-likelihood and optimal estimation. *Int. Stat. Rev.*, **55**,231–44.

Hall, P. and Heyde, C. C. (1980. *Martingale limit theory and its application.* Academic Press, New York.

Hutton, J. E. and Nelson, P. I. (1984). A stable and mixing central limit theorem for continuous time martingales. Kansas State University Technical Report No. 42.

Hutton, J. E. and Nelson, P. I. (1986). Quasi-likelihood estimation for semimartingales. *Stoch. Processes Appl.*, **22**, 245–57.

Jennrich, R. I. (1969). Asymptotic properties of non-linear least squares estimators. *Ann. Math. Stat.*, **40**, 633–43.

Klimko, L. A. and Nelson, P. I. (1978). On conditinal least squares estimation for stochastic processes. *Ann. Stat.*, **6**, 629–42.

McLeish, D. L. (1984). Estimation for aggregate models: the aggregate Markov chain. *Canad. J. Stat.*, **4**, 268–82.

Nelson, P. I. (1980). A note on strong consistency of least squares estimators in regression models with martingale difference errors. *Ann. Stat.*, **5**, 1057–64.

13
Tests based on an optimal estimate
A. Thavaneswaran

ABSTRACT

A general framework for constructing test statistics in stochastic models based on an optimal estimate is developed. Martingale limit theorems are used to prove the asymptotic normality and strong consistency of the optimal estimate. Under general conditions, limiting distribution of the proposed statistics are shown to have χ^2 distribution. Some hypotheses, which are of special interest for time series data, are dealt with in more detail. As in Thompson and Thavaneswaran (1990), a non-parametric problem has been formulated as a parametric one, but with an infinite-dimensional parameter, and the corresponding optimal estimating function is used to define test statistics for testing hypotheses with infinite dimensional parameters.

13.1 Introduction

Consider the situation that X_i are identically distributed real random variables, and Θ is an open subset of $\mathbb{R}^p$. A hypothesis H_0 may be specified by writing θ_i, $i = 1, \ldots, p$ as functions $g_i(\theta_1, \ldots, \theta_k)$ of $\theta \in \mathbb{R}^k$ or by specifying restrictions $R_j(\theta_i) = 0$, for $j = 1, \ldots, r$ $(r + k = p)$. Goodness-of-fit tests for H_0 against the full model classically have been based on the m.l.e. (maximum likelihood estimate), and its associated functions, the score and the likelihood. Let $\hat{\theta}$ denote the m.l.e. for the full model, and θ^* that under the restrictions imposed by hypothesis H_0. Three asymptotically equivalent statistics are used:

(a) the Neyman–Pearson likelihood ratio statistic

$$\lambda_n = 2(l_n(\hat{\theta}) - l_n(\theta^*));$$

(b) Rao's efficient score test statistic

$$Q_n = S^t(\theta_n^*)I^{-1}(\theta_n^*)S(\theta_n^*);$$

(c) Wald's test statistic

$$W_n = nR^t(\hat{\theta}_n)(T^t(\hat{\theta}_n)I^{-1}(\hat{\theta}_n)T(\hat{\theta}_n))^{-1}R(\hat{\theta}_n);$$

where l_n is the usual log likelihood, R is the vector restrictions that define

H_0, $T(\theta) = [\partial R_j/\partial \theta_i]_{q \times r}$, $S(\theta)$ is the (vector) score function (with components $\partial l_h/\partial \theta_i$) and $I(\theta)$ is the Fisher information matrix. Note that Rao's statistic depends only on the m.l.e. for the restricted class of parameters under H_0, while Wald's statistic depends only on the m.l.e. over the whole parameter space. The details of the above may be found in Rao (1973, §.6e). Moreover, the Wald statistic is preferable if the unrestricted estimate has been obtained and shall be tested against a more complex model.

Due to the dependence of these statistics on the m.l.e. and score function, obvious optimal estimation analogues exist. See for example, Godambe (1985) for optimal estimation in stochastic process context. That is, one may write λ_n, Q_n, and W_n, replacing the log-likelihood, score, and Fisher information by the integrated optimal estimating function (w.r.t. the parameter), (i.e. quasi-likelihood function) optimal estimating function, and the appropriate variance matrix of the optimal estimating function respectively.

For stochastic processes, the likelihood ratio test for Markov processes, and the χ^2 test for Markov chains have been studied by Billingsley (1961). The likelihood ratio and related tests are also well known for goodness-of-fit testing in linear stationary time series. Recently there has been a growing interest in non-linear time-series models. It should be mentioned that until recently very little was known about the theoretical properties of the estimation procedures and the corresponding estimates except the results for the random coefficient model (see Nicholls and Quinn (1982)). Tjostheim (1986) considered a wide class of non-linear time-series models and developed a systematic asymptotic theory for estimates and also mentioned about future research on testing.

Note that, given the martingale structure of the optimal estimating function, under regularity conditions, one may readily show that the test statistics introduced at the beginning of this section may be defined for non-linear time-series models with parameter structure as for the identically distributed independent variables case.

Moreover, their properties will be analogous to those in the i.i.d. case. An essential simplification here is that

$$E\{g_n^{*2}|F_{i_{n-1}}^y\} = KE\left[\frac{\partial g_n^*}{\partial \theta}\middle| F_{i_{n-1}}^y\right].$$

where K is a constant and g_n^* is the optimal estimating function as in Godambe (1985). In this paper we are interested in the optimal estimate analogues of the tests above.

In Section 13.2 test statistics based on g_n^* are defined and asymptptic properties are given. Section 13.3 deals with the hypothesis testing problem about an infinite-dimensional parameter (non-parametric problem) in a semimartingale model studied in Thompson and Thavaneswaran (1990).

13.2   Test statistics

In this section first we sketch the construction of the analogous statistics based on optimal estimate and then state and prove our main theorem for their asymptotic behaviour. Let $\{y_t, t \in I\}$ be a discrete-time stochastic process taking values in $\mathbb{R}$ and defined on a probability space (Ω, A, P). The index set I is the set of all positive integers. We assume that observations for $t_1 < t_2 < \cdots < t_n$, (i.e. $y_{t_1}, \ldots, y_{t_n}$) are available and that the parameter $\theta \in \Theta$ an open subset of $\mathbb{R}^p$. This set-up enables us to handle missing observations as well, and when $t_i = i$ our formulation reduces to that of Thavaneswaran and Abraham (1988).

Following Godambe (1985), we say that any $\mathbb{R}^p$ valued function g of the variates $y_{t_1}, \ldots, y_{t_n}$ and the parameter θ, satisfying certain regularity conditions such as square-integrable and differentiable in Θ with non-zero derivative, is called a regular unbiased estimating function if

$$E_\theta[g(y_{t_1}, \ldots, y_{t_n})] = 0, \qquad \theta \in \Theta.$$

Let L be the class of estimating functions g of the form

$$g_n = \sum_{i=1}^n a_{t_{i-1}} h_{t_i}$$

where the function h_{t_i} is such that

$$E[h_{t_i}|F_{t_{i-1}}^y] = 0 \qquad (i = 1, \ldots, n) \tag{13.1}$$

and $a_{t_{i-1}}$ is a function of $y_{t_1}, \ldots, y_{t_{i-1}}$ and θ, for $i = 1, \ldots, n$. The following theorem gives the form of an optimum estimating function.

Theorem 13.1   *In the class L of unbiased estimating functions g, the optimum estimating function g^* is the one which makes the difference $\mathrm{Cov}(b, b) - \mathrm{Cov}(b^*, b^*)$ non-negative definite, where*

$$b = \left[E\left(\frac{\partial g_n}{\partial \theta} \right) \right]^+ g_n, \qquad b^* = \left[E\left(\frac{\partial g_n^*}{\partial \theta} \right) \right]^+ g_n^*,$$

and + denotes the pseudoinverse of a matrix. The optimal estimating function is given by

$$g_n^* = \sum_{i=1}^n a_{t_{i-1}}^* h_{t_i},$$

where

$$a_{i-1}^* = \left[\frac{\partial h_{t_i}}{\partial \theta} \Big| F_{t_{i-1}}^y \right] E[h_{t_i} h_{t_i}^t | F_{t_{i-1}}^y]^+.$$

Proof   See Hutton and Nelson (1986).

For further motivation of the optimality based on efficiency considerations, see Lindsay (1985).

Note Using (13.1) it is easy to show that $\{g_n\}$ is a martingale sequence.

Here, and subsequently $A^{1/2}(A^{T/2})$ denotes a left (the corresponding right) square root of the positive definite matrix A, i.e.

$$A^{1/2}A^{T/2} = A, \qquad A^{-1/2} = (A^{1/2})^{-1}, \qquad A^{-T/2} = (A^{T/2})^{-1}.$$

Note that left (right) square roots are unique up to an orthogonal transformation from the right (from the left). Unique continuous 'versions' of the square root are the Cholesley square root and the well-known symmetric square root. We now state the conditions which assure asymptotic properties of optimal estimates θ^* of θ as well as of the statistics considered below.

Let the unconditional variance $F_n = E g_n^* g_n^{*t}$, and the conditional variance

$$\langle g^* \rangle_n = \operatorname{Cov}(g_n^* | F_{t_{n-1}}^y) = E(g_n^* g_n^{*t} | F_{t_{n-1}}^y).$$

We now state the conditioning which, together with the regularity assumptions on g assure asymptotic properties of the optimal estimate θ_n^* of θ and of the test statistics considered below.

A.   (i) Divergence: $\lambda_{\min}(E g_n^* g_n^{*t}(\theta)) \to \infty$

(here and in the sequel $\lambda_{\min}$ denotes the smallest eigenvalue of a symmetric matrix).

(ii) Convergence and continuity: For any $\theta \in \Theta$ and $\delta > 0$ define a sequence

$$N_n(\delta) = \{\tilde{\theta}: \|F_n^{T/2}(\theta)(\tilde{\theta} - \theta)\| \leqslant \delta\}, \quad n = 1, \ldots, \text{ of neighborhoods of } \theta.$$

Then,

$$\operatorname*{Sup}_{\tilde{\theta} \in N_n(\delta)} \|F_n^{-1/2}(\theta) \langle g_n(\tilde{\theta}) \rangle_n F_n^{-T/2}(\theta) - I\| \xrightarrow{P} 0 \text{ under } P_\theta.$$

(iii) For any $\delta > 0$

$$\operatorname*{Sup}_{\tilde{\theta} \in N_n(\delta)} \|F_n^{-1/2}(\theta) F_n(\tilde{\theta}) \quad F_n^{-T/2}(\theta) - I\| \to 0.$$

The divergence condition (A) (i) is the obvious generalization of a necessary and sufficient condition for consistency of the least-squares estimator in the classical linear regression model with i.i.d. errors.

For further details and interpretation about similar assumptions on the functionals of the likelihood in the i.i.d. set-up see Fahrmeir (1987) and the references therein.

Theorem 13.2   *Under (A) the optimal estimate θ_n^* is consistent, i.e.*

$$\theta_n^* \xrightarrow{\text{a.s.}} \theta \tag{13.2}$$

and asymptotically normal, i.e.

$$F_n^{T/2}(\theta)(\theta_n^* - \theta) \xrightarrow{d} N(0, I) \tag{13.3}$$

for an appropriate right square root $F_n^{T/2}$, for instance, the Cholesley square root.

Proof The optimal estimation function $\{g_n^*\}$ evaluated at θ, is a square-integrable zero-mean martingale. Conditions A(i)–A(iii) are sufficient for the application of the central limit theorem for martingales (e.g. Hall and Heyde 1980, Corollary 3.1). By Taylor expansion (13.2) and (13.3) can be shown with analogous arguments given in Hutton and Nelson (1986).

Most testing problems in stochastic models are tests of linear hypotheses on the unknown parameter θ. Let us consider the problem of testing the value of a subvector of length $r (r \leqslant p)$, say θ_2 of $\theta = (\theta_1, \theta_2)$.

$$H_0: \theta_2 = \theta_{02} \text{ against } H_1: \theta_2 \neq \theta_{02}, \tag{13.4}$$

We partition $g_n^*(\theta)$, $F_n(\theta)$ in conformity with the partitioning of θ as

$$g_n^*(\theta) = \begin{pmatrix} g_{n1}^*(\theta) \\ g_{n2}^*(\theta) \end{pmatrix}, \qquad F_n(\theta) = \begin{pmatrix} F_{n11}(\theta) & F_{n12}(\theta) \\ F_{n21}(\theta) & F_{n22}(\theta) \end{pmatrix}.$$

In the following $\theta_n^* = (\theta_{n1}^*, \theta_{n2}^*)$ denotes the unrestricted optimal estimate, whereas $\tilde{\theta}_n = (\tilde{\theta}_{n1}, \tilde{\theta}_{n2})$ is the optimal estimate under the hypotheses H_0. We propose two test statistics analogous to the ones in Section 13.1.

The Wald statistic analogue

$$W_n = (\theta_{n2}^* - \theta_{02})^t A_{n22}(\theta_{n2}^*)(\theta_{n2}^* - \theta_{02})$$

and the score statistic analogue

$$Q_n = g_{n2}^{*t}(\tilde{\theta}_n) A_{n22}^{-1}(\tilde{\theta}_n) g_{n2}^*(\tilde{\theta}_n),$$

where

$$A_{n22}(\theta) = F_{n22}(\theta) - F_{n21}(\theta) F_{n11}^{-1}(\theta) F_{n12}(\theta)$$

is the inverse of the second diagonal block in the partitioning of $F_n^{-1}(\theta)$.

Let us now consider a general linear hypothesis

$$H_2: C\theta = \beta_0 \quad \text{against} \quad H_3: C\theta \neq \beta_0 \tag{13.5}$$

where C has full rank, say $r \leqslant p$, then the corresponding statistics may be written as

$$\tilde{W}_n = (C\theta_n^* - \beta_0)^t [CF_n^{-1}(\theta_n^*)C^t]^{-1}(C\theta_n^* - \beta_0).$$
$$\tilde{Q}_n = g_n^*(\tilde{\theta}_n) F_n^{-1}(\tilde{\theta}_n) g_n^*(\tilde{\theta}_n).$$

The following theorem gives the limiting central χ^2-distribution of the test statistics under the null hypotheses and is used for computation of appropriate critical values.

Theorem 13.3   *Under (A), the test statistics $W_n(\tilde{W}_n)$ and $Q_n(\tilde{Q}_n)$ are asymptotically equivalent, i.e. the difference between any two among them converges to zero in probability under H_0 and they have the same limiting distributions.*

(a) *$\tilde{W}_n$ and $\tilde{Q}_n \xrightarrow{d} \chi^2(r)$ under H_2*
(b) *W_n and $Q_n \xrightarrow{d} \chi^2(r)$ under H_0.*

Proof $\tilde{Q}_n = g_n^{*t} F_n^{-T/2} F_n^{-1/2} g_n^*.$

According to (13.3) $F_n^{-1/2} g_n^* \xrightarrow{d} N(0, I)$ under H_2. A well-known theorem on quadratic forms and continuity theorem (Billingsley 1968) show that

$$\tilde{Q}_n \xrightarrow{d} \chi^2_{(r)} \qquad \text{under } H_2.$$

To establish the asymptotic of $\tilde{W}_n$ it is enough to show that

$$\tilde{Q}_n = \tilde{W}_n + O_p(1) \tag{13.6}$$

From (A) (ii) (i.e. the probability that the relative difference in conditional and the unconditional variance is arbitrarily small within the neighborhoods $N_n(\delta)$, uniformly in all directions tends to one) and (13.2), the consistency of θ_n^*, we get

$$F_n^{-1/2} g_n^* = F_n^{T/2}(\theta_n^* - \theta) + O_p(1). \tag{13.7}$$

Then using the following appropriate triangular square roots

$$F_n^{T/2} = \begin{pmatrix} F_{n11}^{T/12} & F_{n11}^{-1/2} \\ 0 & A_{n22} \end{pmatrix}, \quad \text{resp.} \quad F_n^{-T/2} = \begin{pmatrix} F_{n11}^{-T/2} & -F_{n11}^{-1} F_{n21}^{T/2} A_{n22}^{-T/2} \\ 0 & A_{n22}^{-T/2} \end{pmatrix}$$

where $F_{n11}^{1/2}$ (resp. $F_{n11}^{T/2}$) and $A_{n22}^{T/2}$ are left (resp. right) square roots of F_{n11} and A_{n22} in (13.7), (13.6) follows. Hence, $\tilde{W}_n \xrightarrow{d} \chi^2_{(r)}$.

Now by an appropriate reparametrization, the hypothesis (13.5) can be reduced to (13.3) and the proof of (b) follows from that of (a).

13.3   Applications

Based on the asymptotic properties of the test statistics of the last section, a great part of the statistical analysis for stochastic models can be handled as for independent observations. Therefore, some problems, which may be treated almost in the same way, are only briefly indicated. Some hypotheses which are of special interest for time-series data (equally spaced as well as unequally spaced, for example, due to missing observations) are dealt with in more detail.

13.3.1 STRUCTURAL CHANGE

Occasionally, some or—as we consume for simplicity—all parameters are supposed to change their values at some known time T, say

$$E[y_s|F^y_{s-1}] = f(F^y_{s-1}, \theta_1), \qquad s = l+1, \ldots, T+1$$
$$E[y_s|F^y_{s-1}] = f(F^y_{s-1}, \theta_2), \qquad s = T+2, \ldots, t.$$

We may wish to test

$$H_0: \theta_1 = \theta_2 = \theta \quad \text{against} \quad H_1: \theta_1 \neq \theta_2.$$

This can be done using the theory of Section 13.3.

13.4 Non-parametric testing

This section extends the result of Section 13.2 on tests based on parametric estimates to tests based on non-parametric estimates.

In Thompson and Thavaneswaran (1989) the non-parametric problem has been formulated as a parametric one, but with infinite-dimensional parameter and under regularity conditions, an optimal estimating equation for the unknown parameter has been obtained.

Here we are concerned with non-parametric testing for a multiplicative intensity model. We propose a test statistic based on the optimal estimating equation to test the hypothesis $H_0: \alpha(t) = \alpha_0(t)$ where $\alpha_0(t)$ is known, in the collection of semimartingale models

$$dX_n(t) = \alpha(t)\, dR_n(t) + dM_n(t), \qquad n = 1, 2, 3, \ldots,$$

where $\{M_n(t)\}$ is a sequence of zero-mean square-integrable martingales. Then the statistic, in analogy with the likelihood ratio statistic in the parametric set-up regarding the likelihood equation as the optimal one, may be defined under H_0 by

$$SS_n(\alpha_0(t)) = G^0_{nt}(\alpha^0) - G^0_{nt}(\alpha_0) = \int_0^t b_{s\alpha}\, dM_n(s),$$

where $G_{n,t}(\alpha) = \int_0^t b_{s\alpha}\, dM_n(t)$ and $G^0_{n,t}(\alpha)$ is the corresponding optimal estimating function as in Thompson and Thavaneswaran (1989), α^0 is the optimal estimate, i.e. $SS_n(\alpha_0(t)) \in M^2(F, P)$, the class of square-integrable martingales.

Now we describe two cases to study the asymptotics:

Case (a): If we consider a sequence of martingales $(X_n(t): t \in [0, T]$ for fixed $T)$ as before, then we may use the asymptotically standard normally distributed statistic

$$U_n(\alpha_0, T) = \frac{SS_n(\alpha_0, T)}{\langle SS_n(T) \rangle^{1/2}}$$

for testing H_0.

Note The proof of the asymptotic normality under suitable regularity conditions follows from the martingale central limit theorem (see for example Liptser and Shiryayev 1980 applied directly to the standardized martingale $U_n(\alpha_0, T)$.

This is the case which is extensively studied in the literature; see for example Anderson *et al.* (1982), Example 4.2.

Case (b): If we consider a single realization of the semimartingale $X = (X_t, F_t)$ in the time interval $[0, T]$, then we may use the asymptotically $(T \to \infty)$ standard normally distributed statistic

$$U(T, \alpha_0) = \frac{SS_n(\alpha_0, T)}{\langle SS_n(\alpha_0, T) \rangle^{1/2}}$$

for testing H_0.

Note The proof of the asymptotic normality follows from the central limit theorem (Feigin 1976) applied directly to the standardized martingale $U(T, \alpha_0)$.

In the literature this case is considered only for the parametric set-up. Hence, the above is a generalization in some sense. Most of the recently suggested test statistics for the one-sample situation are special cases of our proposed statistic, and hence their asymptotic distribution can be found from martingale central limit theorems. This generalization provides a new motivation for an already well-studied technique in the counting-process (especially for Aalen's model) and also provides new test statistics together with their asymptotic distribution for a general semimartingale model. Moreover, the theory of optimal estimating equations suggests a method to motivate test statistics to test values of infinite-dimensional parameters, and is hence applicable in the non-parametric set up.

13.5   Acknowledgements

The author wishes to acknowledge the support of the Natural Sciences and Engineering Research Council of Canada.

References

Anderson, P. K., Borgan, O., Gill, R. D., and Keiding, N. (1982). Linear non-parametric tests for comparison of counting processes, with applications to censored survival data (with discussion). *Int. Stat. Rev.*, **50**, 219–58. Correction **52**, 225.

Billingsley, P. (1968). *Convergence of probability measures.* Wiley, New York.

Billingsley, P. (1961). *Statistical inference for Markov processes.* University of Chicago Press.

Fahremeir, L. (1987). Asymptotic testing theory for generalized linear models. *Math. Operations for Sch. Stat. Ser. Stat.*, **1**, 65–76.

Feigin, P. D. (1976). Maximum likelihood estimation for continuous-time stochastic processes. *Adv. in Prob.*, **8**, 712–36.

Godambe, V. P. (1985). The foundations of finite sample estimation in stochastic processes. *Biometrika*, **72**, 419–28.

Hall, P. and Heyde, C. C. (1980). *Martingale limit theory and its application.* Academic Press, New York.

Hutton, J. E. and Nelson, P. I. (1986). Quasi-likelihood estimation for sem-martingales. *Stoch. Processes Appl.*, **22**, 245–57.

Kaufmann, H. (1987). Regression models for non-stationary categorical time series: Asymptotic estimation theory. *Ann. Stat.*, **15**, 79–98.

Lindsay, B. G. (1985). Using empirical partially Bayes inference for increased efficiency. *Ann. Stat.*, **13** (3), 914–31.

Liptser, R. S. and Shiryayev, A. N. (1980). A functional central limit theorem for semimartingales theory. *Prob. Appl.*, **25**, 667–88.

Nelson, P. I. (1979). Least squares tests for discrete parameter stochastic processes. *Commun. Stat.—Theor. Meth.*, A **8**(3), 283–97.

Nicholls, D. F. and Quinn, B. G. (1982). *Random coefficient autoregressive models: an introduction.* Lecture Notes in Statistics, 11, Springer, New York.

Rao, C. R. (1973). *Linear statistical inference and its applications.* (Second Edition). Wiley, New York.

Thavaneswaran, A. and Abraham, B. (1988). Estimation for nonlinear time series models using estimating equations. *J. Time Series Anal.*, **9**, 99–108.

Thompson, M. E. and Thavaneswaran, A. (1990). Optimal nonparametric estimation for some semimartingale stochatic differential equatiions. *J. Appl. Math. Comp.*, **37**, 169–83.

Tjostheim, D. (1986). Estimation in nonlinear time series models. *Stoch. Proc. Appl.*, **21**, 251–73.

PART 4
Survey sampling

14
Estimating functions in survey sampling: a review

Malay Ghosh

ABSTRACT

The theory of estimating functions has gone a long way in the past three decades, sparked by the stimulating work of Godambe (1960), Kale (1962), and their associates. Unlike many other areas of statistics, the theory of estimating functions seems to be under-utilized in the context of survey sampling. Notable exceptions are the work of Godambe and Thompson (1986a, 1986b). This paper reviews the above work of Godambe and Thompson which discusses the estimation of super-population and survey population parameters, utilizing the theory of linearly optimal unbiased estimating functions. The ideas are extended to the situation when there may be possible non-response.

14.1 Introduction

The theory of estimating functions originated essentially with Godambe (1960) and Durbin (1960). In its earliest version (see Godambe 1960), the theory was designed to provide an optimality property of the score function in a parametric framework.

The past two decades have witnessed a great deal of research activity in this area. A very important message that has emerged out of this research is that the theory of estimating functions provides a general unification of two of the principal methods in the theory of estimation, namely, the least squares and the maximum likelihood. It has been pointed out very effectively by Godambe and Kale in Chapter 1 (see also Heyde 1989) that this unifying theory 'combined the strengths of these two methods eliminating at the same time their weaknesses'.

Unlike some other areas of statistics, the idea of estimating functions seems to be underutilized in the context of survey sampling, although many important estimators of the finite population mean arise naturally as solutions of optimal estimating equations. One such estimator due to Brewer (1963) and Hájek (1971) is a modified version of the Horvitz–Thompson estimator, and rectifies many of the deficiencies associated with the latter.

This modified estimator will arise as a special case of the estimator derived in (14.12) through the estimating-function approach.

Godambe and Thompson (1986a, 1986b) have utilized the theory of unbiased estimating functions to provide a unified approach towards estimation of super population parameters in a model-based procedure, or a finite population function whose form is motivated from the model. The present article is largely a review of the above work of Godambe and Thompson. Section 2 addresses the issue of estimation of superpopulation and survey population parameters when there is complete response, while Section 3 discusses the topic when there may be possible non-response.

14.2   Estimation of super-population and survey population parameters

This section is based on Godambe and Thompson (1986a). Consider a finite population U with units labelled $1, 2, \ldots, N$. Let y_i denote the characteristic of interest associated with the ith population unit. It is assumed that $\mathbf{y} = (y_1, \ldots, y_n)$ is generated from a distribution ξ belonging to a class C. The class C will be referred to as a *super-population model*.

A super-population parameter θ is a real-valued function defined on C. For example, one may consider the model $y_i = \theta x_i + e_i$ $(i = 1, 2, \ldots, N)$, where the x_i are known and the e_i are i.i.d. with zero mean and variance σ^2. In the theory of estimating functions, an unbiased estimation function (UEF) is a function $g(\mathbf{y}, \theta)$ which satisfies

$$\varepsilon_\xi g(\mathbf{y}, \theta) = 0 \qquad \text{for all } \xi \in C, \tag{14.1}$$

where ε_ξ denotes expectation under ξ. We consider only those UEFs g for which $\varepsilon_\xi[\partial g(\mathbf{y}, \theta)/\partial\theta|_{\theta = \theta(\xi)}] \neq 0$ for every $\xi \in C$ and every real $\theta(\xi)$.

Following Godambe (1960), a UEF $g^*(\mathbf{y}, \theta) \in \mathcal{G}$ is said to be optimal within $\mathcal{G}$ if

$$\varepsilon_\xi(g^{*2})/[\varepsilon_\xi(\partial g^*/\partial\theta)|_{\theta = \theta(\xi)}]^2 \leqslant \varepsilon_\xi(g^2)/[\varepsilon_\xi(\partial g/\partial\theta)|_{\theta = \theta(\xi)}]^2 \tag{14.2}$$

for every $\xi \in C$ and every $g \in \mathcal{G}$. The motivation behind this definition of optimality is discussed for example in Godambe (1960) and Chapter 1.

Writing $g_s = g/[\varepsilon_\xi(\partial g/\partial\theta)|_{\theta = \theta(\xi)}]$, a standardized UEF, (14.2) can be alternately expressed as

$$v_\xi(g_s^*) \leqslant v_\xi(g_s) \qquad \text{for every } \xi \in C \text{ and every } g \in \mathcal{G}, \tag{14.3}$$

where v_ξ denotes variance under ξ. It is well known (see e.g., Theorem 1, p. 23 of Heyde 1989)that (14.3) holds if and only if

$$\operatorname{cov}_\xi(g_s^*, g_s) = v_\xi(g_s^*) \qquad \text{for every } \xi \in C \text{ and every } g \in \mathcal{G}. \tag{14.4}$$

It is convenient to appeal to (14.4) for proving optimality of UEFs.

Any optimality property, henceforth referred to, will tacitly assume an underlying class $\mathscr{G}$ of UEFs. If $g^*(\mathbf{y}, \theta)$ is optimal, then the equation $g^*(\mathbf{y}, \theta) = 0$ will be called an *optimal estimating equation*. The corresponding solution for θ will be called the *optimal estimate*.

The solution $\theta_N(\mathbf{y})$ of $g^*(\mathbf{y}, \theta) = 0$ possesses two properties simultaneously.

I. When all the components of $\mathbf{y}$ are known, $\theta_N(\mathbf{y})$ is an estimate of θ.
II. When $\mathbf{y}$ is not completely known, $\theta_N(\mathbf{y})$ is a parameter of the survey population.

We shall illustrate I and II later with examples.

Consider now a super-population model $C = \{\xi\}$ under which $y_1, \ldots, y_N$ are independent. No optimal estimating function will typically exist without extra assumptions. One may then use the more restrictive notion of linearly optimal estimating functions. An estimating function $g(\mathbf{y}, \theta)$ is said to be linear in $\phi_1, \ldots, \phi_N$ if

$$g(\mathbf{y}, \theta) = \sum_{i=1}^{N} \phi_i(y_i, \theta) a_i(\theta), \tag{14.5}$$

where the $a_i(\theta)$ are real differentiable functions of θ, and the p_i are functions satisfying $\varepsilon_\xi \phi_i(y_i, \theta) = 0$ for every $i = 1, \ldots, N$. As an example one may take $\phi_i(y_i, \theta) = y_i - \theta$, where the y_i have the same mean θ. The estimating function g^* is said to be *linearly optimal* if it is linear, and satisfies (14.2) for every g satisfying (14.5). As noticed earlier, a linearly optimal g is defined only up to a constant multiple.

The first main theorem of Godambe and Thompson (1986a) is as follows:

Theorem 14.1 *Consider the super-population model $C = \{\xi\}$ under which $y_1, \ldots, y_N$ are independently distributed, and $\varepsilon_\xi[\phi_i(y_i, \theta)] = 0$ $(1 \leqslant i \leqslant N)$ for every $\xi \in C$. Then an estimating function $g^*(\mathbf{y}, \theta) = \sum_{i=1}^{N} \phi_i(y_i, \theta)$ is linearly optimal (within the class of estimating functions defined in (14.5)) if*

$$\varepsilon_\xi((\partial \phi_i/\partial \theta)|_{\theta = \theta(\xi)}) = k(\theta(\xi))\varepsilon_\xi(\phi_i^2), \qquad 1 \leqslant i \leqslant N. \tag{14.6}$$

If in addition $y_1, \ldots, y_N$ are identically distributed, and we consider the class of linear UEFs with $\phi_1 = \cdots = \phi_N = \phi$ in (14.5), then $g^*(\mathbf{y}, \theta) = \sum_{i=1}^{N} \phi(y_i, \theta)$ is linearly optimal.

Proof Using (14.5),

$$\operatorname{cov}_\xi(g_\mathrm{s}, g_\mathrm{s}^*) = \sum_{i=1}^{N} a_i \varepsilon_\xi(\phi_i^2) \bigg/ \left\{ \sum_{i=1}^{N} \varepsilon_\xi[(\partial \phi_i/\partial \theta)|_{\theta = \theta(\xi)}] a_i(\theta) \right\}$$

$$\times \left\{ \sum_{i=1}^{N} \varepsilon_\xi[(\partial \phi_i/\partial \theta)]_{\theta = \theta(\xi)} \right\}. \tag{14.7}$$

Now, if (14.6) holds, it follows from (14.7) that

$$\text{cov}_\xi(g_s, g_s^*) = \sum_{i=1}^{N} a_i \varepsilon_\xi(\phi_i^2) \Bigg/ \left[k^2(\theta(\xi)) \left\{ \sum_{i=1}^{N} a_i \varepsilon_\xi \phi_i^2 \right\} \left\{ \sum_{i=1}^{N} \varepsilon_\xi \phi_i^2 \right\} \right]$$

$$= \left[k^2(\theta(\xi)) \sum_{i=1}^{N} \varepsilon_\xi(\phi_i^2) \right]^{-1}. \tag{14.8}$$

Also, using (14.6),

$$v_\xi(g_s^*) = \left\{ \sum_{i=1}^{N} \varepsilon_\xi(\phi_i^2) \right\} \Bigg/ \left[\sum_{i=1}^{N} \varepsilon_\xi \{ (\partial \phi_i / \partial \theta)|_{\theta = \theta(\xi)} \} \right]^2 = \left[k^2(\theta(\xi)) \sum_{i=1}^{N} \varepsilon_\xi(\phi_i^2) \right]^{-1}. \tag{14.9}$$

Combining (14.8) and (14.9), one gets (14.4). This proves the linear optimality of g^* if (14.6) holds.

If $\phi_1 = \cdots = \phi_N = \phi$, and $y_1, \ldots, y_N$ are i.i.d., then (14.6) holds, and, hence $g(\mathbf{y}, \theta) = \sum_{i=1}^{N} \phi(y_i, \theta)$ is linearly optimal. The proof of Theorem 14.1 is complete.

Example 14.1   Suppose the class C consists of all distributions ξ for which (i) $y_1, \ldots, y_N$ are independently distributed, (ii) $\varepsilon_\xi(y_i) = \theta(\xi)x_i$, $1 \leqslant i \leqslant N$, the x_i are not all zeros, and (iii) $v_\xi(y_i)$ does not depend on i. Then $g^* = \sum_{i=1}^{N} x_i(y_i - \theta x_i)$ is linearly optimal since (14.6) holds and $\theta_N = \sum_{i=1}^{N} x_i y_i / \sum_{i=1}^{N} x_i^2$ is both an optimal estimate of θ based on $\mathbf{x}$ and $\mathbf{y}$, and also a parameter of the survey population. In the special case when $x_1 = \cdots = x_N = 1$, $\theta_N = N^{-1} \sum_{i=1}^{N} y_i$ is the survey population mean.

As a follow up of Example 14.1, consider the normal regression model where the y_i are independent $N(\theta x_i, \sigma^2)$s. Then the score function equals $\sigma^{-2} \sum_{i=1}^{N} x_i(y_i - \theta x_i)$, and then by Godambe's (1960) result, the linearly optimal UEF g^* becomes the optimal UEF.

An extension of Example 14.1 to a multiple regression model is given in Chapter 16. Also, Chapter 17 considers a slightly general version of Example 14.1, and establishes the optimality of g^* within a bigger class of UEFs.

In practice, it is more important to estimate the survey population parameters θ_N from a sample drawn from the survey population. The following approach due to Godambe and Thompson (1986a) shows how the estimating function theory can be used in this context.

A sample s is a subset of $\{1, \ldots, N\}$. Let $\mathscr{S} = \{s\}$ denote the space of all possible samples. A *sampling design* p is a probability distribution on s such that $p(s) \in [0, 1]$ and $\sum_{s \in \mathscr{S}} p(s) = 1$. The data when a sample s is drawn, and the corresponding y-values are observed will be denoted by $\mathscr{X}_s = \{(i, y_i): i \in s\}$.

Consider now a super-population model C under which conditions of Theorem 14.1 (including (14.6)) hold. Suppose θ_N is the solution of $g^*(\mathbf{y}, \theta) = \sum_{i=1}^{N} \phi_i(y_i, \theta)$ is the linearly optimal estimating function. Let $\pi_i = \sum_{i \in s} p(s)$ denote the inclusion probability of the ith population unit ($i = 1, \ldots, N$).

Then, a function $h(\mathscr{X}_s, \theta)$ is said to satisfy the criterion of design unbiasedness, if

$$E[h(\mathscr{X}_s, \theta)] = \sum_{i=1}^{N} \phi(y_i, \theta) \qquad (14.10)$$

for each population vector $\mathbf{y}$ and θ. In the above E denotes expectation under the design p. The optimal h satisfying (14.10) is one for which

$$\varepsilon_\xi E[h^2(\mathscr{X}_s, \theta)] \Big/ \left\{ \varepsilon_\xi E\left[\frac{\partial h}{\partial \theta}\right]\Big|_{\theta = \theta(\xi)} \right\}^2 \qquad (14.11)$$

is minimized. Since $\varepsilon_\xi E[(\partial h/\partial \theta)|_{\theta = \theta(\xi)}] = \varepsilon_\xi[\partial/\partial \theta \sum_1^N \phi_i]$ does not depend on h, minimization of (14.11) amounts to the minimization of $\varepsilon_\xi E[h^2(\mathscr{X}_s, \theta(\xi))]$ with respect to h subject to $E(h) = \sum_{i=1}^N \phi_i$. The following theorem is proved in Godambe and Thompson (1986a).

Theorem 14.2 *Suppose, under the super-population model $C = \{\xi\}$, $y_1, \ldots, y_N$ are mutually independent with $\varepsilon_\xi[\phi_i(y_i, \theta)] = 0$ for each $i = 1, \ldots, N$ and each $\xi \in C$. Consider estimating functions $h(\mathscr{X}_s, \theta)$ satisfying (14.10). Then, (14.11) is minimized by taking $h = h^*$, where $h^*(\mathscr{X}_s, \theta) = \sum_{i \in s} \pi_i^{-1} \phi_i(y_i, \theta)$. Further, the optimal sampling design within the class of sampling designs of fixed size n is one for which $\pi_i \alpha [\varepsilon_\xi \{\phi_i^2(y_i, \theta(\xi))\}]^{1/2}$.*

The estimator $\hat{\theta}_s$ of the survey population parameter θ_N is obtained by solving $h^*(\mathscr{X}_s, \theta) = 0$. When h^* is an optimal UEF, we say that $\hat{\theta}_s$ is also an optimal estimator, although $\hat{\theta}_s$ is not necessarily an unbiased estimator of θ_N.

Suppose now one is interested in estimating the survey population parameter $\sum_{i=1}^N \psi_i(y_i)$, where $\varepsilon_\xi[\psi_i(y_i)] = \alpha_i(\theta(\xi))$. Now write $\phi_i(y_i) = \psi_i(y_i) - \alpha_i(\theta)$. Then, the optimal estimator of $\sum_{i=1}^N \phi_i(y_i)$ is $\sum_{i \in s} \phi_i(y_i)/\pi_i$. Accordingly, a very intuitively appealing estimator of $\sum_{i=1}^N \psi_i(y_i)$ is $\sum_{i=1}^N \alpha_i(\hat{\theta}_s) + \sum_{i \in s} \{\psi_i(y_i) - \alpha_i(\hat{\theta}_s)\}/\pi_i$. This result is used in Godambe (1989b) for estimating the distribution function in a finite population, that is when $\psi_i(y_i) = I_{[y_i \leqslant t]}$, $i = {}^1 N$, I denoting the usual indicator function.

We illustrate the application of Theorem 14.2 with an example.

Example 14.2 Consider the set-up of Example 14.1 except that now $\varepsilon_\xi(y_i - \theta(\xi)x_i)^2 \alpha \sigma_i^2$. Since (14.6) holds in this case, the linearly optimal UEF is given by $\phi_i(y_i, \theta) = x_i(y_i - \theta x_i)/\sigma_i^2$. The corresponding survey population parameter $\theta_N = \sum_{i=1}^N x_i y_i \sigma_i^{-2} / \sum_{i=1}^N x_i^2 \sigma_i^{-2}$, and its optimal estimator as obtained from Theorem 14.2 is $\hat{\theta}_s = \sum_{i \in s} x_i y_i (\pi_i \sigma_i^2)^{-1} / \sum_{i \in s} x_i^2 (\pi_i \sigma_i^2)^{-1}$. Also, the optimal sampling design within the class of all designs with fixed sample size n is given by $\pi_i \alpha x_i \sigma_i^{-1}$.

It should be noted that even though $\sum_{i=1}^{N} \phi_i(y_i, \theta)$ is not an optimal (or linearly optimal) UEF, $h^*(\mathcal{X}_s, \theta) = \sum_{i \in s} \pi_i^{-1} \phi(y_i, \theta)$ is nevertheless optimal within the class of all functions h satisfying (14.10). The motivation for θ_N is now tenuous, but for a given θ_N, the optimality of $\hat{\theta}_s$ as its estimator as obtained from Theorem 14.2 is still valid. Thus, in Example 14.2, if $\phi_i(y_i, \theta) = (y_i - \theta x_i)\alpha_i$ $(i = 1, \ldots, N)$, unless $\alpha_i \alpha x_i \sigma_i^{-2}$, $\theta_N = \sum_{i=1}^{N} y_i \alpha_i / \sum_{i=1}^{N} x_i \alpha_i$ is not optional. Nevertheless,

$$\hat{\theta}_s = \frac{\sum_{i \in s} \pi_i^{-1} y_i \alpha_i}{\sum_{i \in s} \pi_i^{-1} x_i \alpha_i} \tag{14.12}$$

is an optimal estimator of θ_N. In the special case when $x_1 = \cdots x_N = 1$ and $\alpha_1 = \cdots = \alpha_N = 1$, this reduces to the Brewer–Hájek estimator as mentioned in Section 14.1. Note that such an estimator given in (14.12) does not require the knowledge of the σ_i^2.

Multiparameter analogues of these results are discussed in Godambe and Thompson (1986a). Also as pointed out by these authors (1986b), the optimality of $\hat{\theta}_s$ mentioned in the preceding paragraph holds not only for fixed sample-size designs, but also for random-sample designs. This fact is needed in the next section to find optimal UEFs in the presence of non-response.

14.3   Non-response and optimality

The previous section deals with the situation when there is complete response from the sampled units. In reality, however, most surveys are faced with the problem of non-response from certain units that are selected. The theory of unbiased estimating functions can still be utilized to derive optimal estimators in such cases. To see this suppose a sample s is drawn from a survey population U using a sampling design p. Assume now that because of non-response the y_i are available only for $i \in s' \subset s$. Thus the data are

$$\mathcal{X}_{s, s'} = \{(s, s'): (i, y_i), i \in s'\}. \tag{14.13}$$

The response mechanism assumed throughout this section is as follows:

If the individual i of the survey population U were included in the same s drawn, then i would respond with *known* probability q_i, where $q_i > 0$, $i = 1, \ldots, N$. Write $\mathbf{q} = (q_1, \ldots, q_N)$. \hfill (14.14)

In the following, as before we assume that the parameter θ_N is defined by the equation

$$\tilde{g} = \sum_{i=1}^{N} (y_i - \theta x_i)\alpha_i = 0$$

for some specified numbers of α_i, $i = 1, \ldots, N$. Further

$$h^*(\mathcal{X}_s, \theta) = \sum_{i \in s} (y_i - \theta x_i)\alpha_i/\pi_i.$$

There are two ways to approach the estimation problem:

I. If the complete data $\mathcal{X}_s = \{(i, y_i): i \in s\}$ were available, θ_N is estimated by solving $h^*(\mathcal{X}_s, \theta) = 0$. When the hypothetical data $\mathcal{X}_s$ are replaced by $\mathcal{X}_{s,s'}$ one may try to estimate h^* by $h'(\mathcal{X}_{s,s'})$. Then the class of UEFs h' (for h^* given the sample s) is given by

$$H'(p, \mathbf{q}, s) = \{h': E(h' - h^*|s) = 0 \qquad \text{for all } \mathbf{y} \text{ and } \theta\}, \quad (14.15)$$

E being expectation under the sampling design p employed to draw s. The optimal UEF in H' is defined by h'^*, where $\varepsilon_\xi E(h'^*)^2 \leqslant \varepsilon_\xi E(h')^2$ for all $h' \in H'$ and $\xi \in C$.

II. The second approach is to estimate the survey population parameter θ_N directly using the data $\mathcal{X}_{s,s'}$ without estimating h^* as in I above. Define the class of UEFs $h''(\mathcal{X}_{s,s'})$ such that

$$H''(p, q) = \{h'': E(h'' - \tilde{g}) = 0 \qquad \text{for all } \mathbf{y} \text{ and } \theta\}, \quad (14.16)$$

The 'optimum' UEF in the class H'' is given as before by minimizing $\varepsilon_\xi E[h'']^2$.

The two approaches lead to the same optimal estimating function. The following two theorems of Godambe and Thompson (1986b. 1987) substantiate this.

Theorem 14.3 *Consider the super-population model $C = \{\xi\}$ under which $y_1, \ldots, y_N$ are independent, and $\varepsilon_\xi(y_i) = \theta(\xi)x_i$, $i = 1, \ldots, N$. Then, for any sampling design p for which $\pi_i = \sum_{i \in s} p(s) > 0$ for all $i = 1, \ldots, N$, and any sample s, for every $h' \in H'$, $\varepsilon_\xi E\{(h')^2|s\}$, is minimized for $h' = h'^*$, where*

$$h'^* = \sum_{i \in s'} (y_i - \theta(\xi)x_i)\alpha_i/(\pi_i q_i). \quad (14.17)$$

Theorem 14.4 *Let $\bar{H}''$ be a subclass of H'' for which $h''(\mathcal{X}_{s,s'})$ depends only on s'. Then, for any sampling design p for which $\pi_i > 0$ for all $i = 1, \ldots, N$ under the super-population model described in Theorem 14.3, for every $h'' \in H''$, $\varepsilon_\xi E(h'')^2$ is minimized for $h'' = h''^*$, where $h''^* = \sum_{i \in s'} (y_i - \theta x_i)\alpha_i/(\pi_i q_i)$.*

It is argued in Godambe (1989a) that when there is no non-response, h^* should not be estimated by $h'^* = h'^{**}$. The optimal UEF h'^* or h''^* is useful when there is considerable non-response.

Next assume that the survey population U is divided into k strata U_j of sizes N_j $(j = 1, \ldots, k)$. Assume that

$$q_i = q^{(j)} \qquad \text{for all } i \in U_j, \quad j = 1, \ldots, k. \tag{14.18}$$

Unlike before, we *do not* need to assume the $q^{(j)}$ to be known. Let p_0 denote stratified sampling design consisting of drawing from the stratum U_j a simple random sample without replacement of size n_j $(j = 1, \ldots, k)$. Consider the class of UEFs.

$$H_1(p_0) = \{h_1 : E(h_1 - \tilde{g}) = 0 \qquad \text{for all } y, \theta \text{ and } q^{(j)}, j = 1, \ldots, k\},$$

$$h_1 \equiv h_1(\mathscr{X}_{s,s'}) \tag{14.19}$$

where the $q^{(j)}$ are as in (14.18). Let $s'_j = s' \cap U_j$; and $|s'_j| = n'_j$ $(j = 1, \ldots, k)$. The following theorem is proved in Godambe and Thompson (1986b).

Theorem 14.5 *For the sampling design p_0, in the class of UEFs $H_1(p_0)$ in (14.19), and the super-population model considered in Theorem 14.4, $\varepsilon_\xi E(h_1^2)$ is minimized for $h_1 = h_1^*$, where*

$$h_1^* = \sum_{j=1}^{k} \sum_{i \in s} (y_i - \theta(\xi)x_i)\alpha_i N_j / n'_j. \tag{14.20}$$

Next observe that in Theorem 14.4, the optimality of h''^* was established without specifying the variance function. However, the specification of the variance function in the given super-population model is required to obtain the optimal inclusion probabilities. Assume

$$\varepsilon_\xi(y_i - \theta(\xi)x_i)^2 = \sigma^2 f(x_i), \qquad i = 1, \ldots, N, \tag{14.21}$$

where f is a *known* function, but σ^2 may be unknown. Now, observe that

$$\varepsilon_\xi E(h''^*)^2 = \sum_{i=1}^{N} \varepsilon_\xi(y_i - \theta(\xi)x_i)^2 \alpha_i^2 / (\pi_i q_i) = \sigma^2 \sum_{i=1}^{N} f(x_i)\alpha_i^2 / (\pi_i q_i). \tag{14.22}$$

The objective is to minimize (14.22) subject to

$$(A): \sum_{i=1}^{N} \pi_i = \text{constant} \qquad \text{and} \qquad (B): \sum_{i=1}^{N} \pi_i q_i = \text{constant}. \tag{14.23}$$

In (A) we hold the average sample size fixed, that is $E|s| = \sum_{i=1}^{N} \pi_i$. In (B) we hold the average size of the effective sample s' fixed, that is, $E|s| = \sum_{i=1}^{N} \pi_i q_i$. Since, the q_i are fixed $\varepsilon_\xi E(h''^*)^2$ is minimized under (A) when $\pi_i \alpha (f(x_i)/q_i)^{1/2} \alpha_i$ and under (B) when $\pi_i \alpha (f^{1/2}(x_i)/q_i)\alpha_i$.

References

Brewer, K. R. W. (1963). Ratio estimation and finite populations; some results deducible from the assumption of an underlying stochastic process. *Aust. J. Stat.*, **5**, 93–105.

Durbin, J. (1960). Estimation of parameters in time series regression models. *J. Roy. Stat. Soc., Ser.* B, **22**, 139–53.

Godambe, V. P. (1960). An optimum property of regular maximum likelihood estimation. *Ann. Math. Stat.*, **31**, 1208–12.

Godambe, V. P. (1989a). Quasi-score function, quasi-observed Fisher information and conditioning in survey sampling. Tech. Report No. STAT-89-09. Department of Statistics and Actuarial Science, University of Waterloo.

Godambe, V. P. (1989b). Estimation of cumulative distribution of a survery population. Tech. Report No. STAT-89-17. Department of Statistics and Actuarial Science, University of Waterloo.

Godambe, V. P. and Thompson, M. E. (1986a). Parameters of super-population and survey population: their relationships and estimation. *I.S.I. Rev.*, **54**, 127–38.

Godambe, V. P. and Thompson, M. E. (1986b). Some optimality results in the presence of non-response. *Survey Methodology*, **12**, 29–36.

Godambe, V. P. and Thompson, M. E. (1987). Corrigendum. *Survey Methodology*, **13**, 123.

Hájek, J. (1971). Contribution to discussion of paper by D. Basu. In V. P. Godambe and D. A. Sprott (ed.), *Foundations of Statistical Inference*, p. 236. Holt, Rinehart and Winston, Toronto.

Heyde, C. C. (1989). Quasi-likelihood and optimality for estimating functions: some current unifying themes. *Bull. Int. Stat. Inst.*, Book 1, 19–29.

Kale, B. K. (1962). An extension of Cramér–Rao inequality for statistical estimation functions. *Skand. Actuar.*, **45**, 80–9.

15
Confidence intervals for quantiles
V. P. Godambe

ABSTRACT

In an early paper Woodruff (1952) described a method for computing confidence intervals for quantiles for a survey population. The paper is very remarkable for its clarity and lucidity. The method described in the paper is indeed applicable for many sampling designs. Usually the confidence intervals for large samples are constructed by inverting asymptotic normal distributions of some 'optimal' estimate. Here the variance of the estimate plays a central role. On the other hand, Woodruff obtains confidence intervals by inverting asymptotic distribution of an estimating function. For this, he is required to use the variance of the estimating function itself, not of the resulting estimate. In this note we show how these confidence intervals are a special case of the confidence intervals provided by the theory of optimal estimating functions.

15.1 Introduction

Usually long before scientific *formulation* or *abstraction* of concepts, the concepts are used informally at an intuitive level. A clear 'abstraction' gives a better understanding of the 'concept' and renders it very generally applicable. For instance the concept of sufficiency was abstracted by Fisher (1920). However, long before Fisher, Laplace (Stigler, 1973) informally understood it and used it. The same is true about the concepts of ancillarity, power, robustness and the like. With the concept of estimating functions, it is no different: underlying the confidence intervals described by Woodruff (1952), is a deep intuition for what today we call the *optimal estimating function*. These confidence intervals, according to Woodruff, have been used since 1945 in the Bureau of the Census.

Let the survey population $\mathscr{P}$, under study consist of N individuals i, $\mathscr{P} = \{i: i = 1, \ldots, N\}$. The (real) variate value associated with the individual i, is y_i, $i = 1, \ldots, N$. We define a real function $\psi(y, \theta)$ with arguments (y, θ), θ real, such that $\psi(y, \theta) = 1$ if $y \leqslant \theta$ and $= 0$ otherwise. Now θ_N defines a pth quantile of the survey population if it is a θ-solution of the equation

$$\sum_{i=1}^{N} \{\psi(y_i, \theta) - p\} = 0. \qquad (15.1)$$

Woodruff discusses the special case of $p = \frac{1}{2}$, the median. But as he has stated, his method is applicable for any specified p. Let the population $\mathscr{P}$ be divided into k strata, $\mathscr{P}_j, j = 1, \ldots, k$. A sample s, $s \subset \mathscr{P}$, is drawn with a stratified simple random sampling without replacement design; $s = \bigcup s_j$, $s_j \subset \mathscr{P}_j$, $|s_j| = n_j$, $|\mathscr{P}_j| = N_j$, $j = 1, \ldots, k$. For this sampling design and data $\{(i, y_i): i \in s\}$, Woodruff's estimate $\hat{\theta}$ for the pth quantile is given by a θ-solution of the equation

$$0 = F \qquad \text{where} \quad \sum_{j=1}^{k} \sum_{i \in s_j} \frac{N_j}{n_j} \{\psi(y_i, \theta) - p\} \stackrel{\text{def.}}{=} F. \tag{15.2}$$

A conventional estimate of the variance of F

$$\hat{V}(F) = \sum_{j=1}^{k} N_j^2 \left(\frac{1}{n_j} - \frac{1}{N_j}\right) \cdot \frac{n_j \alpha_j (1 - \alpha_j)}{n_j - 1}, \tag{15.3}$$

where α_j = the proportion of ys in the same s_j which are less than or equal to the pth quantile θ_N. Since θ_N is unknown we replace it by its estimate $\hat{\theta}$ obtained from (15.2). Let $\{\hat{V}(F))\}_{\theta = \hat{\theta}} = \hat{V}_1$. Now for large samples an approximation to the distribution of the function F or rather $F/\{(\hat{V}_1)^{1/2}\}$ in (15.2) and (15.3) is obtained noting that the design expectation of F, $E(F) = 0$ at $\theta = \theta_N$ and its estimated variance is $\hat{V}_N$. An inversion of this approximate distribution 'normal' say, provides Woodruff's confidence intervals for the pth quantile θ_N. Woodruff provides the appropriate 'frequency' rationale underlying the method. For reasons by now apparent, henceforth we will refer to F as an *estimating function*.

An important aspect of the above confidence intervals is that they are *not* based on the inversion of the distribution of the estimate $\hat{\theta}$, obtained by solving the equation (15.2), $F = 0$. On the contrary they utilize the distribution of the estimating function F itself or that of $F/\{(\hat{V}_1)^{1/2}\}$, the latter generally being a near *pivotal quantity* for large samples. Now the use of pivotal quantity to form confidence intervals was recognized right at the inception of the theory of confidence intervals. Yet generally in the literature, confidence intervals based on the distribution of some estimates are recommended, possibly for operational simplicity. But they generally perform badly and are *ad hoc*. One important reason for their bad performance is the fact that convergence to the limiting distribution is much slower for the 'estimate' than for the corresponding estimating function (Mach 1988). These negative features could of course be corrected if for operational simplicity the 'estimate' and the 'estimated variance' are derived from the confidence intervals based on F (Kovar *et al.* 1988). The next section provides a motivation for the confidence intervals based on the estimating function F and a generalization.

15.2 Motivation and generalization

As was shown in Section 15.1, the estimating function F in (15.2) plays a central role in Woodruff's confidence intervals. It so happens that F is also, as a special case, an *optimal estimating function*, within the theoretical framework of Godambe and Thomson (1986): If the survey population is a random sample from an *unspecified* super-population ξ, that is if $y_1, \ldots, y_N$ are i.i.d. as ξ then equation (15.1) performs two roles. (I) It provides the optimal estimation based on $y_1, \ldots, y_N$ for the super-population parameter, pth quantile θ. (II) It defines a survey population parameter, pth quantile θ_N. Given the stratified sampling design and the data $\{(i, y_i): i \in s\}$ as in Section 15.2, the *jointly* optimal estimating equation for both the super-population and survey-population parameters θ and θ_N is given by the equation (15.2). If estimation 'only of the survey population parameter θ_N' is of interest, the 'optimality' of F obtains under a milder super-population structure; the variates $y_1, \ldots, y_N$ are *independent* with a *common* pth quantile θ.

'An optimal estimating function' and a *score function* have many important statistical properties in common. Hence the former is called a *quasi-score function* (Godambe and Heyde 1987; Godambe 1989). These conceptual relationships between a quasi-score function and a score function immediately suggest the following. According to an early work of Wilks (1938), the confidence intervals based on a suitably *standardized* score function tend to be 'shortest' for large samples. The same should be true for the quasi-score function; of course granting appropriate conditions. Using such standardized optimal estimating functions, we give below a general procedure for calculating confidence intervals.

To generalize, we replace in (15.1), $\psi(y_i, \theta) - p$ by $\phi_i(y_i, \theta)$ where the function ϕ_i is constructed in such a way that the solution of the equation

$$\sum_{i=1}^{N} \phi_i(y_i, \theta) = 0, \qquad (15.4)$$

namely θ_N, defines the survey population parameter of interest. For instance if $\phi_i = y_i - \theta$, θ_N = the mean of the survey population. Again if for a covariate x_i, $\phi_i = (y_i - \theta x_i)x_i$, θ_N = the regression coefficient. Obviously different functions ϕ_i can define many parameters of interest including quantiles for the survey population. For simplicity we restrict to the stratified sampling design of Section 15.2. For this sampling design, with the data $\{(i, y_i): i \in s\}$ as before, the optimal estimating function for θ_N, under a very mild super-population structure, is given by

$$F = \sum_{j=1}^{k} \sum_{i \in s_j} \frac{N_j}{n_j} \phi_i(y_i, \theta), \qquad (15.5)$$

(Godambe and Thompson 1986). Note that the function F in (15.5) is obtained from (15.2), by replacing in it $\psi(y_i, \theta) - p$ by $\phi_i(y_i, \theta)$. A 'standardized estimating function' is given by $F/\{(\hat{V}_1)^{1/2}\}$ with F as in (15.5) and

$$\hat{V}_1 = \left[\sum_{j=1}^{k} \sum_{i \in s_j} N_j^2 \left(\frac{1}{n_j} - \frac{1}{N_j} \right) \frac{1}{n_j - 1} \sum_{i \in s_j} (\phi_i - \bar{\phi}_j)^2 \right]_{\theta = \hat{\theta}}, \qquad (15.6)$$

$\hat{\theta}$ being the solution of the equation $F = 0$ and $\bar{\phi}_j = \sum_{i \in s_j} \phi_i/n_j, j = 1, \ldots, k$. Note $\hat{V}_1$ in (15.6) is a generalization of $\hat{V}_1 = \{\hat{V}\}_{\theta = \hat{\theta}}$ in (15.3). Under suitable conditions, for large samples $F/\{\hat{V}_1)^{1/2}\}$ will have an approximate t or $N(0, 1)$ distribution. An inversion of this distribution provides confidence intervals for the survey-population parameter θ_N defined by the equation (15.4).

Although the standardization of the estimating function F using $\hat{V}_1$ in (15.6) for the stratified sampling looks natural, it needs some justification, for there can be many unbiased estimates of variance $V(F)$ of F. If $V(F) = V$ is known (which of course is not the case) then unambiguously the confidence intervals could be based on the distribution of $F/\{(V)^{1/2}\}$. Under some conditions, the functions F and $\hat{V}_1$ would be orthogonal and for large samples $F/\{(V)^{1/2}\}$ and $F/\{(\hat{V}_1)^{1/2}\}$ would be nearly equal. Of course this justification needs further elaboration (Godambe 1991).

For another interesting informal application of estimating functions to survey sampling we refer to Binder (1983).

15.3   Acknowledgement

Thanks are due to D. Murdoch, Boxin Tang, and William Welch for comments.

References

Binder, D. A. (1983). On the variances of asymptotically normal estimators from complex surveys. *I.S.I. Rev.*, **51**, 279–92.

Fisher, R. A. (1920). A mathematical examination of methods of determining the accuracy of an observation by the mean error, and by the mean square error. *Monthly Notices Roy. Astron. Soc.*, **80**, 758–70.

Godambe, V. P. (1989). Quasi-score function, quasi-observed Fisher information and conditioning in survey sampling. University of Waterloo, Tech. Report Ser. Stat-89-09.

Godambe, V. P. (1991). Orthogonal estimating functions and nuisance parameters. *Biometrika*, **78**, 143–51.

Godambe, V. P. and Heyde, C. C. (1987). Quasi-likelihood and optima estimation. *I.S. Rev.*, **55**, 231–44.

Godambe, V. P. and Thompson, M. E. (1986). Parameters of superpopulation and survey population: their relationships and estimation. *I.S. Rev.*, **54**, 127–38.

Kovar, J. G., Rao, J. N. K., and Wu, C. F. J. (1988). Measuring errors in survey estimates. *Canad. J. Stat.*, **16**, 25–45.

Mach, L. (1988). The use of estimating functions for confidence interval construction: The case of population mean. Working Paper No. BSMD-88-028E. Methodology Branch, Statistics Canada.

Stigler, S. M. (1973). Laplace, Fisher and the discovery of the concept of sufficiency. *Biometrika*, **60**, 439–45.

Wilks, S. S. (1938). Shortest average confidence intervals from large samples. *Ann. Math. Stat.*, **9**, 166–75.

Woodruff, R. S. (1952). Confidence intervals for medians and other position measures. *J. Amer. Stat. Assoc.*, **47**, 635–46.

16
Making use of a regression model for inferences about a finite population mean

H. Mantel

ABSTRACT

Godambe and Thompson (1986) define and develop simultaneous optimal estimation of super-population and finite-population parameters based on a super-population model and a survey-sampling design. However, we may be interested in estimating a finite population mean that has no direct relationship to the finite-population parameter of Godambe and Thompson's theory. In this paper we extend the super-population model in such a way that there is a direct relationship between the finite-population mean and the finite-population parameter. The resulting estimator of the finite-population mean is compared to the generalized regression estimator.

16.1 Introduction

The problem discussed in this paper is the estimation of a finite-population mean based on a sample survey. There is also a hypothesized super-population regression model relating the variable of interest to some known covariables. The objective is an estimation procedure which has good properties with respect to both the sampling design and the hypothesized model. The approach here is based on the work of Godambe and Thompson (1986).

We suppose we have a finite population of labelled individuals $\mathbf{P} = \{i: i = 1, \ldots, N\}$. With each individual i is associated an unknown variable y_i and a known vector of covariables, $\mathbf{x}_i$. Letting $\mathscr{E}$ denote expectation with respect to the super-population model, the model assumptions are:

(i) y_i and y_j are independent for $i \neq j$
(ii) $\mathscr{E}(y_i) = \mathbf{x}_i^{\mathsf{T}}\boldsymbol{\beta}$ for some unknown real vector β
(iii) $\mathscr{E}(y_i - \mathbf{x}_i^{\mathsf{T}}\boldsymbol{\beta})^2 = \sigma^2 v_i, i = 1, \ldots, N$, for known v_i and some unknown σ^2.

Following Godambe and Thompson (1986) we define a finite population parameter $\hat{\boldsymbol{\beta}}_N$ as the solution of the linearly optimal estimating equation

$$\mathbf{g}^* = \sum_{i=1}^{N} (y_i - \mathbf{x}_i^{\mathsf{T}}\boldsymbol{\beta})\mathbf{x}_i/v_i = 0, \qquad (16.1)$$

that is,

$$\hat{\boldsymbol{\beta}}_N = (X_N^T V_N^{-1} X_N)^{-1} X_N^T V_N^{-1} \mathbf{y}_N \qquad (16.2)$$

where $\mathbf{y}_N^T = (y_1, \ldots, y_N)$, V_N is a diagonal matrix with entries $v_1, \ldots, v_N$, and X_N is a matrix with N rows, the ith row being $\mathbf{x}_i^T$.

Now $\hat{\boldsymbol{\beta}}_N$ is unknown. Godambe and Thompson (1986) defined and developed simultaneous optimal estimation of $\boldsymbol{\beta}$ and $\hat{\boldsymbol{\beta}}_N$ based on the model and the sampling design. We will denote the data from a sample survey by

$$\chi_s = \{(i, y_i), i \in s\}.$$

For simultaneous estimation of $\boldsymbol{\beta}$ and $\hat{\boldsymbol{\beta}}_N$ we consider estimating functions $\mathbf{h}(\chi_s, \boldsymbol{\beta})$ such that $E(\mathbf{h}) = \mathbf{g}^*$ in (16.1), where E denotes expectation with respect to the sampling design. A function $\mathbf{h}^*$ in this class is called optimal if for all other $\mathbf{h}$ in the class, $\mathscr{E}E\{\mathbf{hh}^T\} - \mathscr{E}\{\mathbf{h}^*\mathbf{h}^{*T}\}$ is non-negative definite. Theorem 1 of Godambe and Thompson (1986) shows that the optimal function $\mathbf{h}^*$ is given by

$$\mathbf{h}^*(\chi_s, \boldsymbol{\beta}) = \sum_{i \in s} (y_i - \mathbf{x}_i^T \boldsymbol{\beta}) \mathbf{x}_i / \pi_i v_i \qquad (16.3)$$

where π_i is the probability under the sampling design that individual i is included in the sample s. We will denote the root of the equation $\mathbf{h}^* = 0$ by $\hat{\boldsymbol{\beta}}_s$, that is,

$$\hat{\boldsymbol{\beta}}_s = (X_s^T \Pi_s^{-1} V_s^{-1} X_s)^{-1} X_s^T \Pi_s^{-1} V_s^{-1} \mathbf{y}_s, \qquad (16.4)$$

where $\mathbf{y}_s$ is the vector of y_i for $i \in s$, Π_s and V_s are diagonal matrices with entries π_i and v_i respectively, $i \in s$, and X_s is the matrix with rows $\mathbf{x}_i^T$, $i \in s$.

So far we have discussed only estimation of $\boldsymbol{\beta}$ or $\hat{\boldsymbol{\beta}}_N$. Our problem was to estimate $\bar{y}_N$, the population mean of the y_i. One possibility is to use a generalized regression estimator.

$$\bar{y}_{\text{greg}} = \bar{\mathbf{x}}_N^T \hat{\boldsymbol{\beta}}_s + \mathbf{1}_s^T \Pi_s^{-1} (\mathbf{y}_s - X_s \hat{\boldsymbol{\beta}}_s) / N \qquad (16.5)$$

where $\mathbf{1}_s$ is a vector of 1s whose length is the size of the sample s. This estimator is discussed, for example, by Cassel et al. (1977). The first part of the estimator gives good model properties while the second part gives good design properties. However, the model and design justifications of $\bar{y}_{\text{greg}}$ in (16.5) do not depend on the particular form of $\hat{\boldsymbol{\beta}}_s$, and there is no immediately apparent reason why $\hat{\boldsymbol{\beta}}_s$ in (16.5) could not be replaced by a purely model-based estimator of $\boldsymbol{\beta}$. The design optimality of $\hat{\boldsymbol{\beta}}_s$ is apparently irrelevant.

The estimator we will propose here more closely integrates the hypothesized model with the finite population parameter $\bar{y}_N$. Since $\hat{\boldsymbol{\beta}}_N$ in (16.2) is optimally estimated by $\hat{\boldsymbol{\beta}}_s$ in (16.4), functions of $\hat{\boldsymbol{\beta}}_N$ are optimally estimated by the same function of $\hat{\boldsymbol{\beta}}_s$. If it should happen that $\bar{y}_N = \mathbf{u}^T \hat{\boldsymbol{\beta}}_N$ for some vector $\mathbf{u}$, then we could estimate $\bar{y}_N$ by $\mathbf{u}^T\hat{\boldsymbol{\beta}}_s$. Such a $\mathbf{u}$ exists if and only if $V_N\mathbf{1}_N$ is in the column space of X_N, and if $V_N\mathbf{1}_N = X_N\mathbf{w}$, then we may take $\mathbf{u} = X_N^T V_N^{-1} X_N \mathbf{w}/N = \bar{\mathbf{x}}_N$. The idea then is that if $V_N\mathbf{1}_N$ is not in the column space of X_N, we will add it. In doing so we lose something of model efficiency, though the augmented model remains valid in light of the original model. We relax model efficiency to gain some sort of finite population relevance. As an interesting special case we note that when the model variances do not depend on i our approach leads to including an arbitrary constant term in the regression model.

16.2   Comparison to the generalized regression estimator

For the discussion of this section we suppose that $V_N\mathbf{1}_N$ is not in the column space of X_N.

Let W_N be the design matrix for the augmented model, that is

$$W_N = (V_N\mathbf{1}_N, X_N). \tag{16.6}$$

Similarly, let W_s be the augmented form of X_s, and γ, $\hat{\gamma}_N$, and $\hat{\gamma}_s$ be the augmented forms of $\boldsymbol{\beta}$, $\hat{\boldsymbol{\beta}}_N$, and $\hat{\boldsymbol{\beta}}_s$ respectively.

For convenience, we will refer to our estimator of the population mean as the augmented regression estimator,

$$\bar{y}_{\text{areg}} = \bar{\mathbf{x}}_N^T \hat{\gamma}_s. \tag{16.7}$$

We first show that $\bar{y}_{\text{areg}}$ is also a type of generalized difference estimator. From (16.6), if $\mathbf{u}$ is a vector of appropriate length with the first entry equal to 1 and the rest 0s, then $W_N\mathbf{u} = V_N\mathbf{1}_N$ and $W_s\mathbf{u} = V_s\mathbf{1}_s$. Then

$$\mathbf{1}_s^T\Pi_s^{-1}W_s\hat{\gamma}_s = \mathbf{u}^T W_s^T V_s^{-1}\Pi_s^{-1}W_s\hat{\gamma}_s = \mathbf{u}^T W_s^T V_s^{-1}\Pi_s^{-1}\mathbf{y}_s = \mathbf{1}_s^T\Pi_s^{-1}\mathbf{y}_s,$$

and it follows that the second part of the generalized regression estimator in (16.5) with $\hat{\boldsymbol{\beta}}_s$ replaced by $\hat{\gamma}_s$ is equal to 0.

Secondly, let us compare $\bar{y}_{\text{areg}}$ in (16.7) to $\bar{y}_{\text{greg}}$ in (16.5). A few tedious calculations give us that

$$\bar{y}_{\text{areg}} = \bar{\mathbf{x}}_N\hat{\boldsymbol{\beta}}_s + (c_1/c_2)\mathbf{1}_s^T\Pi_s^{-1}(\mathbf{y}_s - X_s\hat{\boldsymbol{\beta}}_s)/N,$$

where

$$c_1 = \mathbf{1}_N^T(V_N\mathbf{1}_N - X_N(X_s^T V_s^{-1}\Pi_s^{-1}X_s)^{-1}X_s^T\Pi_s^{-1}\mathbf{1}_s)$$

and

$$c_2 = \mathbf{1}_s^T\Pi_s^{-1}(V_s\mathbf{1}_s - X_s(X_s^T V_s^{-1}\Pi_s^{-1}X_s)^{-1}X_s^T\Pi_s^{-1}\mathbf{1}_s).$$

Written in this way $\bar{y}_{\text{areg}}$ appears very similar to $\bar{y}_{\text{greg}}$ except for an adjusted weight for the second part. It does not seem possible to give a heuristic explanation of the weight (c_1/c_2). However, we note that c_1 is just the population sum of the residuals from a weighted regression of the v_i onto the $\mathbf{x}_i$ based on the sample s, and c_2 looks something like a Horvitz–Thompson estimator of c_1, except that the residuals also depend on the sample s. For large samples from large populations we would expect (c_1/c_2) to be close to 1.

In comparing $\bar{y}_{\text{areg}}$ with $\bar{y}_{\text{greg}}$ we may say that $\bar{y}_{\text{areg}}$ is more design based and $\bar{y}_{\text{greg}}$ is more model based. Of course, $\bar{y}_{\text{greg}}$ is design consistent, but $\bar{y}_{\text{areg}}$ has also a finite sample design justification in that $\hat{\gamma}_s$ is the solution of an estimating equation which is design unbiased for the parameter defining equation of $\hat{\gamma}_N$. Parameter defining equations are discussed by Godambe and Thompson (1984, 1986).

16.3   Variance estimation and confidence intervals

For the discussion in this section we assume that $V_N \mathbf{1}_N$ is already in the column space of X_N.

A method of confidence interval construction which would be consistent with the general philosophy of estimating functions would be to construct as asymptotically multivariate normal pivotal based on $\mathbf{h}^*$ and an estimate of its variance. Approximate confidence regions for $\hat{\boldsymbol{\beta}}_N$ would then correspond to probability regions of the estimated multivariate normal distribution of this approximate pivotal. However, we are interested not in $\hat{\boldsymbol{\beta}}_N$ but in a non-injective function of $\hat{\boldsymbol{\beta}}_N$. We will adopt the more straightforward approach of estimating the variance of $\bar{y}_{\text{areg}}$ directly.

Särndal *et al.* (1989) have investigated variance estimation for $\bar{y}_{\text{greg}}$ in (16.5) for the case that the second part is zero. As we have seen in Section 16.2, our estimator $\bar{y}_{\text{areg}}$ is precisely of that type. Their estimator may be written as

$$\hat{V}_g = \sum_{i \in s} \sum_{j \in s} \tilde{\Delta}_{ij} g_{is} \tilde{e}_{is} g_{js} \tilde{e}_{js}, \tag{16.8}$$

where $\tilde{\Delta}_{ij} = (\pi_{ij} - \pi_i \pi_j)/\pi_{ij}$, π_{ij} is the design probability that both individuals i and j are included in the sample s, $\tilde{e}_{is} = (y_i - \mathbf{x}_i^T \hat{\boldsymbol{\beta}}_s)/\pi_i$, and g_{is} is the ith element of the row vector $\bar{\mathbf{x}}_N^T (X_s^T V_s^{-1} \Pi_s^{-1} X_s)^{-1} X_s^T V_s^{-1}$. See Särndal *et al.* (1989) for a detailed discussion of the model and design properties of $\hat{V}_g$ in (16.8). We note here that our estimator $\bar{y}_{\text{areg}}$ in (16.7) may be written as

$$\bar{y}_{\text{areg}} = \sum_{i \in s} g_{is} y_i / \pi_i$$

and we may also write

$$\bar{y}_{\text{areg}} = \bar{y}_N = \sum_{i \in s} g_{is} \tilde{E}_i = \bar{\mathbf{x}}_N^T (\hat{\boldsymbol{\beta}}_s - \hat{\boldsymbol{\beta}}_N),$$

where $\tilde{E}_i = (y_i - \mathbf{x}_i^T \hat{\boldsymbol{\beta}}_N)/\pi_i$. Now since $V_N \mathbf{1}_N$ is in the column space of X_N, say $V_N \mathbf{1}_N = X_N \mathbf{w}$, we have $\bar{\mathbf{x}}_N^T = \mathbf{1}_N^T V_N V_N^{-1} X_N / N = \mathbf{w}^T X_N^T V_N^{-1} X_N / N$, so that for large samples g_{is} will be near $1/N$ for $i \in s$. The design variance of $\bar{y}_{\text{areg}}$ is then approximately equal to

$$\sum_{i \in \mathbf{P}} \sum_{j \in \mathbf{P}} \Delta_{ij} \tilde{E}_i \tilde{E}_j / N^2$$

where $\Delta_{ij} = (\pi_{ij} - \pi_i \pi_j)$, and this may be estimated by

$$\hat{V}_1 = \sum_{i \in s} \sum_{j \in s} \tilde{\Delta}_{ij} \tilde{e}_{is} \tilde{e}_{js} / N^2. \tag{16.9}$$

$\hat{V}_1$ in (16.9) was considered in early work on the general regression estimator; for example, Särndal (1981, 1982). Now $\hat{V}_g$ is (16.8) may be thought of as a version of $\hat{V}_1$ in (16.9) adjusted for the realized values of g_{is}, $i \in s$. Särndal *et al.* (1989) show that $\hat{V}_g$ in (16.8), as well as being design consistent for the design variance of $\bar{y}_{\text{areg}}$, is often model unbiased or nearly model unbiased for the model mean squared error of $\bar{y}_{\text{areg}}$.

Now approximate confidence intervals for $\bar{y}_N$ could be constructed based on a standard normal approximation to the distribution of

$$(\bar{y}_{\text{areg}} - \bar{y}_N)/\{\tilde{V}_g\}^{1/2}.$$

The justification of this procedure, from both a design and a model point of view, is asymptotic and the question of its appropriateness for particular finite samples must be addressed. One possibility is to compare a set of confidence intervals obtained by this procedure to a set of purely model-based intervals based on a further assumption of normality of the errors and a *t*-statistic. If the two sets of intervals are wildly different there may be reason to doubt the validity of the jointly model- and design-based intervals, but more work is needed before this question can be answered satisfactorily.

References

Cassel, C. M., Särndal, C. E., and Wretman, J. H. (1977). *Foundations of inference in survey sampling*. Wiley, New York.

Godambe, V. P. and Thompson, M. E. (1984). Robust estimation through estimating equations. *Biometrika*, **71**, 115–25.

Godambe, V. P. and Thompson, M. E. (1986). Parameters of superpopulation and survey population: their relationships and estimation. *Int. Stat. Rev.*, **54**, 127–38.

Särndal, C. E. (1981). Frameworks for inference in survey sampling with applications to small area estimation and adjustment for nonresponse. *Bull. Int. Stat. Inst.*, **49**, 494–513.

Särndal, C. E. (1982). Implications of survey design for generalized regression estimation of linear functions. *J. Stat. Plan. Inf.*, **7**, 155–70.

Särndal, C. E., Swennson, B., and Wretman, J. H. (1989). The weighted residual technique for estimating the variance of the general regression estimator. *Biometrika*, **76**, 527–37.

17
Estimating functions in survey sampling: estimation of super-population regression parameters
K. Vijayan

ABSTRACT

It is usual to model the values of a characteristic on units of a finite population by a super-population through regression on some auxiliary information on the population. The parameters of the model are of interest. Using optimal estimating functions suitably defined for the situation we explain how these parameters could be estimated. An alternative method was given by Godambe and Thompson (1986). Some discussion on the logic of the use of estimating functions for estimation is also given.

17.1 Introduction

Survey sampling involves a finite population, say P, which is a collection of identifiable units, a list of which is available beforehand. We are interested in estimating some summary value (like average) of a certain characteristic associated with the population (for each unit of the population the characteristic will assign a unique value). As an illustration the population may be households in a town and our interest may be in the average amount of money spent per week on food by the households. We may number the units as $1, 2, \ldots, N$, where N is the population size and y_i is the value of the characteristic for unit i ($i = 1, 2, \ldots, N$). The y_i values are not known but can be obtained accurately at a certain cost. Because of restrictions on the cost of the survey the summary value needs to be estimated based on a sample of units from the population.

The inference problem involved in the above set-up is different from the inference problems for infinite populations. The parameters here are the values $y_1, \ldots, y_N$ and what we observe through the sample is some of the parameters themselves. The problem could be transformed to a prediction problem by viewing $\mathbf{y} = (y_1, y_2, \ldots, y_N)$ as a realization of an N-dimensional random variable $\mathbf{Y} = (Y_1, \ldots, Y_N)$. Such a formulation is possible since generally we have some information on the magnitude of the y_i-values based on prior surveys of similar nature or other considerations. For the example on expenditure on food, we may know how much each population unit spend on food the previous year from a census and might model the current year's values to have a linear regression on the previous year's values. (In a way we are relating the observed y-values of the sample with the unobserved y_i of the population through the model.)

The problem of the statistician then is to devise a scheme of sample selection and prescribe an estimation procedure for the quantities of interest so that the estimate and the quantity are as close as possible. He may also need to construct appropriate confidence intervals for the quantity. A general approach to the problem is possible through estimating functions. (Godambe and Thompson 1986). Godambe had earlier used estimating functions for inferences from infinite population (Godambe 1960).

In the succeeding sections we examine the logic of estimating functions and how it fits with the general frame work of estimation (both interval and point) in survey sampling. Our emphasis would be on the estimation of parameters in the prediction model.

17.2   Optimal estimating functions

Suppose we have data $\mathbf{y}$ whose distribution depends on a real-valued parameter θ and would like to estimate θ. A function $g(\mathbf{y}, \theta)$ of $\mathbf{y}$ and θ is an estimating function of θ if an estimate of θ could be obtained as a solution of the equation

$$g(\mathbf{y}, \theta) = 0.$$

(For example, in maximum likelihood methods if $L(\mathbf{y}, \theta)$ is the likelihood of θ and is differentiable in θ, then $\partial L(\mathbf{y}, \theta)/\partial \theta$ is an estimating function.)

Let us write $\hat{\theta}_g$ as an estimate of θ using the estimating function $g(\mathbf{y}, \theta)$, i.e.

$$g(\mathbf{y}, \hat{\theta}_g) = 0. \tag{17.1}$$

Now, assuming that Taylor expansion of g w.r.t. θ is valid, we may write

$$g(\mathbf{y}, \theta) = g(\mathbf{y}, \hat{\theta}_g) + (\theta - \hat{\theta})g'(y, \theta^*), \tag{17.2}$$

where θ^* is in the interval $(\hat{\theta}_g, \theta)$. Using (17.1), (17.2) may be rewritten as

$$\theta - \hat{\theta} = \frac{g(\mathbf{y}, \theta)}{g'(\mathbf{y}, \theta^*)}, \qquad \text{assuming } g'(\mathbf{y}, \theta^*) \neq 0.$$

A good estimating function should have the difference $\theta - \hat{\theta}$ small. Hence as a measure of the efficiency of an estimating function we may use the criterion $\lambda_g^{-1}(\theta)$, where

$$\lambda_g(\theta) = \frac{E(g(\mathbf{y}, \theta))^2}{E(g'(\mathbf{y}, \theta)))^2}. \tag{17.3}$$

where E stands for the expectation under the model.

The use of $\lambda_g(\theta)$ as measuring optimality of g makes it natural to demand that the estimating function g be unbiased in the sense that expected value of g is 0. This is so because the estimating equations $g(\mathbf{y}, \theta)$ and $k(\theta)g(\mathbf{y}, \theta)$, where $k(\theta)$ is some known function of θ, provides us with the same estimate for θ and hence should be considered equally efficient, that is, $\lambda_g(\theta)$ and $\lambda_{kg}(\theta)$ are the same, and this means that

$$\frac{k(\theta)E\{g(\mathbf{y}, \theta)\}^2}{[E\{k(\theta)g'(\mathbf{y}, \theta) + g(\mathbf{y}, \theta)k'(\theta)\}]^2} = \frac{E\{g(\mathbf{y}, \theta)\}^2}{(E(g'(\mathbf{y}, \theta))^2},$$

which is possible iff $g(\mathbf{y}, \theta)$ is unbiased. Thus we can restrict our attention to the class G of all unbiased estimating functions g. We may define an estimating function $g* \in G$ as an optimal estimating function for θ if

$$\lambda_{g*}(\theta) \leqslant \lambda_g(\theta) \ \forall \ g \in G. \tag{17.4}$$

It may be noted that in the above definition of optimal estimating function the distribution of the data need not have to be completely specified by θ, i.e. nuisance parameters could be present. This property is important in the survey sampling set up where we have only incomplete information about the distribution of the data.

The definitions of optimal estimating function above can easily be extended to the case when θ is vector valued. If θ has k components, g also will have to be vector valued with k components. We can assume that expectation of g is the zero vector. Suppose

$$\theta' = (\theta_1, \theta_2, \ldots, \theta_k)$$

$$g' = (g_1, g_2, \ldots, g_k)$$

$$c_{ij} = E\left(\frac{\partial}{\partial \theta_i} g_j\right)$$

$$d_{ij} = E(g_i g_j).$$

Then, generalizing (17.3) we can write

$$\Lambda_g^{-1} = CD^{-1}C' \tag{17.5}$$

where Λ and D are (k, k) matrices and

$$C = (c_{ij}), \quad D = (d_{ij}), \qquad i = 1, \ldots, k, \quad j = 1, \ldots, k.$$

The estimating function g^* in the class of all unbiased estimating functions G is optimal if

$$\Lambda_{g^*}^{-1} - \Lambda_g^{-1} \quad \text{is non-negative definite} \quad \forall g \in G. \tag{17.6}$$

17.3   Regression and optimal estimating functions

As mentioned in the introduction, in sample survey situation, we have a finite population of N identifiable units, N known beforehand. The characteristic of interest has a value y_i for the ith unit and is modelled as a realization of a random variable Y_i ($i = 1, 2, \ldots, N$). It is reasonable to assume that the Y_i are uncorrelated.

To model the distribution of the Y_i is generally difficult but the mean and variance could be modelled adequately from the background information available on the population. Let us assume the model to be

$$E(Y_i) = \alpha_i(\theta), \quad \text{Var}(Y_i) = \sigma^2 \tau_i, \quad i = 1, 2, \ldots, N \tag{17.7}$$

where the α_i are known functions of θ and the τ_i are known constants. Before considering the influence of sampling on the optimal estimating function of θ, we may first consider the case where all the N units are observed. To choose an optimal estimating function of θ, we can restrict the class of unbiased estimating functions G to estimating functions of the form

$$g(\mathbf{y}, \theta) = \sum_{i=1}^{N} a_i(\theta)\phi_i, \tag{17.8}$$

where

$$\phi_i = y_i - \alpha_i(\theta), \quad i = 1, \ldots, N. \tag{17.9}$$

Note that the restriction to G is forced on us because we need to compare expected values of $g^2(\mathbf{Y}, \theta)$ and only the first two moments of the Y_i are available. The unbiasedness of g is easily seen by noting that $E(\phi_i) = 0$ and the ϕ_i are uncorrelated. We have the following characterizations of optimal estimates of g in G.

Theorem 17.1 *In the class of all unbiased estimating functions of the form (17.8) g^* is an optimal estimator if and only if it is of the form*

$$g^*(\mathbf{y}, \theta) = a(\theta) \sum_{i=1}^{N} \frac{\alpha_i'(\theta)}{\tau_i} \phi_i. \tag{17.10}$$

Proof   We have

$$
\lambda_g = \frac{E(g(\mathbf{Y}, \theta)^2}{(E(g'(\mathbf{Y}, \theta)))^2}
$$

$$
= \sigma^2 \frac{\sum_{i=1}^{N} a_i^2 \tau_i}{(\sum a_i \alpha_i'(\theta))^2}
$$

$$
= \sigma^2 \frac{\sum_{i=1}^{N} \frac{(\alpha_i'(\theta))^2}{\tau_i} \cdot \left(\dfrac{a_i \tau_i}{\alpha_i'(\theta)}\right)^2}{\left(\sum_{i=1}^{N} \dfrac{(\alpha_i'(\theta))^2}{\tau_i} \dfrac{a_i \tau_i}{\alpha_i'(\theta)}\right)^2}
$$

$$
\geqslant \sigma^2 \frac{1}{\sum_{i=1}^{N} \dfrac{(\alpha_i'(\theta))^2}{\tau_i}} \qquad \text{from the Schwartz inequality.} \qquad (17.11)
$$

The equality in (17.12) is possible if and only if

$$
a_i(\theta)\tau_i = a(\theta)\alpha_i'(\theta) \qquad \text{for some } a(\theta),
$$

that is

$$
a_i(\theta) = a(\theta)\frac{\alpha_i'(\theta)}{\tau_i}.
$$

Hence the theorem.

Note   Godambe and Thompson (1986) had established Theorem 17.1 using a slightly different proof.

The estimating function g^* given in (17.10) can be shown to be optimal in a much wider class than G if the Y_i are assumed to be independent.

Theorem 17.2   *Consider the class G^* of all unbiased estimating functions of θ of the form*

$$
g(\mathbf{y}, \theta) = \sum_i a_i(\mathbf{y}, \theta)\phi_i, \qquad (17.12)
$$

where $a_i(\mathbf{y}, \theta)$ does not depend on y_i. Also assume that the Y_i are independent. Then g^* is an optimal estimating function in G^* if and only if

$$
g^*(\mathbf{y}, \theta) = a(\theta) \sum_{i=1}^{N} \frac{\alpha_i'(\theta)}{\tau_i} \phi_i
$$

for some $a(\theta)$.

Proof   In the proof, for convenience, we write a_i for $a_i(\mathbf{Y}, \theta)$ and α_i for $\alpha_i(\theta)$.

First we observe

$$E\left\{\left[\sum_{i=1}^{N}(a_i - E(a_i))\phi_i\right]\left[\sum_{i=1}^{N}E(a_i)\phi_i\right]\right\}$$

$$= E\left\{\sum_{i=1}^{N}E(a_i)(a_i - E(a_i))\phi_i^2\right\} + E\left\{\sum_{i\neq j}E(a_j)(a_i - E(a_i))\phi_j \cdot \phi_i\right\}$$

$$= \sum_{i=1}^{N}E(a_i)E(a_i - E(a_i))E(\phi_i^2) + \sum_{i\neq j}E(a_j)E((a_i - E(a_i)\phi_j) \cdot E(\phi_i)$$

$$= 0,$$

since a_i does not depend on Y_i. Hence

$$E\left(\sum_{i=1}^{N}a_i\phi_i\right)^2 = E\left(\sum_{i=1}^{N}E(a_i)\phi_i\right)^2 + E\left(\sum_{i=1}^{N}(a_i - E(a_i))\phi_i\right)^2$$

$$\geqslant E\left(\sum_{i=1}^{N}E(a_i)\phi_i\right)^2$$

Also,

$$E\left(\frac{\partial}{\partial\theta}g(\mathbf{Y}, \theta)\right) = -\sum_{i=1}^{N}E(a_i)\alpha_i'.$$

Hence

$$\lambda_g \geqslant \lambda_{g_1}, \tag{17.13}$$

where

$$g_1 = \sum_{i=1}^{N}E(a_i(\mathbf{y}, \theta))\phi_i.$$

Since g_1 belongs to the class G and the equality in (17.13) is possible only when $g = g_1$, the theorem follows from Theorem 17.1.

17.4 Optimal estimating function in survey sampling

Now we may extend the theory of the last section to the case in which only part of the finite population is observed. Let us denote the set of sample units selected by s. The selection is made using some selection procedure which for all practical purposes can be viewed as choosing a sample from a collection of samples S using some random mechanism. We will denote the probability of selecting sample s by p_s.

We pointed out in the last section that a reasonable class of estimating functions from which an optimum one is to be chosen is given by (17.12). As we would observe only some of the y_i-values, we modify $g(\mathbf{y}, \theta)$ appropriately and hence restrict our attention to class of estimating functions $h(\mathbf{y}, s, \theta)$

of the form

$$h(\mathbf{y}, s, \theta) = \sum_i a_{is}(\theta)\phi_i \qquad (17.14)$$

$$\text{where } a_{is}(\theta) = 0 \qquad \text{if } i \notin s$$

and $a_{is}(\theta)$ depends only on unit values other than y_i and in the sample. We may also assume that the Y_i are independent. Here the underlying distribution depends on two things: the regression model for the Y_i as envisaged in previous section; and the sampling design that chooses the probabilities of selection for the different samples in S. To distinguish between the two probability distributions when taking expectations, we would use E_m when the expectation is with respect to the regression model M and E_p when it is with respect to the design P. When it is over the pair (M, P), we use E itself.

We have

$$E(h(\mathbf{Y}, s, \theta)^2) = E_p E_m(h(\mathbf{Y}, s, \theta)^2)$$

$$= E_p E_m \left(\sum_{i=1}^{N} a_{is}(\theta)\phi_i \right)^2$$

$$\geqslant E_p \sum_{i=1}^{N} (E_m^2(a_{is}(\theta)))E_m(\phi_i^2) \qquad (17.15)$$

using appropriate modified version of (17.13) where a_i is changed to $a_{is}(\theta)$. (The modified version can be proved using similar arguments of Theorem 17.2.) Hence

$$E(h(\mathbf{Y}, s, \theta)) \geqslant \sum_{i=1}^{N} E_p(E_m^2(a_{is}(\theta)))(E_m \phi_i^2)$$

$$= \sigma^2 \sum_{i=1}^{N} E_p(E_m^2(a_{is}(\theta)))\tau_i.$$

Also

$$E(h'(\mathbf{Y}, s\, \theta)) = E_p E_m h'(\mathbf{Y}, s, \theta))$$

$$= E_p \sum_{i=1}^{N} E_m(a_{is}(\theta))(-\alpha_i'(\theta))$$

$$= -\sum_{i=1}^{N} E(a_{is}(\theta))\alpha_i'(\theta).$$

Now

$$E_p(E_m^2(a_{is}(\theta))) = E(E_m^2 a_{is}(\theta)|s \supset i)P(s \supset i)$$

and

$$E(a_{is}(\theta)) = E(a_{is}(\theta)|s \supset i)P(s \supset i)$$

and hence, denoting the inclusion probability of the ith unit, $P(s \supset i)$, as π_i, we have

$$\lambda_h \geqslant \sigma^2 \frac{\sum_{i=1}^N E(E_m^2 a_{is}(\theta) | s \supset i) \pi_i \tau_i}{(\sum_{i=1}^N E(a_{is}(\theta) | s \supset i) \pi_i \alpha_i'(\theta))^2}$$

$$\geqslant \sigma^2 \frac{\sum_{i=1}^N (E(a_{is}(\theta) | s \supset i))^2 \pi_i \tau_i}{(\sum_{i=1}^N E(a_{is}(\theta) | s \supset i) \pi_i \alpha_i'(\theta))^2} \qquad (17.16)$$

$$\geqslant \sigma^2 \frac{1}{\sum_{i=1}^N \dfrac{\pi_i (\alpha_i'(\theta))^2}{\tau_i}}, \qquad (17.17)$$

using the Schwartz inequality as in (17.11).

The equality in (17.16) is possible if and only if $a_{is}(\theta)$ is the same for all s that contains i; that is

$$a_{is}(\theta) = E(a_{is}(\theta) | s \supset i) \qquad (17.18)$$

and the equality of (17.17) is possible if and only if

$$E(a_{is}(\theta) | s \supset i) = a(\theta) \frac{\pi_i \alpha_i'(\theta)}{\pi_i \tau_i}$$

$$= a(\theta) \frac{\alpha_i'(\theta)}{\tau_i}, \qquad (17.19)$$

Thus we have the result.

Theorem 17.3 *In the class of all unbiased estimating functions of the form (17.14), h^* is an optimal estimating function if and only if h^* is of the form*

$$h^*(\mathbf{y}, s, \theta) = a(\theta) \sum_{i \in s} \frac{\alpha_i'(\theta)}{\tau_i} (y_i - \alpha_i(\theta)), \qquad (17.20)$$

where $a(\theta)$ is any known function of θ and its efficiency is

$$\lambda_{h^*}^{-1} = \sum_{i=1}^N \pi_i \frac{(\alpha_i'(\theta))^2}{\tau_i} \frac{1}{\sigma^2}. \qquad (17.21)$$

It is interesting to observe that the optimal estimating equation (17.20) does not depend on the sample design, but only on the units in the sample. This is intuitively obvious since there is no information about the model parameters in the way the sample is chosen and so the estimation of the model parameters should not involve the sample design. We may notice that the efficiency of the estimating function as calculated by (17.21) is the expected value of conditional efficiency, conditioning being with respect to the sample chosen, which again is consistent with our intuition.

From (17.21) we observe that the efficiency λ_{h*}^{-1} can be made larger and larger by increasing the inclusion probabilities of units that have $(1/\tau)(\alpha'(\theta))^2$ largest. Hence if n is the size of the sample taken, the optimum strategy is to take the n units that has the largest $(1/\tau_i)(\alpha_i'(\theta))^2$.

In many sampling situations an appropriate regression model is of the form

$$\alpha_i(\theta) = \theta x_i; \qquad \tau_i = \sigma^2 x_i^q, \tag{17.22}$$

where x_i is some size measure based on prior information and g is usually a number lying between 1 and 2 (see Vijayan 1966). Then $(1/\tau_i)(\alpha_i'(\theta))^2$ is proportional to x_i^{2-g} and hence is an increasing function of x_i. Thus the optimum sample, on the assumption that the model given by (17.25) is reasonably accurate, is to take the n units having the largest size, and an optimal estimating function is

$$h(\mathbf{y}, s, \theta) = \sum_{i \in s} \frac{y_i - \theta x_i}{x_i^{g-1}}$$

which gives an estimator for θ as

$$\hat{\theta} = \frac{\sum_{i \in s} y_i / x_i^{g-1}}{\sum_{i \in s} x_i^{2-g}}.$$

17.5   Extension to the multiparameter situation

Theorems 17.1 and 17.2 can be extended easily when the regression model depends on more than one parameter. We might model the regression as

$$E(Y_i) = \alpha_i(\boldsymbol{\theta}), \qquad \text{Var}(Y_i) = \sigma^2 \tau_i, \tag{17.23}$$

where $\boldsymbol{\theta}$ has k components. Obviously we would need k independent (unbiased) estimating functions, i.e. an estimating vector function $\mathbf{g}$ of k components. We restrict the class $\mathbf{G}$ of estimating vector functions $\mathbf{g}$ to vectors of the form

$$\mathbf{g}(\mathbf{y}, \theta) = \sum_{i=1}^{N} \mathbf{a}_i(\boldsymbol{\theta})\phi_i \tag{17.24}$$

$$= A\boldsymbol{\phi}$$

where

$$\phi_i = y_i - \alpha_i(\boldsymbol{\theta}), \qquad i = 1, 2, \ldots, N, \tag{17.25}$$

and A is a matrix whose ith column is $\mathbf{a}_i$, $i = 1, 2, \ldots, N$. Theorem 17.1 can now be extended as follows.

Theorem 17.4 *In the class of all estimating vector functions of the form (17.24), g^* is an optimal estimating vector function if and only if it is of the form*

$$\mathbf{g}^* = H(\boldsymbol{\theta})J\Sigma^{-1}\boldsymbol{\phi} \tag{17.26}$$

where H is any (k, k) non-singular matrix whose components depend only on $\boldsymbol{\theta}$ and

$$J = \frac{\partial}{\partial\theta}\boldsymbol{\phi}^{\mathrm{T}}$$

and

$$\Sigma = \operatorname{diag}(\tau_1, \tau_2, \ldots, \tau_n).$$

where superscript T *is used for the transpose of the matrix.*

Proof We easily notice that the matrices C and D of (17.5) are given as

$$C = JA^{\mathrm{T}}$$

and

$$D = A\Sigma A^{\mathrm{T}}\sigma^2.$$

For Λ as defined in (17.5), it follows that

$$\Lambda_g^{-1} = JA^{\mathrm{T}}(A\Sigma A^{\mathrm{T}})^{-1}AJ^{\mathrm{T}}\cdot\frac{1}{\sigma^2}$$

and

$$\Lambda_{g^*}^{-1} = J\Sigma^{-1}J^{\mathrm{T}}H^{\mathrm{T}}(HJ\Sigma^{-1}J^{\mathrm{T}}H^{\mathrm{T}})^{-1}HJ\Sigma^{-1}J^{\mathrm{T}}\cdot\frac{1}{\sigma^2}$$

$$= J\Sigma^{-1}J^{\mathrm{T}}H^{\mathrm{T}}(H^{\mathrm{T}})^{-1}(J\Sigma^{-1}J^{\mathrm{T}})^{-1}H^{-1}HJ\Sigma^{-1}J^{\mathrm{T}}\cdot\frac{1}{\sigma^2}$$

$$= J\Sigma^{-1}J^{\mathrm{T}}\cdot\frac{1}{\sigma^2}.$$

Hence

$$\Lambda_{g^*}^{-1} - \Lambda_g^{-1} = J(\Sigma^{-1} - A^{\mathrm{T}}(A\Sigma A^{\mathrm{T}})^{-1}A)J^{\mathrm{T}}\cdot\frac{1}{\sigma^2}$$

$$= J\Sigma^{-1/2}F\Sigma^{-1/2}J^{\mathrm{T}}\cdot\frac{1}{\sigma^2} \tag{17.27}$$

where

$$F = I - \Sigma^{1/2}A^{\mathrm{T}}(A\Sigma A^{\mathrm{T}})^{-1}A\Sigma^{1/2}. \tag{17.28}$$

Now F is idempotent and hence non-negative definite. This would imply that (17.27) is also non-negative definite satisfying (17.6). Thus g^* as defined by (17.26) is an optimal estimating function.

Now we wish to show that all optimal estimating functions must be of the form (17.26). For a particular choice of A, if the corresponding g is optimal, then using (17.6) again we would need $\Lambda_g^{-1} - \Lambda_{g*}^{-1}$ to be non-negative definite and since $\Lambda_{g*}^{-1} - \Lambda_g^{-1}$ also is non-negative definite we have

$$\Lambda_{g*}^{-1} - \Lambda_g^{-1} = 0$$

i.e. $J\Sigma^{-1/2} . F\Sigma^{-1/2}J^{\mathrm{T}} = 0$, which would then imply that

$$F\Sigma^{-1/2}J^{\mathrm{T}} = 0.$$

Now F is an idempotent matrix and hence the orthogonal space of the row space of F is the column space of $I - F$, where I is the identity matrix of order n. Columns of $\Sigma^{-1/2}J$ are in the column space of

$$I - F = \Sigma^{1/2}A^{\mathrm{T}}(A\Sigma^{1/2}\Sigma^{1/2}A^{\mathrm{T}})^{-1}A\Sigma^{1/2} \qquad (17.29)$$

From (17.29) we observe that the column space of $I - F$ is the same as that of $\Sigma^{1/2}A^{\mathrm{T}}$ and hence we can write

$$\Sigma^{-1/2}J^{\mathrm{T}} = \Sigma^{1/2}A^{\mathrm{T}}K, \qquad (17.30)$$

where K is a non-singular matrix of order k. From (17.30) we get

$$A = (K^{-1})^{\mathrm{T}}J\Sigma^{-1},$$

which would then produce an estimating function of the form (17.26).

Theorem 17.2 also can be extended to the multiparameter case following the same line of arguments there and we will write down the general version as the following theorem.

Theorem 17.5 *When the Y_i are independent, in the class of all unbiased estimating functions of the form*

$$\mathbf{g}(\mathbf{y}, \boldsymbol{\theta}) = \sum_{i=1}^{N} a_i(\mathbf{y}, \theta) . \phi_i,$$

where $\mathbf{a}_i(\mathbf{y}, \boldsymbol{\theta})$ does not depend on y_i, $i = 1, \ldots, N$, $\mathbf{g}^*(\mathbf{y}, \boldsymbol{\theta})$ is an optimal estimating function if and only if we can write $\mathbf{g}^*$ in the form (17.26).

The extension to the sampling situation also is exactly along the same lines as in Section 17.4 and we omit the details.

17.6 The confidence interval for θ and σ^2

Having obtained an estimate of $\hat{\theta}$ of θ using an optimal estimating function, one might like to obtain a confidence interval for θ. An approach we adopt is to get two functions $h_1(\mathbf{y}, s, \theta)$ and $h_2(\mathbf{y}, s, \theta)$, where h_1 and h_2 are obtained from the optimal estimating function h^*, such that the interval $(h_1(\mathbf{y}, s, \theta), h_2(\mathbf{y}, s, \theta))$ covers zero with probability $1 - \alpha$ where $1 - \alpha$ is the level of confidence needed and the confidence limits of θ are taken as the solutions of the equations $h_1(\mathbf{y}, s, \theta) = 0$ and $h_2(\mathbf{y}, s, \theta) = 0$.

Recall that the optimal estimating function h^* is a linear function of the ϕ_i, which are uncorrelated. If

$$\phi_i^* = \phi_i / \sqrt{\tau_i},$$

then ϕ_i^* has mean zero and variance σ^2 and the ϕ_i^* are uncorrelated. A natural choice of estimating functions for σ^2 is one of the form

$$\sum_i b_{is}(\sigma^2)((\phi_i^*)^2 - \sigma^2, \tag{17.31}$$

where $b_{is} = 0$ if $i \notin s$. The efficiency function of (17.31) depends on the fourth central moments of the Y_i and hence needs further assumptions on the model. In the absence of any further assumptions, a natural choice would be to take $b_{is}(\sigma^2)$ to be same for all $i \in s$. Hence σ^2 is estimated from the estimating equation

$$\sum_{i \in s} (\phi_i^{*2} - \sigma^2) = 0$$

$$\hat{\sigma}^2 = \frac{1}{n} \sum_{i \in s} \phi_i^2 / \tau_i.$$

Now,

$$h^*(\mathbf{y}, s, \theta) = \sum_{i \in s} b_i(\theta)\phi_i^*$$

$$b_i(\theta) = \alpha_i'(\theta) / \sqrt{\tau_i}.$$

Obviously,

$$\text{Var}(h^*) = E_p \sum_{i \in s} b_i^2(\theta)\sigma^2,$$

which can be estimated by

$$\hat{v}^2 = \frac{1}{n} \sum_{i \in s} b_i^2(\theta) \sum_{i \in s} \phi_i^2 / \tau_i$$

$$= \frac{1}{n} \sum_{i \in s} (\alpha_i'(\theta))^2 / \tau_i \sum_{i \in s} \phi_i^2 / \tau_i. \tag{17.32}$$

Now assuming n is large we may assume that $h^*/\hat{v}$ is approximately normally distributed with mean zero and variance 1. Then a $(1 - \alpha)$ level confidence interval is chosen as $(\hat{\theta}_1, \hat{\theta}_2)$, where $\hat{\theta}_1$ and $\hat{\theta}_2$ are respectively solutions of

$$h_1(\mathbf{y}, s, \theta) = \sum_{i\varepsilon s} \frac{\alpha_i'(\theta)}{\tau_i} \phi_i - z_{\alpha/2} \hat{v} = 0$$

and

$$h_2(\mathbf{y}, s, \theta) = \sum_{i\varepsilon s} \frac{\alpha_i'(\theta)}{\tau_i} \phi_i + z_{\alpha/2} \hat{v} = 0. \qquad (17.33)$$

Example 17.1 Consider the situation when $\alpha_i(\theta) = \theta x_i$. We have $\alpha_i'(\theta) = x_i$ and the two equations (17.33) become, after substituting for ϕ_i from (17.9),

$$\sum_{i\varepsilon s} \frac{x_i}{\tau_i} (y_i - \theta x_i) \pm z_{\alpha/2} \left(\frac{1}{n} \sum_{i\varepsilon s} \frac{x_i^2}{\tau_i} \cdot \sum_{i\varepsilon s} \frac{(y_i - \theta x_i)^2}{\tau_i} \right)^{1/2} = 0.$$

This can be simplified as

$$(\hat{\theta} - \theta) \sum_{i\varepsilon s} \frac{x_i^2}{\tau_i} \pm z_{\alpha/2} \left(\frac{1}{n} \sum_{i\varepsilon s} \frac{x_i^2}{\tau_i} \left(\sum \frac{1}{\tau_i} (y_i - \hat{\theta} x_i)^2 + (\hat{\theta} - \theta)^2 \sum \frac{x_i^2}{\tau_i} \right) \right)^{1/2} = 0,$$

where $\hat{\theta}$ is the estimate obtained from the optimal estimating function, that is

$$\hat{\theta} = \frac{\sum x_i y_i / \tau_i}{\sum x_i^2 / \tau_i}. \qquad (17.34)$$

We can then write down the confidence interval as

$$\hat{\theta} \pm \frac{\left(z_{\alpha/2} \sqrt{\sum_{i\varepsilon s} \frac{y_i^2}{\tau_i} \sum_{i\varepsilon s} \frac{x_i^2}{\tau_i} - (\sum_{i\varepsilon s} x_i y_i / \tau_i)^2} \right)}{\sqrt{(n - z_{\alpha/2}^2)} \cdot \sum_{i\varepsilon s} x_i^2 / \tau_i}.$$

It is interesting to note that in this example the estimate of θ coincides with the best linear unbiased estimator of θ while the confidence interval would have been the same if $n - z_{\alpha/2}^2$ is replaced by $(n - 1)$.

17.7 Acknowledgement

I am thankful to Dr Harold Mantel and Mr Boxin Tang and the referees for their suggestions which helped tremendously in the improvement of the paper.

References

Godambe, V. P. (1960). An optimum property of maximum likelihood estimation. *Ann. Math. Stat.*, **31**, 1208–12.

Godambe, V. P. and Thompson, M. E. (1986). Parameters of superpopulation and survey population: their relationship and estimation. *Int. Stat. Rev.*, **54**, 127–38.

Vijayan, K. (1966). On Horvitz and Thompson and Des Raj estimators. *Sankhya* A, **28**, 87–93.

Comments on Vijayan's chapter

V. P. Godambe

Theorem 17.3 of Vijayan's paper has a very important implication: in the context of survey sampling, if estimation '*exclusively* of a super-population parameter θ' is of interest, the 'optimum estimating function' must be 'independent of the sampling design' used to draw the sample.

Of course mathematically the theorem is correct granting the underlying super-population model. In practice, even if one is only interested in estimating a super-population parameter θ, one may doubt some aspects of the model. Suppose the optimum estimating function is based in an essential manner on these 'doubtful aspects' of the model. In such a situation, even for estimating the super-population parameter θ, one may not like to use the above-mentioned design-free 'optimum estimating function'. Some design-based estimating functions can be more appropriate. This is illustrated below.

Simplifying Vijayan's super-population model somewhat, we assume $Y_1, \ldots, Y_N$ to be independent, $E(Y_i) = \theta$ and $\text{Var}(Y_i) = \tau_i, i = 1, \ldots, N$. Hence according to Theorem 17.3, the optimum estimating function

$$h^* = \sum_{i \in s} (y_i - \theta)/\tau_i, \tag{1}$$

which is independent of the sampling design. Now suppose one is doubtful about the specification of super-population variances τ_i, $i = 1, \ldots, N$. For the allowable variations of τ_i, $i = 1, \ldots, N$, h^* may vary so much as to be practically *useless*.

In the following, for notational convenience, we assume the survey population $\mathcal{P} = \{i: i = 1, \ldots, N\}$ to be divided into two strata $\mathcal{P}_1 = \{i = 1, \ldots, N_1\}$ and $\mathcal{P}_2 = \{i: N_1 + 1, \ldots, N_1 + N_2\}$. For the super-population $E(Y_i) = \theta$, $i = 1, \ldots, N$ and $\text{Var}(Y_i) = \tau_1$, $i = 1, \ldots, N_1$ and $\text{Var}(Y_i) = \tau_2$, $i = N_1 + 1, \ldots, N_1 + N_2$. We suppose the true values τ_1 and τ_2 are unknown but some guessed values τ_1' and τ_2' are available. In terms of the guessed values τ_1' and τ_2' we define a survey population parameter θ_N by the equation

$$0 = \sum_{i=1}^{N_1} \frac{y_i - \theta_N}{\tau_1'} + \sum_{i=N_1+1}^{N_1+N_2} \frac{y_i - \theta_N}{\tau_2'}. \tag{2}$$

We note two points:

(A) For any allowable departures $|\tau_1 - \tau_1'|$ and $|\tau_2 - \tau_2'|$ for sufficiently large N_1 and N_2, $\theta_N \simeq \theta$. That is, for sufficiently large strata, for all practical purpose the survey population parameter θ_N can be assumed to be equal to the super-population parameter θ.

(B) For a stratified sampling design, with strata samples s_1 and s_2, $|s_1| = n_1$ and $|s_2| = n_2$ respectively, the 'optimum estimating function' for estimating the survey population parameter θ_N in (2) is given by

$$\tilde{h} = \sum_{i \in s_1} \frac{y_i - \theta}{\tau_1'} \cdot \frac{N_1}{n_1} + \sum_{i \in s_2} \frac{y_i - \theta}{\tau_2'} \cdot \frac{N_2}{n_2} \tag{3}$$

(Godambe and Thompson 1986). In the present case from (1)

$$h^* = \sum_{i \in s_1} \frac{y_i - \theta}{\tau_1} + \sum_{i \in s_2} \frac{y_i - \theta}{\tau_2}. \tag{4}$$

The estimating function $\tilde{h}$ in (3) is *design dependent* but is *independent* of the true variances τ_1 and τ_2. The situation is *reversed* with h^*.

CONCLUSION

If the true variances τ_1 and τ_2 are not known sufficiently accurately, but the strata sizes are sufficiently large, in view of (A) and (B) above, even to estimate *exclusively* the super-population parameter θ, the 'design-based' estimating function $\tilde{h}$ in (3) is preferable to 'design-independent' h^* in (4). Of course the appropriate allocation of the samples sizes, n_1 to n_2, to different strata is to be obtained using the guessed values τ_1' and τ_2' of the true variances τ_1 and τ_2.

A reply to Godambe's comments

K. Vijayan

I completely agree with the main thrust of Godambe's Comment. The super-population variances are very rarely known exactly and only reasonable guesses of the variances could be made. Similarly other aspects of the specification of the model may not be completely true. To guard against such failures it is absolutely essential that the sample is taken using some appropriate sample design and if diagnostic checks indicate that the assumed model is not appropriate one can then fall back on a design-based estimation procedure suggested by Godambe and Thompson (1986).

For the example Godambe is using, I would still prefer the estimating function given by (4), after replacing τ_1 and τ_2 by τ_1' and τ_2' respectively, for the following reason.

Solving (2), we have

$$\theta_N = \frac{(N_1/\tau_1')\bar{Y}_1 + (N_2/\tau_2')\bar{Y}_2}{N_1/\tau_1' + N_2/\tau_2'}, \tag{5}$$

where $\bar{Y}_i$ is the ith stratum mean $(i = 1, 2)$ and is an estimate of θ. The estimate of θ_N from (3) is the same as one obtained from (5) on replacing $\bar{Y}_i$ by its corresponding sample mean, say $\bar{y}_i$ $(i = 1, 2)$. Now if n_i is small, $|\bar{y}_i - \bar{Y}_i|$ is expected to be comparatively large and hence it would be better to reduce the weight on $\bar{y}_i$. This is achieved if we use the estimating function h^* given by (4) (after the suggested modification) since in this case the estimate of θ is given by

$$\hat{\theta} = \frac{(n_1/\tau_1')\bar{y}_1 + (n_2/\tau_2')\bar{y}_2}{n_1/\tau_1' + n_2/\tau_2'}.$$

PART 5

Theory (foundations)

18
Sufficiency, ancillarity, and information in estimating functions
V. P. Bhapkar

ABSTRACT

The optimality of the score function as an estimating function is naturally related to the Fisher information for the parameter. For any estimating function, a sufficient statistic can be used to derive a possibly more informative version of the given estimating function. Similarly, the use of the conditional distribution given the observed value of an ancillary statistic leads to a possibly more informative estimating function. The concepts of sufficiency, or ancillarity, for the parameter of interest in the presence of nuisance parameter are termed *p*-sufficiency, or *p*-ancillarity, in this paper. The role of such *p*-sufficient, or *p*-ancillary, statistics are studied in producing optimal estimating functions for parameters of interest. Suitable generalizations of Fisher information function for parameter of interest are considered here in the presence of nuisance parameters.

18.1 Introduction

Suppose x is the observed value of random variable X with probability density function (p.d.f.) involving an unknown real-valued parameter θ. Consider an estimation procedure which uses a solution of the equation

$$g(x, \theta) = 0$$

as an estimate of θ. g is said to be *unbiased* if $E_\theta g(X, \theta) = 0$ for all θ. Under some regularity assumptions for estimating functions g and the p.d.f. $p(x, \theta)$ of X, it was shown by Godambe (1960) that the usual maximum likelihood estimating function, i.e. the *score function*,

$$l(x, \theta) \equiv \frac{\partial \log p(x, \theta)}{\partial \theta} \tag{18.1}$$

enjoys a certain optimal property.

In view of this result Bhapkar (1972) defined the *information in the estimating function g* as

$$i_g(\theta) = \frac{[E_\theta(\partial g(X, \theta)/\partial \theta)]^2}{E_\theta g^2(X, \theta)}. \tag{18.2}$$

The basic result due to Godambe (1960) may be then presented as

$$i_g(\theta) \leqslant I(\theta) = i_l(\theta), \tag{18.3}$$

where $I(\theta)$ is the *Fisher information function* for the distribution of X, and l is the score function which is the maximum likelihood estimating function (m.l.e.f.).

More generally, suppose the p.d.f. p involves the real-valued parameter of interest θ_1 along with some nuisance parameters θ_2, and consider the estimating function g for θ_1, viz. $g = g(x, \theta_1)$, which does not involve θ_2. Under regularity assumptions for g and p, it was shown by Godambe (1976) that the *conditional likelihood* estimating function, l_c, was optimal in some sense, provided conditioning was with respect to a statistic which is *ancillary* for θ_1 in some specific sense, when θ_2 is unknown.

On the other hand, it has been shown by Lloyd (1987) that the *marginal likelihood* estimating function, l_m, for θ_1 is optimal when the marginal distribution of a suitable statistic, if available, is used. Such a statistic has been termed *partially sufficient* for θ_1. This paper brings together results concerning optimality of estimating functions, sufficiency and partial sufficiency, ancillarity and partial ancillarity, and finally Fisher information function generalizations.

In Section 18.2 we state the regularity assumptions and consider the use of sufficient statistic, or ancillary statistic, in the context of estimating functions and Fisher information functions. Section 18.3 briefly deals with the case of vector parameter $\mathbf{\theta}$. Sections 18.4 and 18.5 consider the case of estimating functions for θ_1 when nuisance parameters θ_2 are unknown. The role of partially ancillary statistics for θ_1, when they exist, is examined in producing l_c as the optimal estimating function for θ_1. A similar study is made concerning partially sufficient statistics, when they exist, in producing l_m as the optimal estimating function for θ_1. Finally, if such statistics are not available, the resulting information loss is examined.

18.2   Sufficiency and ancillarity for θ

Assume first that the probability distribution P_θ of X is indexed by a real-valued θ in an open interval Ω of the real line. We assume that the p.d.f. of P_θ with respect to some σ-finite measure μ is $p(x, \theta)$.

The regularity assumptions for p are denoted as R, given below.

R: (a) $\partial \log p/\partial\theta$ and $\partial^2 \log p/\partial\theta^2$ exist for all $\theta \in \Omega$ and for almost all $(\mu)x$;

(b) $\int p\,d\mu$ and $\int (\partial \log p/\partial\theta)p\,d\mu$ are differentiable with respect to θ under the integral sign at all $\theta \in \Omega$;

(c) $I(\theta) \equiv E_\theta[\partial \log p(X, \theta)/\partial\theta]^2 > 0$ for all $\theta \in \Omega$.

The regularity assumptions for the unbiased estimating function $g \in G$ are denoted as R_G, given below.

R_G: (i) $\partial g/\partial \theta$ exists for all $\theta \in \Omega$ and almost all $(\mu)x$;
   (ii) $\int gp \, d\mu$ is differentiable with respect to θ under the integral sign at all $\theta \in \Omega$;
   (iii) $E_\theta[\partial g(X, \theta)/\partial \theta]$ exists and is non-zero for all $\theta \in \Omega$;
   (iv) $0 < E_\theta g^2(X, \theta) < \infty$ for all $\theta \in \Omega$.

Let now the statistic $S = S(X)$ be sufficient for the family $\{P_\theta, \theta \in \Omega\}$ and $f(s, \theta)$ be the p.d.f. of S with respect to the induced measure v.

Define now

$$g^*(s, \theta) = E_\theta g(X, \theta | s), \tag{18.4}$$

where $s = S(x)$ is the observed value of statistic S. It was shown by Bhapkar (1972) that

$$i_g(\theta) \leqslant i_{g*}(\theta), \qquad \text{all } \theta \in \Omega, \tag{18.5}$$

with equality iff $g(x, \theta) = g^*(S(x), \theta)$ a.e. (P_θ), where the information function i_g of g is defined by (18.2). The relative optimality property (18.5) of estimating functions g^*, based in S, in relation to g, based on X, provides a more general version in terms of estimating functions of the well-known Rao–Blackwell theorem; the result for estimators $t^*(S)$ and $t(X)$ follows by writing

$$g(x, \theta) = t(x) - \theta, \qquad g^*(s, \theta) = t^*(s) - \theta.$$

Godambe's basic result (18.3) for the optimality of score function $l(x, \theta)$, given by (18.1), may be interpreted in terms of sufficient statistic S as follows.

Let $T = T(X)$ be any statistic and suppose $q(t, \theta)$ is its p.d.f. Denote by

$$l^{(T)}(t, \theta) = \frac{\partial \log q(t, \theta)}{\partial \theta}$$

the score function based on T, and

$$I^{(T)}(\theta) = E_\theta[l^{(T)}(T, \theta)]^2,$$

the Fisher information in T, i.e. in its distribution. Then it is well known (see, e.g. Rao 1965) that

$$l^{(T)}(t, \theta) = E_\theta[l(X, \theta)|t]$$

for almost all $(\mu)t$, and

$$I^{(T)}(\theta) \leqslant I(\theta), \qquad \text{all } \theta \in \Omega. \tag{18.6}$$

Furthermore, $l(x, \theta) = l^{(S)}(s, \theta)$ for almost all x and $I^{(S)}(\theta) = I(\theta)$ iff $S = S(X)$ is sufficient. Thus

$$i_g(\theta) \leqslant i_{g*}(\theta) \leqslant I(\theta) = I^{(S)}(\theta) = i_l(\theta) = i_{l*}(\theta)$$

when S is sufficient.

An ancillary statistic can also be used for improving the relative performance of the estimating function. Suppose $U = U(X)$ is ancillary for θ in the sense that its distribution does not involve θ. Let

$$k(u, \theta) = E_\theta[g(X, \theta)|u], \tag{18.7}$$

for $u = U(x)$, and define

$$g_*(x, \theta) = g(x, \theta) - k(u, \theta). \tag{18.8}$$

Then it can be shown that

$$i_g(\theta) \leqslant i_{g_*}(\theta) \qquad \text{all } \theta \in \Omega. \tag{18.9}$$

The conditional distribution of X, given $u = U(x)$, is useful in constructing a more efficient estimating function g_* relative to g, when U is an ancillary statistic for θ. Note that for any ancillary statistic U, $l_*(x, \theta) = l(x, \theta)$.

These situations correspond to the forms of 'attainment' of Fisher information for given random variable X by using a particular statistic $T = T(X)$ provided it has desirable properties. Generally, for any arbitrary statistic T,

$$I(\theta) = I^{(T)}(\theta) + E_\theta[I(\theta|T)]; \tag{18.10}$$

the first component is the Fisher information in the marginal distribution of T, while the second component is the average information in the conditional distribution of X, given $t = T(x)$, averaged with respect to distribution of T. If U happens to be ancillary for θ, from (18.10) we have

$$I(\theta) = E_\theta[I(\theta|U)]; \tag{18.11}$$

there is no 'loss of information' in fixing $u = U(x)$ in the sense that marginal distribution of U has no information for θ. On the other hand, the inequality (18.6) for information in the marginal distribution of T is replaced by the equality

$$I(\theta) = I^{(S)}(\theta), \tag{18.12}$$

when S is sufficient; then there is no 'loss of information' in using S alone, rather than X.

Example 18.1   Let $X = (X_1, \ldots, X_n)$ be a random sample from $N(\theta, k\theta^2)$ where $k > 0$ is known. Then $I(\theta) = (2k + 1)n/k\theta^2$. For statistic $T = \bar{X}$ we have $I^{(T)}(\theta) = (2k + n)/k\theta^2$. Then $I(\theta|t) = 2(n - 1)/\theta^2$ for every t; furthermore, for $T_1 = \Sigma(X_i - \bar{X})^2$, $I^{(T_1)}(\theta) = 2(n - 1)/\theta^2$. Thus, neither T nor T_1 is fully informative, and the score function depends on both components of the sufficient statistic $S = (T, T_1)$.

Example 18.2 Suppose $X = (X_1, X_2, X_3)$ has multinomial distribution for known sample size n in four cells so that $X_4 = n - X_1 - X_2 - X_3$, with probabilities $\pi_1 = (1 - \theta)/6$, $\pi_2 = (2 - \theta)/6$. $\pi_3 = (1 + \theta)/6$ and $\pi_4 = (2 + \theta)/6$, for $-1 < \theta < 1$. Then

$$I(\theta) = \frac{n}{3(1 - \theta^2)} + \frac{2n}{3(4 - \theta^2)}.$$

For $T = (X_1, X_2)$, we have

$$I^{(T)}(\theta) = \frac{n}{6(1 - \theta)} + \frac{n}{6(2 - \theta)} + \frac{4n}{6(3 + 2\theta)},$$

$$I(\theta|t) = \frac{n - x_1 - x_2}{(1 + \theta)(2 + \theta)(3 + 2\theta)^2},$$

which depends on t. On the other hand, for ancillary statistics $U_1 = X_1 + X_3$, $U_2 = X_1 + X_4$, we have

$$I(\theta|u_i) = \frac{u_1}{(1 - \theta^2)} + \frac{n - u_1}{(4 - \theta^2)}$$

$$I(\theta|u_2) = \frac{u_2}{(1 - \theta)(2 + \theta)} + \frac{n - u_2}{(2 - \theta)(1 + \theta)}.$$

We note here that $I(\theta) = E_\theta[I(\theta|U_1)] = E_\theta[I(\theta|U_2)]$ and the relation (18.10) for $T = (X_1, X_2)$.

18.3 Estimating functions for vector $\boldsymbol{\theta}$

For the vector case suppose Ω is an open interval in the d-dimensional space, and G is the class of regular unbiased estimating functions $\mathbf{g} = \mathbf{g}(\mathbf{x}, \boldsymbol{\theta})$, $\mathbf{g}$ being d-dimensional. For the generalization of regularity conditions R_G in Section 18.2 we now need, in addition to condition of unbiasedness, differentiability under the integral sign with respect to $\boldsymbol{\theta}$ in $\int g p \, d\mu$. Furthermore, the $d \times d$ matrix

$$\mathbf{G}(\boldsymbol{\theta}) = E_{\boldsymbol{\theta}} \left[\frac{\partial g_i(X, \boldsymbol{\theta})}{\partial \theta_j} \right]_{d \times d} \tag{18.13}$$

is assumed to be nonsingular for all $\boldsymbol{\theta} \in \Omega$. Finally, we assume that

$$\Sigma_{\mathbf{g}}(\boldsymbol{\theta}) \equiv E_{\boldsymbol{\theta}}[\mathbf{g}(X, \boldsymbol{\theta})\mathbf{g}'(X, \boldsymbol{\theta})]_{d \times d} \tag{18.14}$$

is positive definite (p.d.) for all $\boldsymbol{\theta} \in \Omega$.

For the generalization of regularity conditions R on p.d.f. $p(x, \theta)$, we now need in addition to conditions of type (a) and (b) in Section 2, the assumption that the Fisher information matrix

$$I(\theta) \equiv E_\theta[l(X, \theta)l'(X, \theta)] \qquad (18.15)$$

is p.d. for all $\theta \in \Omega$, where l is the score function vector given by

$$l'(x, \theta) = \left[\frac{\partial \log p(x, \theta)}{\partial \theta_j} \right]_{1 \times d}. \qquad (18.16)$$

Define the information matrix of **g** by

$$I_g(\theta) = G'(\theta)\Sigma_g^{-1}(\theta)G(\theta), \qquad (18.17)$$

thus generalizing (18.2). It was then proved by Bhapkar (1972) that

$$I_g(\theta) \leqslant I(\theta) = I_l(\theta), \qquad \text{all } \theta \in \Omega, \qquad (18.18)$$

for all $g \in G$. Here $A \leqslant B$ means that $B - A$ is non-negative definite (n.n.d.). We note that $I(\theta)$ is, in fact, the information matrix of score function l as the estimating function. Thus we have in (18.18) the optimality of the score function with respect to the partial order for information matrices I_g for $g \in G$ as a vector generalization of Godambe's basic result.

If now $S = S(X)$ is sufficient for $\{P_\theta, \theta \in \Omega\}$, and $g^*(s, \theta) = E_\theta g(X, \theta|s)$, then

$$I_g(\theta) \leqslant I_{g^*}(\theta), \qquad \text{all } \theta \in \Omega. \qquad (18.19)$$

Also, if $U = U(X)$ is ancillary for θ and $g_*(x, \theta) = g(x, \theta) - E_\theta[g|u]$, then

$$I_g(\theta) \leqslant I_{g^*}(\theta), \qquad \text{all } \theta \in \Omega. \qquad (18.20)$$

As in the one-dimensional case, the relative improvement of g^* (or g_*), with respect to g, may be interpreted as 'no loss of information' in using the marginal distribution of S (or the conditional distribution given $u = U(x)$), in view of the relation

$$I(\theta) = I^{(T)}(\theta) + E_\theta[I(\theta|T)], \qquad (18.21)$$

for arbitrary statistic $T = T(X)$, and it special forms $I(\theta) = I^{(S)}(\theta)$ for sufficient statistic S, and $I(\theta) = E_\theta[I(\theta|U)]$ for ancillary statistic U.

18.4   Estimating functions in the presence of nuisance parameters

Suppose now that the parameter θ indexing the distribution P_θ of X is of the form $\theta' = (\theta_1', \theta_2')$ where θ_1 is the *parameter of interest* and θ_2 is the *nuisance parameter*. We assume that the parametric space Ω of θ is the Cartesian product $\Omega = \Omega_1 \times \Omega_2$ of spaces Ω_i of θ_i, Ω_i being an open subset of the d_i-dimensional Euclidean space, i = 1, 2.

Assume that the p.d.f. $p(x, \boldsymbol{\theta})$ satisfies the regularity conditions $\boldsymbol{R}$ as generalized in Section 18.3. Now we consider estimating functions $\mathbf{g}_1$ for $\boldsymbol{\theta}_1$, i.e. $\mathbf{g}_1 = \mathbf{g}_1(x, \boldsymbol{\theta}_1)$, satisfying vector generalizations of conditions in Godambe and Thompson (1974). Specifically, G_1 is the class of regular unbiased estimating functions $\mathbf{g}_1$ for $\boldsymbol{\theta}_1$ satisfying conditions:

$\boldsymbol{R}_{G_1}$: (i) $\partial \mathbf{g}_1/\partial \boldsymbol{\theta}_1$ exists for all $\boldsymbol{\theta}_1 \in \Omega_1$ and almost all x;

(ii) $\int g_1 p \, d\mu$ is differentiable with respect to $\boldsymbol{\theta}$ under the integral sign at all $\boldsymbol{\theta} \in \Omega$;

(iii) the $d_1 \times d_1$ matrix

$$G_1(\boldsymbol{\theta}) = E_{\boldsymbol{\theta}}\left[\frac{\partial \mathbf{g}_1(X, \boldsymbol{\theta}_1)}{\partial \boldsymbol{\theta}_1}\right] \tag{18.22}$$

is nonsingular for all $\boldsymbol{\theta} \in \Omega$.

(iv) The $d_1 \times d_1$ matrix $\Sigma_{\mathbf{g}_1}(\boldsymbol{\theta}) \equiv E_{\boldsymbol{\theta}}[\mathbf{g}_1(X, \boldsymbol{\theta}_1)\mathbf{g}_1'(X, \boldsymbol{\theta}_1)]$ is p.d. for all $\boldsymbol{\theta} \in \Omega$.

The information in the estimating function $\mathbf{g}_1$ for $\boldsymbol{\theta}_1$ is now defined as the $d_1 \times d_1$ matrix

$$\mathbf{I}_{\mathbf{g}_1}(\boldsymbol{\theta}_1; \boldsymbol{\theta}) = G_1'(\boldsymbol{\theta})\Sigma_{g_1}^{-1}(\boldsymbol{\theta})G_1(\boldsymbol{\theta}). \tag{18.23}$$

The expression (18.23) formally resembles (18.17); however, the latter matrix is $d \times d$, corresponding to estimating function $\mathbf{g}$ for the whole parameter $\boldsymbol{\theta}$.

Let $U = U(X)$ be a statistic which can be used for conditional inference concerning $\boldsymbol{\theta}_1$ after eliminating the nuisance parameter $\boldsymbol{\theta}_2$. Specifically, assume that the statistic (S, U) is jointly sufficient for the family $\{P_{\boldsymbol{\theta}}, \boldsymbol{\theta} \in \Omega\}$ and, furthermore, U satisfies condition C:

C: The conditional distribution of S, given $u = U(x)$, depends on $\boldsymbol{\theta}$ only through $\boldsymbol{\theta}_1$, for almost all u.

Denote by $h(s; \theta_1|u)$ the conditional p.d.f. of $S = S(X)$, given u with respect to some suitable measure η_u.

Consider the first case $d_1 = 1$ of real-valued θ_1. It was shown by Godambe (1976) that the information (18.23) is maximized by the *conditional score function*

$$l_c(x, \theta_1) \equiv \frac{\partial \log h(s, \theta_1|u)}{\partial \theta_1},$$

provided the statistic $U = U(X)$ satisfies, in addition to the basic *conditioning requirement C*, one of the following two conditions:

1. The conditional distribution (of S), given $u = U(x)$, meets Assumptions 3.1–3 in Godambe (1976);
2. The family of probability distributions $\{P_{\boldsymbol{\theta}}^{(U)}, \boldsymbol{\theta}_2 \in \Omega_2\}$, for fixed θ_1, of statistic U is complete for every $\theta_1 \in \Omega_1$.

In the light of Godambe's results and some previous work by Andersen (1970) and Liang (1983), we consider two definitions of *partial ancillarity* of statistic $U = U(X)$ for parameter θ_1, which is a d_1-dimensional vector, with θ_2 as nuisance parameter; the relationship of these definitions to the optimality of the conditional score function

$$l'_c(x, \theta_1) = \frac{\partial \log h(s, \theta_1 | u)}{\partial \theta_1} \tag{18.24}$$

would be pointed out.

Definition 18.1   A statistic $U = U(X)$ is said to be partially ancillary for θ_1 in the complete sense if (i) U satisfies requirement C, and (ii) the family $\{P_\theta^{(U)}, \theta_2 \in \Omega_2\}$ of distributions of U for fixed θ_1 is complete for every $\theta_1 \in \Omega_1$.

Definition 18.2   A statistic $U = U(X)$ is partially ancillary for θ_1 at least in the weak sense if (i) U meets requirement C, and (ii) the marginal distribution of U depends on θ only through a parametric function $\delta = \delta(\theta)$ such that (θ'_1, δ') is a one-to-one function of θ.

If $\delta = \theta_2$, U is said to be *strongly p-ancillary* for θ_1 and if $\delta \neq \theta_2$, U is said to be a *weak p-ancillary* statistic for θ_1. We further assume that $\delta(\theta)$ is differentiable.

It may be noted now that the statistic U satisfying Godambe's Assumptions 3.1–3 in the (1976) paper meet the more general Definition 18.2 of weak p-ancillarity for θ_1, if $\delta(\theta)$ is differentiable.

It was shown by Bhapkar (1989a, Theorem 4.3) that if $U = U(X)$ is p-ancillary for θ_1, either in the complete sense of Definition 18.1 or at least in the weak sense of Definition 18.2, then the conditional scores function l_c, given by (18.24), is optimal for θ_1 in the sense that

$$\mathbf{I}_{g_1}(\theta_1; \theta) \leqslant \mathbf{I}_{l_c}(\theta_1; \theta), \qquad \text{all } \theta, \tag{18.25}$$

for all $g_1 \in G_1$.

Suppose that $g_{1.}(x, \theta_1) = g_1(x, \theta_1) - k(u, \theta_1)$, as in (18.8), with $k(u, \theta_1)$ defined as in (18.7). For any statistic U satisfying condition C, $g_{1.} \in G_1$. It is then easy to see that

$$\mathbf{I}_{g_1}(\theta_1; \theta) \leqslant \mathbf{I}_{g_{1.}}(\theta_1; \theta) \qquad \text{for all } \theta$$

as in (18.9). We have then $l_c = l_{c_*}$ for every such conditional score function, given $u = U(x)$. Among all such statistics U' satisfying condition C, the information for the conditional score function is *maximized* for a particular U if it happens to be p-ancillary for θ_1 in the complete sense or at least in the weak sense, in view of the result (18.25).

Example 18.3   Let $X = (X_1, \ldots, X_n)$ be a random sample from $N(\mu, \sigma^2)$. For $\theta_1 = \sigma^2$, $U = \bar{X}$ is p-ancillary in the *complete* sense; however, U is not p-ancillary in the weak sense. The conditional score function given u is the optimal estimating function for σ^2.

Example 18.4   Let X_1, X_2 be independent binomial variables with known number n_1, n_2 of trials and probabilities π_1, π_2, respectively. If we take $\theta_1 = \ln[\pi_1(1 - \pi_2)/\pi_2(1 - \pi_1)]$ and $\theta_2 = \ln \pi_2/(1 - \pi_2)$, then $U = X_1 + X_2$ is p-ancillary in the complete sense for estimating θ_1, the log-odds-ratio. The conditional score functions given u is the optimal estimating function for θ_1.

Example 18.5   Suppose $X_1, \ldots, X_n$ are independent Poisson variables with means μ_i, $i = 1, \ldots, n$, respectively, satisfying the log-linear model

$$\ln \mu_i = \theta_2 + \theta_1 y_i,$$

where $\{y_i\}$ are known constants, not all equal. Then $U = \Sigma X_i$ is p-ancillary for θ_1 in the weak sense, and the conditional score function given u is the optimal estimating function for θ_1.

One would also like to consider the dual problem as to when the marginal distribution of statistic $S = S(X)$ is adequate. More specifically, when is the *marginal score function*

$$l'_m(x, \boldsymbol{\theta}_1) \equiv \frac{\partial \log f(s; \boldsymbol{\theta}_1)}{\partial \theta_1}, \tag{18.26}$$

optimal for $\boldsymbol{\theta}_1$? Obviously, for l_m to be an estimating function for θ_1 alone, the marginalization requirement M has to be met, viz.

M The distribution of statistic $S = S(X)$ depends on $\boldsymbol{\theta}$ only through $\boldsymbol{\theta}_1$.

Taking into account Lloyd's (1987) work, we consider the following definition.

Definition 18.3   The statistic $S = S(X)$ is said to be *partially sufficient* for $\boldsymbol{\theta}_1$ in the *complete* sense if (i) S satisfies requirement M, and (ii) given $s = S(x)$, the family $\{P_{\boldsymbol{\theta}}^{(U|s)}, \boldsymbol{\theta}_2 \in \Omega_2\}$ of conditional distributions of U for fixed $\boldsymbol{\theta}_1$ is complete for almost all s and for every $\boldsymbol{\theta}_1 \in \Omega_1$.

It was shown by Lloyd (1987) that if $S = S(X)$ is p-sufficient for $\boldsymbol{\theta}_1$ in the complete sense, then the marginal score function l_m, given by (18.26), is optimal for $\boldsymbol{\theta}_1$, i.e.

$$\mathbf{I}_{\mathbf{g}_1}(\boldsymbol{\theta}_1; \boldsymbol{\theta}) \leq \mathbf{I}_{l_m}(\boldsymbol{\theta}_1; \boldsymbol{\theta}), \qquad \boldsymbol{\theta} \in \Omega, \tag{18.27}$$

for all $\mathbf{g}_1 \in G_1$, for the case $d_1 = 1$.

Similarly, it has been shown by Bhapkar (1989b) that the marginal score function is optimal for θ_1, in the sense of (18.27), if $S = S(X)$ is p-sufficient for θ_1 at least in the weak sense according to the following definition.

Definition 18.4   The statistic $S = S(X)$ is p-sufficient for θ_1 at least in the weak sense if (i) S satisfies condition M, and (ii) the conditional distribution of U, given $s = S(x)$, depends on θ only through $\delta(\theta)$, where (θ'_1, δ') is a one-to-one function of θ, for almost all s.

In fact, as shown by Bhapkar (1989c), the optimality of I_m holds under a somewhat broader condition as well, where the parameter δ in Definition 18.4, for given s, depends on s, as in Example 18.6 below.

Example 18.3 (continued)   For $\theta_1 = \sigma^2$, $S = \Sigma(X_i - \bar{X})^2$ is p-sufficient in the complete sense (with $U = \bar{X}$). The marginal score function in terms of S is the optimal estimating function for θ_1. Here it coincides with the conditional score function given u.

Example 18.6   Let $X = (X_1, \ldots, X_n)$ be a random sample of i.i.d. variables $X_i = (Y_i, Z_i)$, where Y_i, Z_i are independent exponential random variables with means θ_2 and $\theta_1\theta_2$, respectively. If we let $Y = \Sigma Y_i$, $Z = \Sigma Z_i$, $S = Z/Y$ and $U = Y$, then (S, U) is minimal sufficient for θ. Furthermore, the marginal distribution of S has p.d.f.

$$f(s, \theta_1) = \frac{s^{n-1}\theta_1^n}{\beta(n, n)(\theta_1 + s)^{2n}}, \qquad 0 < s < \infty,$$

and the conditional distribution of U, given s, has p.d.f.

$$m(u, \theta|s) = \frac{u^{2n-1}e^{-u/\delta(s)}}{\Gamma(2n)[\delta(s)]^{2n}}, \qquad 0 < u < \infty,$$

where $\delta(s) = \theta_1\theta_2/(\theta_1 + s)$. Thus S is p-sufficient in the complete sense and the marginal score function based on S is optimal for θ_1.

18.5   Fisher information generalizations

We have noted the relations (18.3) or (18.18) between information $i_g(\theta)$ or $\mathbf{I}_g(\theta)$ on one hand, and Fisher information $I(\theta)$ or $\mathbf{I}(\theta)$ on the other hand in the case of estimating function g for θ (or $\mathbf{g}$ for θ). One could then wonder what modification is needed in $I(\theta)$ (or $\mathbf{I}(\theta)$) for similar relations to hold in the case of estimating functions $\mathbf{g}_1$ for $\boldsymbol{\theta}_1$ in the presence of nuisance parameters $\boldsymbol{\theta}_2$.

For the case $d_1 = 1$ of real-valued θ_1, Liang (1983) investigated the generalization used earlier by Efron (1977). The Fisher information concerning θ_1, when the parameter is $\boldsymbol{\theta}$, in the distribution of X is defined by

$$I(\theta_1; \boldsymbol{\theta}) = \inf_{\mathbf{n}} E_{\boldsymbol{\theta}} \left[\frac{\partial \log p(X, \boldsymbol{\theta})}{\partial \theta_1} - \frac{\partial \log p(X, \boldsymbol{\theta})}{\partial \theta_2} \mathbf{n} \right]^2$$

where $\mathbf{n}$ is any $d_2 \times 1$ vector possibly depending on $\boldsymbol{\theta}$.

Godambe (1984) considered a further generalization. Let $\mathscr{U}$ be the class of functions $u = u(X, \boldsymbol{\theta})$ with finite variance which are uncorrelated with all estimating functions $g_1 = g_1(X, \theta_1)$ for θ_1, with $g_1 \in G_1$; thus

$$E_{\boldsymbol{\theta}} u(X, \boldsymbol{\theta}) g_1(X, \theta_1) = 0, \qquad \text{all } \boldsymbol{\theta} \in \Omega.$$

Notice that $(\partial \log p / \partial \boldsymbol{\theta}_2)\mathbf{n}$ can serve as $u \in \mathscr{U}$ for every $\mathbf{n}$. Define

$$I_G(\theta_1; \boldsymbol{\theta}) = \inf_{u \in \mathscr{U}} E_{\boldsymbol{\theta}} \left[\frac{\partial \log p(X, \boldsymbol{\theta})}{\partial \theta_1} - u(X, \boldsymbol{\theta}) \right]^2.$$

It follows that $I_G(\theta_1; \boldsymbol{\theta}) \leqslant I(\theta_1; \boldsymbol{\theta})$ for all $\boldsymbol{\theta}$.

It was shown by Godambe (1984) that if there exists a statistic $U = U(X)$, which satisfies requirement C in Section 18.4, and further satisfies either condition 1 or 2 in Section 18.4, then

$$i_{g_1}(\theta_1, \boldsymbol{\theta}) \leqslant I_G(\theta_1; \boldsymbol{\theta}) = i_{l_c}(\theta_1; \boldsymbol{\theta}), \tag{18.28}$$

where $i_{g_1}(\theta_1; \boldsymbol{\theta})$ is the scalar form of (18.22) for estimating function g_1 of θ_1 alone, and l_c is the conditional score function given $u = U(x)$. Obviously, (18.28) is a generalization of inequality (18.3). More generally, it has been shown by Bhapkar (1989a) that for estimating functions $\mathbf{g}_1$ of $\boldsymbol{\theta}_1$ alone we have

$$\mathbf{I}_{\mathbf{g}_1}(\boldsymbol{\theta}_1; \boldsymbol{\theta}) \leqslant \mathbf{I}_G(\boldsymbol{\theta}_1; \boldsymbol{\theta}) = \mathbf{I}_{l_c}(\boldsymbol{\theta}_1, \boldsymbol{\theta}) \tag{18.29}$$

for all $\boldsymbol{\theta} \in \Omega$ and $\mathbf{g}_1 \in \mathbf{G}_1$, when there exists a statistic $\mathbf{U} = \mathbf{U}(\mathbf{X})$ which is p-ancillary for $\boldsymbol{\theta}_1$ either in the complete sense (Definition 18.1) or at least in the weak sense (Definition 18.2). Here

$$\mathbf{I}_G(\boldsymbol{\theta}_1; \boldsymbol{\theta}) = \underset{\mathbf{u}}{\text{Min}} \ E_{\boldsymbol{\theta}} \left[\frac{\partial \log \mathbf{p}(\mathbf{X}, \boldsymbol{\theta})}{\partial \boldsymbol{\theta}_1} - \mathbf{u}' \right]' \left[\frac{\partial \log \mathbf{p}(\mathbf{X}, \boldsymbol{\theta})}{\partial \boldsymbol{\theta}_1} - \mathbf{u}' \right].$$

In the above expression, Min denotes the minimal matrix, say $\mathbf{M}^*$, in the class $\mathscr{M}$ of all n.n.d. matrices $\mathbf{M} \in \mathscr{M}$ such that $\mathbf{M} \geqslant \mathbf{M}^*$, i.e. $\mathbf{M} - \mathbf{M}^*$ is n.n.d; also $\mathbf{u}$ denotes a $\mathbf{d}_1$-dimensional vector such that every element of $\mathbf{u}$ is in $\mathscr{U}$. Although such a minimal matrix might not necessarily exist for every distribution $\mathbf{P}_{\boldsymbol{\theta}}$ of $\mathbf{X}$, it does exist when a p-ancillary statistic $\mathbf{U}$ is available for $\boldsymbol{\theta}_1$. Thus (18.29) is seen as the generalization of (18.18) for the case where there are nuisance parameters $\boldsymbol{\theta}_2$, and a p-ancillary statistic $\mathbf{U} = \mathbf{U}(\mathbf{X})$ exists for $\boldsymbol{\theta}_1$ either in the complete sense or at least in the weak sense.

A similar generalization is available for (18.18) in another direction; here suppose a p-sufficient statistic $\mathbf{S} = \mathbf{S}(\mathbf{X})$ is available for $\boldsymbol{\theta}_1$, satisfying Definition 18.4. Then it has been shown by Bhapkar (1989b) that

$$\mathbf{I}_{\mathbf{g}_1}(\boldsymbol{\theta}_1; \boldsymbol{\theta}) \leqslant \mathbf{I}_G(\boldsymbol{\theta}_1; \boldsymbol{\theta}) = \mathbf{I}_{l_m}(\boldsymbol{\theta}_1; \boldsymbol{\theta}), \tag{18.30}$$

where $\mathbf{l}_m$ is the marginal score function, defined by (18.26), using such a p-sufficient statistic S. The same relation (18.30) holds if S is p-sufficient for $\boldsymbol{\theta}_1$ in the complete sense (i.e., S satisfies Definition 18.3) provided the distribution of X is such that $\mathbf{g}_1^*(s, \boldsymbol{\theta}) \equiv E_{\boldsymbol{\theta}} \mathbf{g}_1(X, \boldsymbol{\theta} | s)$ depends on $\boldsymbol{\theta}$ only through $\boldsymbol{\theta}_1$ for every $\mathbf{g}_1 \in G_1$ (Bhapkar 1989c). In particular, the property (18.30) holds when S satisfies the sufficiency property as in Godambe (1980) for $\boldsymbol{\theta}_1$ in the conditional distribution of X, given $u = U(x)$, where U is p-ancillary for $\boldsymbol{\theta}_1$ in the complete sense.

Suppose $T = T(X)$ is any statistic satisfying the conditioning prerequisite C for $\boldsymbol{\theta}_1$. Thus T is one candidate for choice as a conditioning vehicle for inference concerning $\boldsymbol{\theta}_1$ eliminating θ_2. It has been shown (Bhapkar 1989a) that we have under the regularity assumptions

$$\mathbf{I}(\boldsymbol{\theta}_1; \boldsymbol{\theta}) = \mathbf{I}^{(T)}(\boldsymbol{\theta}_1; \boldsymbol{\theta}) + \mathbf{E}_{\boldsymbol{\theta}} \mathbf{I}(\boldsymbol{\theta}_1 | T). \tag{18.31}$$

Note here that the second term on the right-hand side of (18.31) is the expectation of the usual Fisher information for $\boldsymbol{\theta}_1$ in the conditional distribution given t and, thus, the relation (18.31) is a generalization of (18.21), when T satisfies C. However, if T further satisfies Definition 18.2 of p-ancillarity for $\boldsymbol{\theta}_1$ at least in the weak sense, then the relation (18.31) simplifies to

$$\mathbf{I}(\boldsymbol{\theta}_1; \boldsymbol{\theta}) = \mathbf{E}_{\boldsymbol{\theta}} \mathbf{I}(\boldsymbol{\theta}_1 | T), \tag{18.32}$$

which is a generalization of (18.11), when $\mathbf{T}$ is p-ancillary for $\boldsymbol{\theta}_1$ at least in the weak sense. As noted earlier, the information bound $\mathbf{I}_G(\boldsymbol{\theta}_1; \boldsymbol{\theta})$ for $\mathbf{I}_{\mathbf{g}_1}(\boldsymbol{\theta}_1; \boldsymbol{\theta})$ is attained by the conditional score function $\mathbf{l}_c$ in view of the relation

$$\mathbf{I}(\boldsymbol{\theta}_1; \boldsymbol{\theta}) = \mathbf{I}_G(\boldsymbol{\theta}_1; \boldsymbol{\theta}) = \mathbf{I}_{\mathbf{l}_c}(\boldsymbol{\theta}_1; \boldsymbol{\theta}) \tag{18.33}$$

that now follows from (18.29) and (18.32).

A relation similar to (18.31) also holds, viz.

$$\mathbf{I}_G(\boldsymbol{\theta}_1; \boldsymbol{\theta}) = \mathbf{I}_G^{(T)}(\boldsymbol{\theta}_1; \boldsymbol{\theta}) + \dot{\mathbf{E}}_{\boldsymbol{\theta}} \mathbf{I}(\boldsymbol{\theta}_1 | \mathbf{T}), \tag{18.34}$$

for $\mathbf{I}_G(\boldsymbol{\theta}_1; \boldsymbol{\theta})$ when T satisfies the prerequisite C for conditioning (Bhapkar 1989a). This is also a generalization of (18.21). If now T satisfies further conditions for p-ancillarity for $\boldsymbol{\theta}_1$, either in the complete sense or at least in the weak sense, then (18.34) simplifies to

$$\mathbf{I}_G(\boldsymbol{\theta}_1; \boldsymbol{\theta}) = \mathbf{E}_{\boldsymbol{\theta}} \mathbf{I}(\boldsymbol{\theta}_1 | T), \tag{18.35}$$

as an analogue of (18.32) and thus, a generalization of (18.11). In view of (18.29), we then have

$$I_G(\theta_1; \theta) = E_\theta(I(\theta|T)) = I_{l_c}(\theta_1; \theta).$$

Thus, optimality of conditional score function, as an estimating function for θ_1, with θ_2 as nuisance parameters, is seen to be closely related to Fisher information generalizations $I(\theta_1; \theta)$ and $I_G(\theta_1; \theta)$, especially the latter.

A similar relationship is seen between the optimality of marginal score function as an estimating function for θ_1 on one side and $I(\theta_1; \theta)$, $I_G(\theta_1; \theta)$ on the other side, especially the latter, when $S = S(X)$ happens to be p-sufficient for θ_1.

If $S = S(X)$ satisfies the marginalization prerequisite M for inference concerning θ_1, eliminating θ_2, then the analogues of (18.31) and (18.34) are

$$I(\theta_1; \theta) \geqslant I^{(S)}(\theta_1) + E_\theta I(\theta_1; \theta|S)$$

$$I_G(\theta_1; \theta) \geqslant I^{(S)}(\theta_1) + E_\theta I_G(\theta_1; \theta|S) \tag{18.36}$$

as shown by Bhapkar (1989c).

However, if S is p-sufficient for θ_1 at least in the weak sense, we have

$$I(\theta_1; \theta) = I^{(S)}(\theta_1) = I_{l_m}(\theta_1; \theta); \tag{18.37}$$

furthermore,

$$I_G(\theta_1; \theta) = I^{(S)}(\theta_1) = I_{l_m}(\theta_1; \theta), \tag{18.38}$$

where S is p-sufficient for θ_1 at least in the weak sense. Here l_m is the marginal score function as the estimating function for θ_1. The property (18.38) holds also if S is p-sufficient in the complete sense with the same additional qualification as was needed for (18.30) to hold. Thus, the relations (18.37), (18.38) establish the connection between optimality of l_m and Fisher information generalizations $I(\theta_1, \theta)$ and $I_G(\theta_1; \theta)$, especially the latter. As a matter of fact, as indicated in the remark after Definition 18.4, this connection with I_G holds under a somewhat broader condition than just p-sufficiency in the weak sense.

The equations of type (18.31) and (18.34) in fact provide some idea about 'loss of information' involved in conditioning on $T(x) = t$, when T meets the prerequisite C but does not satisfy any p-ancillarity property for θ_1. Similarly, inequalities (18.36) give the extent of possible loss of information when S is used, instead of X, for inference concerning θ_1, but it does not satisfy any p-sufficiency property for θ_1.

18.6   Acknowledgement

This work was supported in part by the NSF Grant MCS-8806233.

References

Andersen, E. B. (1970). Asymptotic properties of conditional maximum likelihood estimators. *J. Roy. Stat. Soc.*, **B32**, 283–301.

Bhapkar, V. P. (1972). On a measure of efficiency of an estimating equation. *Sankhya*, A **34**, 467–72.

Bhapkar, V. P. (1989a). Conditioning on ancillary statistics and loss of information in the presence of nuisance parameters. *J. Stat. Plan. Inf.*, **21**, 139–60.

Bhapkar, V. P. (1989b). On optimality of marginal estimation equations. Department of Statistics, Technical Report No. 285, University of Kentucky.

Bhapkar, V. P. (1989c). Loss of information in the presence of nuisance parameters and partial sufficiency. Department of Statistics, Technical Report No. 286, University of Kentucky. *J. Stat. Plan. Inf.* (In press.)

Efron, B. (1977). The efficiency of Cox's likelihood function for censored data. *J. Amer. Stat. Assoc.*, **72**, 557–65.

Godambe, V. P. (1960). An optimum property of a regular maximum likelihood estimation. *Ann. Math. Stat.*, **31**, 1208–12.

Godambe, V. P. (1976). Conditional likelihood and unconditional optimum estimating equations. *Biometrika*, **63**, 277–84.

Godambe, V. P. (1980). On sufficiency and ancillarity in the presence of a nuisance parameter. *Biometrika*, **67**, 155–62.

Godambe, V. P. (1984). On ancillarity and Fisher information in the presence of a nuisance parameter. *Biometrika*, **71**, 626–29.

Godambe, V. P. and Thompson, M. E. (1974). Estimating equations in the presence of a nuisance parameter. *Ann. Stat.*, **2**, 568–71.

Liang, K. (1983). On information and ancillarity in the presence of a nuisance parameter. *Biometrika*, **70**, 607–12.

Lloyd, C. J. (1987). Optimality of marginal likelihood estimating equations. *Comm. Stat., Theory and Meth.*, **16**, 1733–41.

Rao, C. R. (1965). *Linear statistical inference*, Wiley, New York.

19
Inferential estimation, likelihood, and maximum likelihood linear estimation functions

S. R. Chamberlin and D. A. Sprott

ABSTRACT

The reduction of data $\mathbf{X}$ to a function $u(\mathbf{X}; \theta)$ is examined where u has a distribution $G(u; \theta)$ depending perhaps on an ancillary statistic $\mathbf{A} = \mathbf{A}(\mathbf{X})$ and perhaps on θ. If u is linear in θ it is possible to associate with u a well-defined likelihood function of θ. If this approximates closely the likelihood function of θ based on $\mathbf{X}$, we say that u has high fidelity. If 'likelihood-confidence sets' derived from $G(u; \theta)$ have accurately fixed coverage frequency properties, we say that u has high accuracy. Fidelity and accuracy do not necessarily go together. This approach can yield simple solutions to inferential estimation problems in cases where the distribution of an estimate which has high fidelity is not readily available.

19.1 Introduction

Attention in this paper is focused on estimation problems arising from scientific, that is, *repeatable*, experiments involving an unknown scalar parameter θ. The observations x_i are thus regarded as coming from a hypothetical *infinite* population of possible observations. The problem of estimation is to estimate some one-to-one function $\mu(\theta)$ based on the observations $\mathbf{X} = (x_1, \ldots, x_n)$ having a density function $f(\mathbf{X}; \theta)$. How this is done depends on what is meant by estimation.

The purpose of 'inferential' estimation, as contrasted with 'point' estimation, is to make estimation statements. An estimation statement is a quantitative statement of uncertainty about μ using all the information supplied by $\mathbf{X}$. Such an estimation statement typically takes the form of a nested family of confidence sets that converges to the maximum likelihood (ML) estimate (or estimates, if the likelihood is multimodel) $\hat{\mu}$, any specified interval including the most plausible values of μ at the given level of confidence. These may be called a complete family of likelihood-confidence sets. They are efficient in the sense of reproducing the observed likelihood function based on **X**. This general notion of efficiency as likelihood reproduction is termed fidelity, and is discussed in Section 19.4 and exemplified in Sections 19.5 and 19.6.

This form of solution to the problem of inferential estimation rests in the reduction of the data to an estimating function, $\mathbf{X} \to u(\mathbf{X}; \theta)$, where u has a distribution $G(u; \theta)$ which may depend on θ, but on no other unknown parameter. Also, G can be a conditional distribution depending on an ancillary statistic $\mathbf{A} = \mathbf{A}(\mathbf{X})$, but in what follows we suppress this dependence in the notation. The purpose of this paper is to examine this form of data reduction and related estimation procedures. These will be assessed in terms of their fidelity and accuracy. This requires the association of a well-defined likelihood function of θ with $G(u; \theta)$. This in turn leads to *linear* estimating functions and ML estimation.

19.2   The synthesis of efficiency and likelihood: fidelity

The likelihood function based upon **X** supplies all the material for estimating θ (Fisher 1973, p. 165). Suitable forms of inference about θ can therefore only be known after the properties of the observed likelihood function $L(\theta; \mathbf{X_o})$ are known, where $\mathbf{X_o}$ is the observed value of **X**. The family of possible likelihood functions $L(\theta; \mathbf{X})$ plays a role by utilizing the sample space, allowing a repeated sampling interpretation of the observed likelihood function in terms, for example, of likelihood-confidence sets.

Fisher (1925) derived a measure of efficiency, based on local properties of the likelihood function about its maximum, to assess the loss of information due to the reduction of data to an estimate, $\mathbf{X} \to T(\mathbf{X})$. More generally, the *fidelity* of the distribution of T is the extent, however measured, to which the likelihood $L(\theta; T)$, deduced from the distribution of T, reproduces $L(\theta; \mathbf{X})$. The term fidelity is used to distinguish measures of efficiency based globally on likelihood from those based only locally on likelihood, and more importantly, from others based on variance. See, for example, Rao (1973, pp. 346–52), who examines and compares various definitions of efficiency. As illustrated in Section 19.6, a global evaluation of fidelity can give an assessment of data reduction different from Fisher's measure of efficiency.

If the statistic $\mathbf{T}(\mathbf{X})$ is sufficient for θ, then by definition $L(\theta; \mathbf{T}) = L(\theta; \mathbf{X})$ and the distribution of $\mathbf{T}(\mathbf{X})$ has perfect fidelity. Usually, a reduction of this form cannot be found. A more general form of data reduction is $\mathbf{X} \to u(\mathbf{X}; \theta)$, where u is a one-dimensional function. If u has a distribution $G(u)$ independent of θ, then u is a pivotal quantity (Fisher 1945). Barnard (1977) emphasized that a pivotal quantity should be 'fully efficient'. This means that its distribution should have perfect fidelity, which is termed pivotal sufficiency. The Gauss linear model with an arbitrary error distribution provides a wide set of classical parametric models for which the reduction to a linear pivotal quantity, together with its distribution, conditional on a maximal ancillary **A**, provides the solution to the problem of inferential estimation, Barnard (1977), Chamberlin and Sprott (1989).

ML estimation, as applied by Sprott and Viveros (1984), provides a technique to extend this domain of application of u by its use of functions $u(\mathbf{X}; \theta)$ that are approximately pivotal, that is, have a distribution $G(u; \theta)$ which depends weakly on θ. From the examples dealt with by Sprott and Viveros (1984), Viveros and Sprott (1986), and Viveros (1985), the class of functions $u(\mathbf{X}; \phi)$ that emerges operationally is

$$u_\phi = u(\mathbf{X}; \phi) = [\phi(\hat{\theta}) - \phi(\theta)]I(\mathbf{X}; \hat{\phi})^{1/2}$$
$$= (\hat{\phi} - \phi)I(\mathbf{X}; \hat{\theta})^{1/2}|d\hat{\theta}/d\hat{\phi}|, \qquad (19.1)$$

where ϕ is any one-to-one function, and $I(\mathbf{X}; \hat{\theta})$ is the observed Fisher information under the parametrization θ,

$$I(\mathbf{X}; \theta) = -\partial^2 \log f(\mathbf{X}; \theta)/\partial\theta^2, \qquad (19.2)$$

calculated at the ML estimate $\theta = \hat{\theta}$. The functions (19.1) will be called ML linear estimating functions.

To assess the fidelity of the distribution $G(u; \theta)$ of a general function $u(\mathbf{X}; \theta)$ requires the unambiguous association of a likelihood function of θ with $G(u; \theta)$. There are difficulties in doing this. When u is a pivotal this leads to the identification of a *linear* pivotal. A well-defined likelihood function can be associated with the distribution $G(u)$ of the pivotal linear in $\phi = \phi(\theta)$, $u(\mathbf{X}; \theta) = (T_1 - \phi)/T_2$, where $T_1 = T_1(\mathbf{X})$ and $T_2 = T_2(\mathbf{X}) > 0$ are functionally independent statistics. This likelihood function is proportional to the density $g(u)$ expressed as a function of θ. The details are given in Sprott (1990).

We now proceed to show that if attention is further restricted to the ML linear estimating functions (19.1), a well-defined likelihood function of θ can be associated with the distribution $G(u; \theta)$ even when G depends on θ.

19.3 Likelihoods and ML linear estimating functions

Let $\mathbf{X} = (X_1, \ldots, X_n)$ be a vector of $n \geqslant 2$ observations with probability density function $f(\mathbf{X}; \theta)$, where θ is a scalar parameter and $\hat{\theta}$ and $I = I(\mathbf{X}; \hat{\theta})$ are functionally independent statistics. That is, $I(\mathbf{X}; \hat{\theta})$ cannot be expressed as $I(\hat{\theta})$, a function of $\hat{\theta}$ alone. We restrict attention to the reduction $\mathbf{X} \to (\hat{\theta}, I)$. Chamberlin (1989, Chapter 6) discusses the more general mapping $\mathbf{X} \leftrightarrow (\hat{\theta}, I, \mathbf{A})$, where $\mathbf{A}$ is chosen to make the transformation one-to-one, and I and/or $\mathbf{A}$ are chosen to be ancillary when this is feasible.

The pair of statistics $(\hat{\theta}, I)$ have a joint density $f(\hat{\theta}, I; \theta)$. The likelihood function based on $(\hat{\theta}, I)$ is proportonal to $f_0(\hat{\theta}, I; \theta)$, where the suffix o means that the statistics within the following brackets are to be replaced by their observed values. The one-to-one transformation,

$$(\hat{\theta}, I) \leftrightarrow ((\hat{\theta} - \theta)I^{1/2}, I) = (u(\mathbf{X}; \theta), I)$$

has Jacobian $I^{-1/2}$, so that if we write $\mathbf{Q}(\mathbf{X}; \theta) = (u(X; \theta), I)$, the density of $\mathbf{Q}$ is $g(\mathbf{Q}; \theta) = I^{-1/2}f(\hat{\theta}, I; \theta)$. Since I does not involve θ, $g_0(\mathbf{Q}; \theta)$ is proportional to $f_0(\hat{\theta}, I; \theta)$. The likelihood function of θ based on the distribution of $\mathbf{Q}$ is thereby defined to be proportional as a function of θ to $g_0(\mathbf{Q}; \theta)$ with $\mathbf{Q}$ replaced by $((\hat{\theta} - \theta)I^{1/2}, I)$.

The density of $\mathbf{Q}$ can be factored in two ways to yield

$$g(\mathbf{Q}; \theta) = g(u; \theta|I)g(I; \theta). \tag{19.3a}$$

$$= g(u; \theta)g(I; \theta|u), \tag{19.3b}$$

In (19.3a) $g_0(I; \theta)$ is proportional to the likelihood of θ based upon the density $g(I; \theta)$. Define the likelihood function of θ associated with each of the densities $g(u; \theta|I)$, $g(u; \theta)$, and $g(I; \theta|u)$, to be proportional, as a function of θ, to each of their corresponding g_0, with u replaced by $(\hat{\theta} - \theta)I^{1/2}$.

This definition implies that each of the factorizations (19.3a) and (19.3b) of the density of $\mathbf{Q}$ corresponds to a particular factoring of the likelihood function of θ based upon $(\hat{\theta}, I)$. That is, the pairs of likelihoods of θ, obtained from the densities on the right-hand side of (19.3a) and (19.3b) respectively, multiply as they should to give the likelihood based on $(\hat{\theta}, I)$.

The likelihood of θ based on $g(u; \theta|I)$ is proportional to the likelihood based on the distribution of the statistic $\hat{\theta}|I$ and is thus well defined. Since I is a statistic and $g(I; \theta|u) = g(I; \theta|v)$, where $v = v(u, \theta)$ is a one-to-one function of u for each θ, the likelihood function based upon $g(I; \theta|u)$ is also well defined.

The marginal distribution $g(u; \theta)$ of u is not equivalent to the distribution of any estimate $\tilde{\theta}$. This is indeed what gives it its unique role in inferential estimation, as illustrated in Example 19.2. It is shown in the Appendix that the likelihood function based on $g(u; \theta)$ is also well defined. That is, any other member v of (19.1) that, for each θ, is a one-to-one function $v = v(u, \theta)$ of u produces the same likelihood.

19.4   Assessing the properties of an estimation procedure

If $u(\mathbf{X}; \theta)$ has generic (conditional) density function $g(u; \theta)$ specified by (19.3a) or (19.3b), then a likelihood function of θ can be associated with this particular distribution of u. Its fidelity can be assessed graphically by comparing the relevant likelihood obtained from (19.3) with $L(\theta; \mathbf{X})$.

In practice, the exact distribution of u, and hence, $g(u; \theta)$ may be complicated, as in Example 19.2, so that approximations must be made. Suppose the approximation to $g(u; \theta)$ employed is $h(u; \theta)$.

19.4.1 FIDELITY OF AN ESTIMATION PROCEDURE

If $h(u; \theta)$ is *assumed* to be the true density function, a likelihood function of θ can be deduced from h in the manner of Section 19.3. The fidelity of this likelihood to the likelihood function of θ given $\mathbf{X}$ can also be assessed graphically. This is the fidelity of the *estimation procedure* based on $h(u; \theta)$. It is not the fidelity of the true distribution of u, since $h(u; \theta)$ may be different from $g(u; \theta)$.

19.4.2 ACCURACY OF AN ESTIMATION PROCEDURE

For the estimation procedure based upon $h(u; \theta)$ a second criterion is required. This is the extent, however assessed, to which $h(u; \theta)$ is an accurate representation of $g(u; \theta)$, and is the *accuracy* of the estimation procedure based on $h(u; \theta)$. The evaluation of the accuracy of $h(u; \theta)$ is similar to the evaluation of any other assumption of the analysis. It can be examined analytically, or empirically by simulations.

An estimation procedure can have high fidelity, yet be inaccurate. Its high fidelity implies that the likelihood intervals deduced from $g(u; \theta)$ reflect those of the sample likelihood of θ. The inaccuracy results in the confidence coefficients assigned to these likelihood intervals by $h(u; \theta)$ being inaccurate. See Section 19.5, Example 19.2. More importantly, and probably more commonly, an estimation procedure can have high accuracy yet low fidelity. This is a more serious defect of an estimation procedure. It implies that the confidence coefficients, although accurate, are irrelevant to the observed sample. See Section 19.6.

19.5 Examples

Example 19.1 *An example of the problem of the Nile* Fisher (1973, pp. 169–75) considered n independent pairs (x_i, y_i) with joint density $f(x_i, y_i; \theta) = \exp(-x_i\theta - y_i/\theta)$. The resulting likelihood of θ can be highly skewed. This skewness can be completely removed by the use of $\phi = \log \theta$, the odd derivatives of the log likelihood of ϕ evaluated at $\hat{\phi}$ being identically zero. The relative likelihood of ϕ based on a sample $(\mathbf{X}, \mathbf{Y})$ of n pairs is $R(\phi; \mathbf{X}, \mathbf{Y}) = f(\mathbf{X}, \mathbf{Y}; \phi)/f(\mathbf{X}, \mathbf{Y}; \hat{\phi})$, where $f(\mathbf{X}, \mathbf{Y}; \phi)$ is the joint density function of $\mathbf{X}, \mathbf{Y}$.

Let $S = \sum x_i$, $T = \sum y_i$. Then

$$R(\phi; \mathbf{X}, \mathbf{Y}) = R(\phi; \hat{\phi}, I) = R(u; I)$$
$$= \exp\{(I/2)[2 - \exp(u/I^{1/2}) - \exp(-u/I^{1/2})]\}, \qquad (19.4)$$

where u is in the class of ML linear estimating functions (19.1)

$$u = u(\hat{\phi}, I; \phi) = (\hat{\phi} - \phi)I^{1/2} \qquad (19.5)$$
$$\hat{\phi} = \log \hat{\theta} = \tfrac{1}{2} \log(T/S),$$

and from (19.2)

$$I = I(\mathbf{X}, \mathbf{Y}; \hat{\phi}) = 2(ST)^{1/2}.$$

Expanding $\exp(\pm u/I^{1/2})$ in (19.4) it can be seen that, if $u^2/12I$ is negligible in the range $(-3 < u < 3)$ of non-negligible likelihood, (19.4) is approximately a $N(\hat{\phi}, 1/I)$ likelihood of ϕ as defined by Viveros and Sprott (1987):

$$R(u; I) \doteq \exp[-(\hat{\phi} - \phi)^2 I/2] = \exp(-u^2/2) = R_N(u). \qquad (19.6)$$

The approximation (19.6) suggests the estimation procedure of using u as a $N(0, 1)$ pivotal.

Accuracy of the estimation procedure: $u \sim N(0, 1)$. The accuracy of this estimation procedure can be examined analytically. The joint density of u, I is

$$g(u, I) \propto [I^{2n - 3/2} \exp(-I)]R(u; I), \qquad (19.7)$$

from which the marginal distribution $g(u)$ can be obtained. From (19.6), (19.7) is approximated by $[I^{2n-3/2} \exp(-I)] \exp(-u/2)]$. Thus, to this order of accuracy u and I are statistically independent and $u \sim N(0, 1)$. For example, using u with $n = 5$ the approximate normal 0.90, 0.95, and 0.99 symmetric confidence intervals have exact confidence bounds $\Pr\{|u| < 1.645, 1.960, 2.576\} = 0.9098, 0.9569$, and 0.9930, respectively. With larger values of n this normal approximation will of course be more accurate.

Fidelity: $n = 5$. The likelihood of ϕ based on the marginal distribution of u is proportional to $g(u; n)$ obtained from (19.7), expressed as a function of ϕ. Since this attains its maximum at $\phi = \hat{\phi}$, or equivalently, $u = 0$, the corresponding relative likelihood of ϕ is $R_u(\phi) = g(u; n)/g(0; n)$ expressed as a function of ϕ. Figure 19.1 gives plots of $R(\phi; \mathbf{X}, \mathbf{Y})$, $R_u(\phi)$, and $R_N(\phi)$ for $I = 5, 8.5$ (the most probable value of I), and 20. These plots show that little fidelity is lost by using the marginal distribution of u. Since the probability that I is outside of the range $(5, 20)$ is only 0.05, the graphs demonstrate that u is marginally almost pivotally sufficient for ϕ, even when n is as small as 5. The graphs also show that the fidelity of the estimation procedure $u \sim N(0, 1)$ is also high, although slightly inferior to that of $g(u; n)$.

The fidelity of the marginal distribution of u and the corresponding $N(0, 1)$

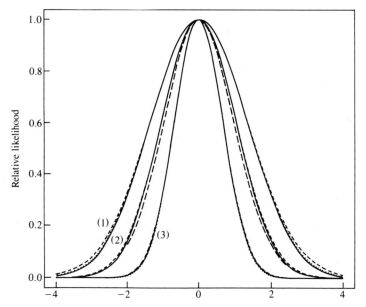

Fig. 19.1. Relative likelihood functions. Section 19.6: $R_w(\phi)$ – – –. Example 19.1 (1): $I = 5$; (2): $I = 8.5$; (3): $I = 20$. $R(\phi; X, Y)$ ——; $R_u(\phi)$ ·····; $R_N(\phi)$ ------.

estimation procedure will of course improve as n increases. Thus these provide a reasonable method to achieve the aim of inferential estimation in this example.

In this example perfect fidelity can be achieved by using the *conditional* distribution of $\hat{\phi}$ given I (Fisher 1973). Thus the use of the pivotal u is not essential. The problem can be reduced to an estimate $\hat{\phi}$ and its conditional distribution. The next example is a case where this does not appear to be true. The solution must be expressed in terms of an approximate pivotal.

Example 19.2 *Linear autoregression* Consider the model $y_i = \theta y_{i-1} + \sigma e_i$, $i = 1, \ldots, n$ where e_i are independent $N(0, 1)$ variates, and σ is assumed known. The initial observation y_0 is regarded as fixed. Here the relative likelihood function $R(\theta; \mathbf{Y})$ based on $\mathbf{Y} = (y_0, y_1, \ldots, y_n)$ is

$$R(\theta; \mathbf{Y}) = R(u) = \exp(-u^2/2), \tag{19.8a}$$

where u is a member of class (19.1)

$$u = u_\theta = (\hat{\theta} - \theta)I^{1/2},$$

$$\hat{\theta} = \sum_{i=1}^{n} y_i y_{i-1} \bigg/ \sum_{i=1}^{n} y_{i-1}^2, \qquad I = I(\mathbf{Y}; \hat{\theta}) = \sum_{i=1}^{n} y_{i-1}^2/\sigma^2. \tag{19.8b}$$

Table 19.1 N autoregression simulations $n = 10$, $\sigma = 1$, $N = 5000$

θ	$y_0 = 0$			$y_0 = 10$		
	0.90	0.95	0.99	0.90	0.95	0.99
0	0.927	0.963	0.994	0.905	0.955	0.992
0.5	0.918	0.961	0.994	0.902	0.953	0.992
1	0.898	0.947	0.990			
2	0.894	0.948	0.987			

This is a normal likelihood (19.6). Thus the likelihood is *exactly* normal for samples of any size.

As before, (19.8a) suggests the estimation procedure $u \sim N(0, 1)$. From (19.8), this estimation procedure has perfect fidelity. Its accuracy must be examined by simulations because exact distributions are difficult to obtain. For analytical details, see Chamberlin (1989, Chapter 6, Section 6.7.4). Consider the likelihood intervals $\theta = \hat{\theta} + u/I^{1/2}$, $u = 1.645$, 1.960, 2.576.

The simulations in Table 19.1 show that the estimation procedure $u \sim N(0, 1)$ yields reasonably accurate confidence coefficients for these likelihood intervals. The accuracy is greatest in the non-ergodic case $\theta > 1$ or when y_0/σ is not small.

Hinkley (1983) also considered the $N(0, 1)$ use of (19.8b) in this example with much the same conclusions. However, he implied that the conditional distribution of $\hat{\theta}|I$ can be obtained as $N(\theta, 1/I)$ from the $N(0, 1)$ marginal distribution of u, at least for the case $\theta = 1$, $y_0 = 0$. This is true only if the conditional distribution of $u|I$ is $N(0, 1)$ for all I. The simulations in Table 19.2 indicate that this is apparently not true in general. These simulations are those of Table 19.1 in the row $\theta = 1$ subdivided into five different intervals of I each containing $N = 1000$ simulations. From these results it appears

Table 19.2   Conditional distribution of $u|I \in R$ based on the above $N = 5000$ simulations with $\theta = 1$, $y_0 = 0$, $n = 10$.

R		N	$|u| < 1.96|I \in R$
0.65	10.32	1000	0.885
10.33	19.06	1000	0.929
19.06	34.69	1000	0.980
34.69	68.00	1000	0.990
68.00	570.90	1000	0.951

that the approximate $N(0, 1)$ distribution of u is not equivalent to the distribution of $\hat{\theta}|I$. Like Example 1 of Sprott (1990), here is another example where $\hat{\theta}|I$ appears to offer no escape from the use of an approximate pivotal like u of (19.1).

19.6   Adjustments to the ML linear estimating function

The purpose of ML inferential estimation is to obtain a member of the class (19.1) that is an approximate pivotal with high fidelity (Sprott and Viveros 1984). Unfortunately, its purpose is usually thought to be to obtain an estimate that is asymptotically unbiased with minimum variance. For this reason $1/I$ is usually interpreted as an estimate of Var $\hat{\phi}$. This often leads to attempts to improve upon the resulting ML approximate pivotal by adjusting it for the bias and variance of the ML estimate. This leads to replacing (19.1) by $w = [\hat{\phi} - E(\hat{\phi})]/[\text{Var } \hat{\phi}]^{1/2}$.

In Example 19.1 the effect of this adjustment can be evaluated analytically. It can easily be shown that $\hat{\phi}$ is unbiased and, for $n = 5$, Var $\hat{\phi} = 0.11066$. This leads to the estimation procedure of using the marginal distribution of $w = (\hat{\phi} - \phi)/0.3326$. The relative likelihood function $R_w(\phi)$ based on the marginal distribution of w, or equivalently, on $\hat{\phi}$, is $R_w(\phi) = \{[\exp(\hat{\phi} - \phi) + \exp(\phi - \hat{\phi})]/2\}^{-2n}$. Figure 19.1 shows a plot of $R_w(\phi)$. $R_w(\phi)$ appears only in the central group of graphs ($I = 8.5$) since it is independent of I, and $1/\text{Var } \hat{\phi} \doteq 9$. The fidelity of the marginal distribution of w is generally poor, as shown in Fig. 19.1. This poor fidelity is due to the fact that the marginal distribution of w cannot take the observed precision I into account. The marginal distribution of u does, resulting in its higher fidelity. This poor fidelity also affects confidence levels. For example, the 0.950 symmetric confidence interval obtained from the marginal distribution of w has a confidence coefficient of 0.871 and 0.990 obtained from the marginal distribution of u when $I = 5, 15$ respectively. All three confidence coefficients are accurate, since they are based on the exact marginal distributions of w and of u. But that obtained from w is irrelevant for the observed sample, since it ignores the precision specified by the observed I.

The fidelity of the marginal distribution of w improves as the sample size n increases. But the rate of improvement is extremely slow since it depends on the stochastic fluctuations of I becoming negligible. Thus the marginal use of w does not achieve the aim of inferential estimation. The estimation procedure based on adjusting u to obtain $w \sim N(0, 1)$ is inappropriate to achieve this aim since it is based upon approximating a form of the marginal distribution of the ML estimate. This distribution is appropriate only if the ML estimate is sufficient for θ, which is usually not the case.

19.7 Discussion

Although perfect fidelity can be achieved in Example 19.1 by using the conditional distribution of $\hat{\phi}$, or equivalently w, given I, this method is not available in general, as Example 19.2 shows. The purpose of Section 19.6 is to illustrate that the adjustments commonly applied to the appropriate ML quantity (19.1) usually impair its use in inferential estimation by lowering its fidelity. See Sprott and Viveros (1984) and Sprott (1990) for other examples of how such adjustments also impair the accuracy of the $N(0, 1)$ approximation.

In assessing the information lost by the marginal distribution of $\hat{\phi}$ in Example 1, Fisher (1973) remarked that this was less than that supplied by the sample by a single observation in $2n + 1$. In light of Section 19.6, this assessment seems overly optimistic. For finite samples, more global measures of fidelity, which take into account the variablity of the entire shape of the likelihood function, are required.

Although estimation in the presence of 'nuisance' parameters is beyond the scope of this paper, if σ is assumed unknown in Example 19.2, the simulations can be used to show that using $t = u/(s/\sigma)$ as a Student t variate with 9 d.f., where $s^2 = \sum (y_i - \hat{\theta}y_{i-1})^2/(n-1)$ has the same accuracy as exhibited in Table 19.1. The likelihood function of θ in the absence of knowledge of σ is a $t_{(n-1)}$ likelihood centred at $\hat{\theta}$ and scaled by s as defined by Viveros and Sprott (1987). This likelihood function is also well defined. The pivotal t is also not equivalent to any estimate. Hence ML linear estimating functions such as u and t in Example 19.2 exhibit the special role that the class (19.1) can play when seeking a solution to a problem in inferential estimation. Further examples can be found in Chamberlin (1989), Chamberlin and Sprott (1989), and Sprott (1990).

19.8 Acknowledgements

We should like to thank Professor G. A. Barnard for many constructive suggestions. This work was partially supported by grants from the Natural Sciences and Engineering Research Council of Canada.

References

Barnard, G. A. (1977). On ridge regression and the general principles of estimation. *Utilitas Math.*, **11**, 299–311.

Chamberlin, S. R. (1989). The foundation and application of inferential estimation. Ph.D. thesis, Department of Statistics and Actuarial Science, University of Waterloo.

Chamberlin, S. R. and Sprott, D. A. (1989). Linear systems of pivotals and associated pivotal likelihoods with applications. *Biometrika*, **76**, 685–91.

Fisher, R. A. (1925). Theory of statistical estimation. *Proc. Camb. Phil. Soc.*, **22**, 309–68.

Fisher, R. A. (1945). The logical inversion of the notion of the random variable. *Sankhya*, **7**, 129–32.

Fisher, R. A. (1973). *Statistical methods and scientific inference*. Hafner Press, New York.

Hinkley, D. V. (1983). Can frequentist inferences be very wrong? A conditional 'yes'. In G. E. P. Box, T. Leonard, and C. F. J. Wu (ed.), *Scientific inference, data analysis, and robustness*, pp. 85–103. Academic Press, New York.

Rao, C. R. (1973). *Linear statistical inference and its applications* (2nd edn). Wiley, New York.

Sprott, D. A. (1990). Inferential estimation, likelihood, and linear pivotals (with discussion). *Canad. J. Stat.*, **18**, 1–15.

Sprott, D. A. and Viveros, R. (1984). The interpretation of maximum likelihood estimation. *Canad. J. Stat.*, **12**, 27–38.

Viveros, R. (1985). Estimation in small samples. Ph.D. thesis, Department of Statistics and Actuarial Science, University of Waterloo.

Viveros, R. and Sprott, D. A. (1986). Conditional inference and maximum likelihood in a capture–recapture model. *Comm. Statist. Theory-Meth.*, 15, 1035–46.

Viveros, R. and Sprott, D. A. (1987). Allowance for skewness in maximum likelihood estimation with application to the location-scale model. *Canad. J. Stat.*, **15**, 349–61.

Appendix

Theorem   Let u_θ and u_ϕ be members of the class (19.1). If

$$u_\phi = h(u_\theta; \theta) \tag{19.A1}$$

is a one-to-one function of u_θ for every θ, and $T = I(\mathbf{X}; \hat{\theta})^{1/2}$, $\hat{\theta}$ and θ are functionally independent, then $u_\phi = \pm u_\theta$.

Proof   It follows that T, u_θ, and θ are functionally independent. Differentiating (19.A1) twice with respect to ϕ gives an equation of the form

$$0 = T^2(\partial^2 h/\partial u_\theta^2)(\mathrm{d}\theta/\mathrm{d}\phi)^2 + Ta(u_\theta; \theta) + b(u_\theta, \theta),$$

where $a(u_\theta; \theta)$ and $b(u_\theta; \theta)$ are complicated functions independent of T whose functional forms are not required in subsequent arguments. Since T is arbitrary and $\mathrm{d}\theta/\mathrm{d}\phi < 0$ or $\mathrm{d}\theta/\mathrm{d}\phi > 0$, it follows that $\partial^2 h/\partial u_\theta^2 = 0$. Noting that $u_\theta = 0$ if and only if $u_\phi = 0$ gives

$$u_\phi = A(\theta)u_\theta,$$

so that from (19.1) $(\hat{\phi} - \phi)|d\hat{\theta}/d\hat{\phi}| = A(\theta)(\hat{\theta} - \theta)$. Differentiating with respect to θ and setting $\theta = \hat{\theta}$ gives $(d\hat{\phi}/d\hat{\theta})|d\hat{\theta}/d\hat{\phi}| = \pm 1 = A(\hat{\theta})$. Since $\hat{\theta}$ is arbitrary this implies $A(\theta) = \pm 1$ for all θ, so that $u_\phi = \pm u_\theta$, as required.

20
Geometrical aspects of efficiency criteria for spaces of estimating functions
Christopher G. Small and D. L. McLeish

ABSTRACT

A concept of the efficiency of an estimating function was formally proposed by Godambe (1960). In that paper, it was shown that within the class of estimating functions satisfying certain mild regularity conditions the score function maximizes this measure of efficiency.

In this paper, we shall develop a geometrical interpretation of such efficiency criteria for estimating functions. In particular, we show how the existence and continuity of linear functionals on Hilbert spaces of estimating functions relate to the necessary regularity that is fundamental to Godambe's results and its extensions. This leads to the concepts of *E-ancillary* subspaces and *complete E-sufficient* subspaces of inference functions. Applications will be given to problems involving the inferential separation of parameters.

20.1   Introduction

While the formal theory of estimating functions is of fairly recent development, its roots go back to early work by Pearson (1894) on the method of moments, which was an important stepping stone in the realization that estimators are naturally obtained as the solution to a set of equations which may not be capable of explicit solution. The controversy that subsequently raged between Pearson and Fisher over the merits of the method of moments and maximum likelihood estimation can be naturally set today in the context of the modern theory of estimating functions, where both the method of moments and the score function are examples of unbiased estimating functions. The term 'unbiased' as applied to estimating functions appears in Kendall (1951). That paper makes the important distinction between the unbiasedness of an estimating function and the unbiasedness of the estimator defined as the root of that function. The idea of an optimal estimating function appears in the restricted context of linear estimating functions in the paper by Durbin (1960) and was generalized to a wider class of functions by Godambe (1960), who demonstrated the optimality of the score function within that wider class. In this paper we examine a geometrical interpretation of Godambe's efficiency criterion and use this interpretation to examine the properties of models that fail to have sufficient regularity to admit the score function as an unbiased square-integrable estimating function. It will be shown that the failure of this regularity is the failure of a linear functional to be continuous. This will raise the possibility of extending the functional to a wider class, among which are to be found continuous functionals. Such extensions lead to the notions of *E-ancillarity* and *E-sufficiency* and to methods for the inferential separation of nuisance parameters from parameters of interest by decomposing estimating functions into *E*-ancillary and *E*-sufficient components.

Throughout this paper we shall denote the entire data set by X and the parameter by θ. The parameter space Θ will typically be some subset of Euclidean space $\mathbf{R}^k$. However, we do not rule out the possibility that Θ is infinite dimensional at this stage. In such cases, we shall assume that Θ is endowed with some appropriate metric. By an estimating function $\psi(\theta; X)$ we mean a real-valued function of both the data and the parameter that is measurable in the former for each value of θ. Such functions can be added pointwise in θ and X. By a scalar multiple of $\psi(\theta; X)$ we shall mean a function of the form $k(\theta)\psi(\theta; X)$, where $k(\theta)$ does not depend upon X. An inference function $\psi(\theta; X)$ is said to be unbiased if $E_\theta\psi(\theta; X) = 0$ for all $\theta \in \Theta$ and is said to be square integrable if $E_\theta\psi^2(\theta; X) < \infty$ for all $\theta \in \Theta$. Let Ψ be a collection of unbiased square-integrable inference functions that is a vector space in the sense that it is closed under the operations of pointwise addition and scalar multiplication. The vector space Ψ can be endowed with a family of inner products and norms as well. Let $\langle \cdot, \cdot \rangle_\theta$ be defined by $\langle \psi, \phi \rangle_\theta = E_\theta\psi(\theta; X)\phi(\theta; X)$ for each ψ and ϕ in Ψ. Also let $\|\psi\|_\theta^2 = \langle \psi, \psi \rangle_\theta$. Finally, we assume that Ψ is closed under limits in the following sense. Given any $\theta \in \Theta$ and any sequence of functions $\psi_n \in \Psi$ such that

$$\lim_{mn} \|\psi_m - \psi_n\|_\theta = 0 \tag{20.1}$$

there exists a function ψ in Ψ such that $\|\psi_n - \psi\|_\theta \to 0$. Note that the choice of ψ will in general depend on θ.

The term *estimating function* is in widespread usage, but should in no way exclude the possibility that a function may be chosen from Ψ for the purposes of testing rather than estimation. For this general context, the term *inference function* is perhaps more accurate. If a test statistic for θ is desired, then an appropriately chosen function ψ can serve the purpose for testing $H\colon \theta = \theta_0$ using $\psi(\theta_0)$. For the purposes of estimation, one would typically construct a set $\{\psi_\alpha\}_{\alpha \in A}$ from Ψ such that $\hat{\theta}$ is obtained as the simultaneous solution to $\psi_\alpha(\hat{\theta}; X) = 0$ for all $\alpha \in A$. Typically, when θ is a k-dimensional parameter, then A will have cardinality k. For example, a k-dimensional parameter can be estimated by the k estimating functions which are the k components of the score vector.

A function $h: \Psi \times \Theta \to \mathbf{R}^p$ is said (by abuse of terminology) to be a linear functional on Ψ if $h(\cdot, \theta)$ is a linear functional on the Hilbert space $\Psi_\theta = \{\psi(\theta): \psi \in \Psi\}$ for every $\theta \in \Theta$. As an example of such a functional we note in particular the *score functional* which is defined by

$$h[\psi, \theta] = \frac{\partial}{\partial \eta} E_\eta \psi(\theta)|_{\eta = \theta} \qquad (20.2)$$

where $\partial/\partial \eta$ represents the k-dimensional gradient of the function. We shall write $\nabla \psi$ for the application of this functional to ψ and shall typically suppress the θ in the notation. Note here that the score functional exists in much greater generality that does the score vector, regarded as a k-tuple of unbiased square integrable inference functions. However, under standard regularity assumptions we can write

$$\nabla \psi = -E_\theta \frac{\partial \psi}{\partial \theta} = (\langle \psi, S_1 \rangle_\theta, \ldots, \langle \psi, S_k \rangle_\theta) \qquad (20.3)$$

where $S_i = \partial/\partial \theta_i \log f(x; \theta)$ is the ith component of the score vector.

The invocation of regularity conditions necessary in order to interchange derivatives and integrals is required here, and has little in the way of an intuitive interpretation as stated. There is, however, a geometric interpretation of the required regularity which we now describe. Recall that Hilbert spaces are self-dual. Therefore, all continuous linear functionals can be represented as inner products. So in particular if $\nabla: \Psi_\theta \to \mathbf{R}^k$ is continuous, then we can find a set of inference functions $\psi_1, \psi_2, \ldots, \psi_k$ such that $\nabla \psi = (\langle \psi, \psi_1 \rangle_\theta, \ldots, \langle \psi, \psi_k \rangle_\theta)$. It remains to determine these functions ψ_i. From (20.3) we see that ψ_i is the projection of S_i into Ψ. So the required regularity is that which ensures that the score functional is continuous as a mapping into $\mathbf{R}^k$.

Now such linear functionals are continuous if and only if they are bounded (i.e. have finite norm). The norm of the linear transformation ∇ is defined as

$$\|\nabla\|_\theta = \sup_{\psi \in \Psi} \frac{\|\nabla \psi\|_\theta}{\|\psi\|_\theta}. \qquad (20.4)$$

20.2 One parameter case

Hereafter we will restrict to the one-parameter case ($k = 1$). Using equation (20.3), the ratio on the right-hand side of (20.4) is also recognizable as the square root of the efficiency of ψ in the sense defined by Godambe (1960) for the one parameter case. *Thus we deduce that the score functional is continuous if and only if Godambe's efficiency criterion is bounded away from infinity. It can then be seen from (20.3) that the projection of the score function into* Ψ *is a* ψ *which attains this supremum.*

In some cases, regularity fails and the score function is not unbiased and square integrable. See, for example, McLeish and Small (1988, pp. 7–8). Clearly, a more general theory is needed for such cases. The obvious approach is suggested by the geometrical interpretation of the score functional as a continuous linear functional when regularity holds. If the functional is not continuous, then appropriately chosen continuous alternatives are desirable. Even if regularity does hold, alternative continuous functionals need to be considered, as the score functional is a measure of sensitivity only to infinitesimal shifts in parameter values. Clearly the class of all continuous linear functionals on Ψ is too rich as there are many such functionals which are irrelevant to assessing the sensitivity of an inference function to parameter changes. One way to restrict the class of continuous linear functionals is to specify some subspace of appropriately 'insensitive' inference functions and to demand that the functionals under consideration annihilate that subspace. A sensible choice of such inference functions is the class of $\phi \in \Psi$ such that

$$E_\eta \phi(\theta) = 0 \qquad (20.5)$$

for all η and for all θ in the parameter space Θ. As we shall only be considering continuous linear functionals, we can close this subspace without changing the annihilating class of linear functionals. See Small and McLeish (1988, p. 537) for details of the closure.

Definition 20.1   The closed subspace in which functions which satisfy (20.5) are dense shall be called the *subspace of E-ancillary functions* of Ψ. Let $\mathscr{A}$ be the *E-ancillary subspace*.

The continuous functionals which annihilate $\mathscr{A}$ are naturally isomorphic to the orthogonal complement of $\mathscr{A}$, i.e. those functions ψ for which $\langle \psi, \phi \rangle_\theta = 0$ for all θ and for all ϕ satisfying equation (20.5).

Definition 20.2   Let $\mathscr{S}$ be the subspace of all functions ψ which are orthogonal as above to the elements of $\mathscr{A}$. We shall call $\mathscr{S}$ the *complete E-sufficient subspace of* Ψ.

The natural isomorphism between an element ψ in $\mathscr{S}$ and a continuous functional Λ which annihilates $\mathscr{A}$ is such that $\Lambda \phi = \langle \phi, \psi \rangle_\theta$ and

$$\|\Lambda\|_\theta = \frac{|(\Lambda \psi)(\theta)|}{\|\psi\|_\theta}. \qquad (20.6)$$

Of course, any multiple of ψ also satisfies this latter property.

Consider two functions ψ_1 and ψ_2 lying in the complete E-sufficient subspace, and suppose that $E_\eta \psi_1(\theta) = E_\eta \psi_2(\theta)$ for all η and θ in Θ. Then $\psi_1 - \psi_2$ is E-ancillary and therefore $\psi_1 = \psi_2$. Thus an inference function in $\mathcal{S}$ with a given mean function is unique. For each $\eta \in \Theta$, we define a functional ∇_η by $(\nabla_\eta)\psi(\theta) = E_\eta\psi(\theta)$, and suppose that these functionals are continuous. If there is a function $\psi \in \Psi$ with a given mean function $\nabla_\eta \psi(\theta) = E_\eta \psi(\theta)$ for all $\eta, \theta \in \Theta$, then there is a unique $\psi^* \in \mathcal{S}$ having the same mean function, and this is the projection of ψ onto $\mathcal{S}$.

Just as the score functional, when continuous, determines the score function using (20.3), so the functional ∇_η, when continuous, determines an element ψ_η in the complete E-sufficient subspace. More explicitly, there exists a function ψ_η such that $E_\eta\psi(\theta) = \langle \psi, \psi_\eta \rangle_\theta$ for all ψ, θ. In our context it is natural to require that these functions have a power-series expansion in the sense that

$$\psi_\eta(\theta) = \sum_j \frac{(\eta - \theta)^j}{j!} \frac{\partial^j \psi_\eta}{\partial \eta^j}\bigg|_{\eta = \theta} \tag{20.7}$$

where the series on the right side is assumed to be $L_2(\theta)$ convergent.

We may approximate the behaviour of the mean functions of the form $\nabla_\eta \psi(\theta)$ by using the local behaviour of the function at $\eta = \theta$. Then functionals

$$\nabla^{(j)}\psi(\theta) = \frac{\partial^j}{\partial \eta^j} E_\eta\psi(\theta)|_{\eta=\theta}, \qquad j = 1, 2, \ldots$$

have a representation of the form $\nabla^{(j)}\psi(\theta) = \langle \psi, \psi^{(j)} \rangle_\theta$ for all j, θ, where $\psi^{(j)}(\theta) = \partial^j \psi_\eta / \partial \eta^j|_{\eta = \theta}$. Using this notation, we observe that $\nabla = \nabla^{(1)}$ and that $\psi^{(1)}(\theta)$ is a projection of $S(\theta)$ into Ψ. An inference function is termed rth-order E-ancillary if it is a limit of functions ψ with mean functions satisfying

$$\nabla_\eta \psi(\theta) = o(\eta - \theta)^r \tag{20.8}$$

as $\eta \to \theta$. See McLeish and Small (1988). Analogously to the definition of the complete E-sufficient subspace, the rth-order complete E-sufficient subspace is the orthogonal complement of the subspace of rth-order E-ancillary functions and is generated by $\{\psi^{(1)}, \ldots, \psi^{(r)}\}$.

Proposition 20.1 *Under regularity conditions* (20.7) *above, the functions* $\psi^{(1)}, \psi^{(2)}, \ldots$, *form a basis for the complete E-sufficient subspace.*

Proof First note that all $\psi^{(i)}$ are elements of $\mathcal{S}$ since every function satisfying (20.5) and hence every function in $\mathcal{A}$ is orthogonal to all $\psi^{(i)}$. Conversely, if a function ϕ satisfies $\langle \phi, \psi^{(j)} \rangle_\theta = 0$ for all $j = 1, 2, \ldots$, its mean function satisfies

$$\frac{\partial^j}{\partial \eta_j} \nabla_\eta \phi(\theta)|_{\eta = \theta} = 0 \qquad \text{for all } j, \theta.$$

However, under the conditions (20.7), $\nabla_\eta \phi(\theta)$ is analytic and therefore this implies $\nabla_\eta \phi(\theta) = 0$ for all η, θ. So $\phi \in \mathscr{A}$.

Projection on the rth-order complete E-sufficient subspace, i.e. projection onto the subspace spanned by $\psi^{(1)}, \psi^{(2)}, \ldots, \psi^{(r)}$, is the natural local analogue for estimating functions of a sufficiency reduction for statistics.

Consider now a finite-dimensional space Ψ spanned by n linearly independent unbiased square integrable inference functions $\beta_1, \ldots, \beta_n$. For convenience suppose there is an $r \times n$ matrix $\mathbf{M}(\theta)$ of rank r whose rows span the same (r-dimensional) space as do the vectors

$$\bigcup \{(E_\eta \beta_1(\theta), E_\eta \beta_2(\theta), \ldots, E_\eta \beta_n(\theta)); \eta \in \Theta\}.$$

Denote by $\Sigma(\theta)$ that $n \times n$ matrix with components $\sum_{ij}(\theta) = \langle \beta_i, \beta_j \rangle_\theta$ and assume this matrix is non-singular. We begin by finding the projection ψ^* of an arbitrary function $\psi(\theta) = \sum_i a_i(\theta)\beta_i(\theta)$ onto the complete E-sufficient subspace. Let $\mathbf{a}(\theta) = (a_1(\theta), \ldots, a_n(\theta))$ be the coefficient vector. Then $\psi^*(\theta)$ has coefficient vector

$$\mathbf{a}^* = \mathbf{a}M^{\mathrm{T}}[M\Sigma^{-1}M^{\mathrm{T}}]^{-1}M\Sigma^{-1}. \qquad (20.9)$$

The same formula provides the projection of ψ onto the locally rth-order complete E-sufficient subspace if $\mathbf{M}(\theta)$ has linearly independent rows where

$$M_{ij}(\theta) = \frac{\partial^i}{\partial\eta^i} E_\eta \beta_j(\theta)|_{\eta=\theta}. \qquad (20.10)$$

Typically an inference function space Ψ is of infinite but countable dimension. In this case, we may attempt to carry out projections by projecting on a space of finite dimension n and then allowing n to increase. Any choice of countably many basis vectors $\beta_1, \ldots, \beta_n, \ldots$ will then result in a limit which is the projection on the complete E-sufficient subspace.

Proposition 20.2   *Let $\beta_1, \beta_2, \ldots$ be a basis for the inference function space Ψ. Consider an arbitrary function $\psi(\theta) = \sum_{i=1}^{\infty} a_i(\theta)\beta_i(\theta)$. Define $\mathbf{M}_n(\theta)$ to be an $r \times n$ matrix with components*

$$\mathbf{M}_{nij}(\theta) = \frac{\partial^i}{\partial\eta^i} E_\eta \beta_j(\theta)|_{\eta=\theta}. \qquad (20.11)$$

Let $\boldsymbol{\beta} = (\beta_1, \ldots, \beta_n)$ and $\mathbf{a}_n = (a_1(\theta), a_2(\theta), \ldots, a_n(\theta))$. As before, define $\Sigma_n(\theta)$ to be the $n \times n$ matrix with i, jth component $\langle \beta_i, \beta_j \rangle_\theta$. Then the sequence of inference functions

$$\psi_n^* = \mathbf{a}_n \mathbf{M}_n^{\mathrm{T}}[\mathbf{M}_n \Sigma_n^{-1} \mathbf{M}_n^{\mathrm{T}}]^{-1} \mathbf{M}_n \Sigma_n^{-1} \boldsymbol{\beta}^{\mathrm{T}} \qquad (20.12)$$

converges to the projection of ψ onto the rth order complete E-sufficient subspace as $n \to \infty$. If furthermore $r \to \infty$ and $n \to \infty$, then the limit is the projection onto the complete E-sufficient space.

Proof Denote by Ψ_n the subspace spanned by the first n basis vectors and put $\psi_n = \sum_{i=1}^n a_i \beta_i$. Let Υ_n denote the rth-order complete E-sufficient subspace of Ψ_n and by Υ the rth order complete E-sufficient subspace of Ψ. Denote by $\Pi[\psi|\Upsilon]$ the projection of the function ψ onto the subspace Υ. Then

$$\|\Pi[\psi_n|\Upsilon_n] - \Pi[\psi|\Upsilon]\|_\theta \leqslant \|\Pi[\psi_n|\Upsilon_n] - \Pi[\psi|\Upsilon_n]\|_\theta + \|\Pi[\psi|\Upsilon_n] - \Pi[\psi|\Upsilon]\|_\theta.$$

Now the square of the first term on the right side above is bounded by $\sum_{i=n+1}^\infty a_i^2(\theta)\|\beta_i\|_\theta^2$ and this goes to zero. The second goes to zero since Υ is the closure of $\bigcup_{n=1}^\infty \Upsilon_n$.

The second part of the proposition follows easily from Proposition 20.1.

Now suppose we have no information on a particular inference function ψ and the underlying model other than the mean function $\nabla_\eta \psi(\theta)$ and the covariance structure of some basis for Ψ. The above proposition permits approximation of the complete E-sufficient projection of ψ. It is natural to terminate the addition of basis vectors when $\|\psi_{n+1}^* - \psi_n^*\|$ is sufficiently small. The particular choice of the basis is obviously important in reaching a reasonable approximation quickly, but does not otherwise affect the limit. Thus, if we work in a computer environment which permits symbolic computation such as MAPLE or MATHEMATICA, we may add basis-estimating functions from any rich class such as polynomials, or built-in special functions, until the projection ceases to change or changes very little, at which point we have a close approximation to the desired projection. In a sense, this indicates that the choice of spanning basis vectors is not as important as we might first suppose. Indeed, they might be selected sequentially from suitable class.

20.3 Nuisance parameters

Small and McLeish (1989) recommended projection on the E-ancillary subspace for the purpose of eliminating or limiting the effects of nuisance parameters. In particular, for estimation or for testing a hypothesis concering a parameter of interest θ in the presence of a nuisance parameter ξ we project a given inference function ψ onto the rth order E-ancillary subspace for the nuisance parameter. Related work is that of Godambe and Thompson (1974).

Let us return to a finite-dimensional inference function space Ψ spanned by n unbiased inference functions $\beta_1(\theta, \xi), \ldots, \beta_n(\theta, \xi)$. Suppose we have a function $\psi(\theta, \xi) \in \Psi$ in mind as a possible estimating function but we would like to limit the effect of the estimation of ξ on that of θ. To this end, define the matrix with entries

$$M_{ij}(\theta, \xi) = \frac{\partial^i}{\partial \eta^i} E_{\theta, \eta} \beta_j(\theta, \xi)|^{\eta = \xi}, \qquad i = 1, \ldots, r, \quad j = 1, \ldots, n. \qquad (20.13)$$

Then if we project an arbitrary function $\psi(\theta, \xi) = \sum_i a_i(\theta, \xi)\beta_i(\theta, \xi)$ onto the rth-order E-ancillary subspace for the nuisance parameter, the coefficient vector for the projection is

$$\mathbf{a}(\theta, \xi)[\mathbf{I} - \mathbf{M}^{\mathrm{T}}(\theta, \xi)(\mathbf{M}(\theta, \xi)\Sigma^{-1}(\theta, \xi)\mathbf{M}^{\mathrm{T}}(\theta, \xi))^{-1}\mathbf{M}(\theta, \xi)\Sigma^{-1}(\theta, \xi)]. \qquad (20.14)$$

Example 20.1   We now consider a general process X_t; $t \in [a, b]$. Assume that each X_t is square integrable and that the process is *continuous in mean*, that is that $E_{\theta, \xi}|X_s - X_t| \to 0$ as $s \to t$. Suppose $E_{\theta, \xi}(X_t) = \xi m_\theta(t)$ where $m_\theta(t)$ belongs to a parametric family of functions indexed by a parameter of interest θ and ξ is a nuisance parameter. Suppose the process $X_.$ is a random element of the space of right continuous functions with left-hand limits endowed with the Skorokhod topology. We also assume that the functions $m_\theta(t)$ are members of this space. Let $\mathcal{L}$ be the space of continuous linear functionals on the above space. Now consider a space Ψ of inference functions of the form

$$\psi(\theta, \xi) = L[X_. - \xi m_\theta(\cdot)] \qquad (20.15)$$

where $L \in \mathcal{L}$. Suppose that there exists an inference function $\gamma \in \Psi$ which satisfies the set of *normal equations* for fixed θ

$$E_{\theta, \xi}[\gamma(\theta, \xi)X_s] = m_\theta(s) \qquad (20.16)$$

for all $a \leqslant s \leqslant b$ and for all ξ. It can be shown (cf. McLeish and Small 1988, pp. 91–2) that γ spans the complete E-sufficient subspace for ξ. We now consider the special case where the covariance function $R(s, t) = \mathrm{Cov}_{\theta, \xi}(X_s, X_t)$ may depend on θ but is independent of the parameter ξ. Denote by ϕ_j, λ_j the orthonormal eigenfunctions, eigenvalues respectively of R, satisfying

$$\lambda_j \phi_j(s) = \int_a^b R(s, t)\phi_j(t) \, \mathrm{d}t, \qquad j = 1, 2, \ldots. \qquad (20.17)$$

Then the Karhunen–Loeve expansion (cf. Grenander 1981) is

$$R(s, t) = \sum_{j=1}^\infty \lambda_j \phi_j(s)\bar{\phi}_j(t) \qquad (20.18)$$

with convergence on the right side occurring absolutely and uniformly.

Denote the coefficients in the expansion of $m_\theta(t)$ by $\eta_j = \int_a^b m_\theta(t)\bar{\phi}_j(t)\,dt$. We will show using criterion (20.16) that the complete E-sufficient subspace is spanned by the function $\gamma(\theta, \xi) = \int_a^b f^*(t)[X_t - \xi m_\theta(t)]\,dt$, where $f^*(t) = \sum_{k=1}^\infty (\eta_k/\lambda_k)\phi_k(t)$ if this series converges uniformly and absolutely and the resulting function γ lies in Ψ. Note that

$$E_{\theta,\xi}[\gamma(\theta, \xi)X_s] = E_{\theta,\xi}\left\{X_s \int_a^b f^*(t)[X_t - \xi m_\theta(t)]\,dt\right\} = \int_a^b f^*(t)R(s, t)\,dt$$

$$= \int_a^b \left\{\sum_{k=1}^\infty \frac{\eta_k}{\lambda_k}\phi_k(t)\right\} R(s, t)\,dt = \sum_{k=1}^\infty \frac{\eta_k}{\lambda_k} \int_a^b \phi_k(t)R(s, t)\,dt$$

$$= \sum_{k=1}^\infty \eta_k \phi_k(s) = m_\theta(s).$$

Therefore the normal equations hold.

Consider an unbiased inference function $\psi(\theta, \xi)$ for inference about the parameter of interest θ. We reduce the effect of the nuisance parameter by projecting this function on the E-ancillary subspace for ξ. This becomes

$$\psi - \psi^* = \psi - \frac{\text{Cov}_\theta(\psi, \gamma)}{\text{Var}_\theta(\gamma)}\gamma. \tag{20.19}$$

From the E-ancillarity of this function with respect to ξ we observe that $(L - L^*)[m_\theta(t)] = 0$, where L^* is the linear functional associated with ψ^*. Thus it can be seen that the E-ancillary function (20.19) is not dependent on the nuisance parameter ξ. In fact if

$$\psi(\theta, \xi) = \int f(t; \theta)(X_t - \xi m_\theta(t))\,dt, \tag{20.20}$$

then

$$(\psi - \psi^*)(\theta) = \int (f - f^*)(t; \theta)X_t\,dt \tag{20.21}$$

is a weighting of the process which eliminates the functional dependence on the nuisance parameter.

The methods discussed in this paper facilitate the selection of a function or class of functions useful for inference on a parameter. If we wish to test a hypothesis or construct interval estimates, frequency properties of these functions can be determined by classical techniques such as simulation and the bootstrap.

References

Durbin, J. (1960). Estimation of parameters in time-series regression models. *J. Roy. Stat. Soc. B* **22**, 139–53.

Godambe, V. P. (1960). An optimum property of regular maximum likelihood estimation. *Ann. Math. Stat.*, **31**, 1208–11.

Godambe, V. P. and Thompson, M. E. (1974). Estimating equations in the presence of a nuisance parameter. *Ann. Stat.*, **2**, 568–71.

Grenander, U. (1981). *Abstract inference.* Wiley, New York.

Kendall, M. G. (1951). Regression, structure and functional relationship I. *Biometrika*, **38**, 11–25.

McLeish, D. L. and Small, C. G. (1988). *The theory and applications of statistical inference functions.* Lecture Notes in Statistics 44, Springer-Verlag, New York.

Pearson, K. (1894). Contributions to the mathematical theory of evolution. *Phil. Trans. Roy. Soc.*, Ser. A, **185**, 71–110.

Small, C. G. and McLeish, D. L. (1988). Generalizations of ancillarity, completeness and sufficiency in an inference function space. *Ann. Stat.*, **16**, 534–51.

Small, C. G. and McLeish, D. L. (1989). Projection as a method for increasing sensitivity and eliminating nuisance parameters. *Biometrika*, **73**, 693–703.

PART 6

Theory (general methods)

21
Estimating equations from modified profile likelihood
H. Ferguson, N. Reid, and D. R. Cox

ABSTRACT

The score equation derived from the modified profile likelihood of Cox and Reid (1987) is examined by means of its stochastic asymptotic expansion. The bias and variance can then be evaluated explicitly, and it is easy to see that the estimating equation is unbiased to higher order than the score equation from the profile likelihood, a result also obtained by Liang (1987). The relation between the derivative of the score function and its variance is also considered, as is the role of parameter orthogonality, the 'usual' relation being recovered for certain exponential family problems. Parameter orthogonality is shown to be crucial.

21.1 Introduction

Let Y be an $n \times 1$ random vector with density $f_Y(y; \psi, \lambda)$, where ψ is a scalar parameter and λ is a vector of length p. We assume that it is desired to construct inference for ψ, in the absence of knowledge of λ, and that the components of the vector Y are independent. Writing

$$l(\psi, \lambda) = \log f_Y(y; \psi, \lambda) \quad \text{and} \quad nj_{\lambda\lambda}(\psi, \lambda) = -\partial^2 l(\psi, \lambda)/\partial\lambda\, \partial\lambda^T,$$

we can construct the *profile log-likelihood function* for ψ

$$l_P(\psi) = l(\psi, \hat{\lambda}_\psi) \tag{21.1}$$

and the *modified profile log-likelihood function* for ψ

$$l_M(\psi) = l(\psi, \hat{\lambda}_\psi) - \tfrac{1}{2}\log \det nj_{\lambda\lambda}(\psi, \hat{\lambda}_\psi), \tag{21.2}$$

where $\hat{\lambda}_\psi$ is the maximum likelihood estimate of λ for fixed ψ.

Although the profile likelihood function is easy to use for inference about ψ, and has some of the properties of a true likelihood function (Barndorff-Nielsen 1988, p. 30) it can generally give misleading inference for ψ if the dimension of λ is of the same order of magnitude as the sample size or amount of information (Neyman and Scott 1948; Kalbfleisch and Sprott 1970; Bartlett 1936). In many such problems a conditional or marginal likelihood is more suitable for inference about ψ. The modification to profile likelihood indicated in (21.2) was suggested in Cox and Reid (1987) as a first-order adjustment to $l_P(\psi)$ to take account of the estimation of λ. It was derived in that paper as an approximation to a conditional likelihood, based on the distribution of Y, given $\hat{\lambda}_\psi$, although as pointed out by Barndorff-Nielsen (1987) it can approximate a marginal likelihood as well. The function l_M is not the same as the modified profile likelihood function of Barndorff-Nielsen (1983, 1987) which differs from (21.2) by an additional term often expressible as $J(\psi) = \log|\partial\hat{\lambda}/\partial\hat{\lambda}_\psi|$, where $\hat{\lambda}$ is the unrestricted maximum likelihood estimate of λ. Inclusion of the additional term ensures that the resulting function is invariant to reparametrization of the nuisance parameter λ; (21.2) does not have this invariance. The precise definition of J is elaborated on in Barndorff-Nielsen (1986) and Fraser and Reid (1989).

As in Cox and Reid (1987), we require for (21.2) that the nuisance parameter λ be orthogonal to ψ, with respect to the expected Fisher information measure; i.e. writing $I_{\psi\lambda} = E(-\partial^2 l/\partial\psi\,\partial\lambda)/n$, we require that $I_{\psi\lambda} = 0$. This ensures that the lack of parameter invariance of (21.2) affects only terms of $O_p(n^{-1})$, whereas l_P is $O_p(n)$, and the correction term $j_{\lambda\lambda}$ is $O_p(1)$, at least in $\sqrt{n}$-neighbourhoods of the true value of ψ. Choice of a version of λ to further reduce this lack of invariance is discussed in Cox and Reid (1989). In principle it is always possible to express a given nuisance parameter ϕ as $\phi = \phi(\psi, \lambda)$, with λ orthogonal to ψ, by solving a set of differential equations, given as (4) in Cox and Reid (1987). Approximate solution of the differential equations is discussed in work as yet unpublished by Cox and Reid.

Treating $l_M(\psi)$ as a likelihood function provides a point estimate of ψ, $\hat{\psi}_M$, say, as well as a likelihood ratio statistic $w_M(\psi) = 2\{l_M(\hat{\psi}_M) - l_M(\psi)\}$ and a score statistic $U_M(\psi) = l_M'(\psi)$. To first order $\hat{\psi}_M$ and $w_M(\psi)$ have the same distribution as $\hat{\psi}$ and $w(\psi)$, the usual maximum likelihood estimate and likelihood ratio statistic. Second-order properties of $\hat{\psi}_M$ are investigated in detail in Ferguson (1989), using stochastic asymptotic expansions of $\hat{\psi}_M$ obtained by inverting the series expansion for the likelihood equation. Among other things, Ferguson's results show that $\hat{\psi}_M - \hat{\psi} = O_p(n^{-1})$, that the constant in the leading term of the remainder is the non-random quantity $I_{\psi\lambda\lambda}/2I_{\psi\psi}I_{\lambda\lambda}$, and that the $O(n^{-1})$ term in the bias of $\hat{\psi}_M$ does not include terms due to the estimation of λ, in contrast to the $O(n^{-1})$ term in the bias for $\hat{\psi}$.

In this paper we consider properties of the estimating equation

$$l_M'(\psi) = U_M(\psi) = 0. \tag{21.3}$$

The equation is investigated using a stochastic asymptotic expansion for $U_M(\psi)$, presented in Section 21.2. Equation (21.3) was also considered in Liang (1987), and his results are closely related to ours. For ease of notation, the scalar λ and vector λ cases are considered separately. Using orthogonal parameters in more general types of estimating equations is discussed in Firth (1987).

21.2   Expansion of the score statistic

To establish the notation used for the expansions, it is convenient to assume the components of Y are identically distributed, with $l_i = \log f(y_i; \psi, \lambda)$, although this assumption and that of independence could be weakened. We then define

$$Z_{\psi\lambda} = \frac{1}{\sqrt{n}} \frac{\partial^2 l(\psi, \lambda)}{\partial\psi \, \partial\lambda} = \frac{1}{\sqrt{n}} \sum_i \frac{\partial^2 l_i(\psi, \lambda)}{\partial\psi \, \partial\lambda} \qquad (21.4)$$

and

$$Z_{\psi\psi\lambda} = \frac{1}{\sqrt{n}} \frac{\partial^3 l(\psi, \lambda)}{\partial\psi^2 \, \partial\lambda} - \sqrt{n} \, I_{\psi\psi\lambda},$$

etc., where we assume in this section that λ is a scalar. The random variables Z are $O_p(1)$ and have mean zero. (We maintain the special notation $nj_{\lambda\lambda}$ used in (21.2) for $-(\sqrt{n} \, Z_{\lambda\lambda} - nI_{\lambda\lambda})$.) The $O(1)$ Is represent the expected values of log-likelihood derivatives. In the special case of only two subscripts, the Is have been defined as the negative expected value of the corresponding log-likelihood derivative. For example,

$$I_{\psi\psi} = -\frac{1}{n} E\left(\frac{\partial^2}{\partial\psi^2} l(\psi, \lambda)\right)$$

and

$$I_{\psi\lambda\lambda} = \frac{1}{n} E\left(\frac{\partial^3}{\partial\psi \, \partial\lambda^2} l(\psi, \lambda)\right).$$

It will also be necessary to use quantities involving Is with subscripts separated by commas. These items occur when taking expected values of products of Zs, and the commas indicate products of log-likelihood derivatives included within the expected values. Some examples include

$$I_{\lambda, \lambda} = \frac{1}{n} E\left(\frac{\partial l}{\partial\lambda}\right)^2$$

and

$$I_{\lambda, \psi\lambda} = \frac{1}{n} E\left(\frac{\partial l(\psi, \lambda)}{\partial\lambda}\right)\left(\frac{\partial^2 l(\psi, \lambda)}{\partial\psi \, \partial\lambda}\right).$$

Since $l'_M(\psi) = l'_P(\psi) - \frac{1}{2} \partial\{\log|j_{\lambda\lambda}(\psi, \hat{\lambda}_\psi)|\}/\partial\psi$, it will be convenient to expand $l'_P(\psi)$ separately. By expanding $l(\psi, \hat{\lambda}_\psi)$ in a Taylor series about λ and differentiating the resulting series with respect to ψ we obtain

$$l'_P(\psi) = \sqrt{n}\, a_{11} + a_{12} + \frac{1}{\sqrt{n}} a_{13} + \frac{1}{n} a_{14} + O_p(n^{-3/2})$$

where the a_{ij} are polynomials of order j in Z with coefficients that are functions of I. The first two terms are

$$a_{11} = Z_\psi \tag{21.5}$$

$$a_{12} = \frac{Z_{\psi\lambda}Z_\lambda}{I_{\lambda\lambda}} + \frac{1}{2}\frac{Z_\lambda^2 I_{\psi\lambda\lambda}}{I_{\lambda\lambda}^2}$$

and expansions for a_{13} and a_{14} are given in the Appendix as (A1). Combining (21.5) and (A1) gives

$$El'_P(\psi) = -\frac{I_{\psi\lambda\lambda}}{2I_{\lambda\lambda}} + \frac{1}{n}A_{14} + O(n^{-2}), \tag{21.6}$$

where $A_{14} = E(\sqrt{n}\, a_{13} + a_{14})$ is given by expression (A2). Note from (21.6) that the profile score is $O_p(\sqrt{n})$ and its bias is $O(1)$.

Differentiating $\log|nj_{\lambda\lambda}(\psi, \hat{\lambda}_\psi)|$ with respect to ψ and performing Taylor series expansions on the resulting numerator and denominator separately, we have

$$\frac{\partial}{\partial\psi}\log|nj_{\lambda\lambda}(\psi, \hat{\lambda}_\psi)| = a_{20} + \frac{1}{\sqrt{n}}a_{21} + \frac{1}{n}a_{22} + O_p(n^{-3/2}) \tag{21.7}$$

with again a_{2j} a polynomial of order j in Z with coefficients that are functions of I, and $Ea_{21} = 0$. In this case a_{20} is in fact non-random:

$$a_{20} = -\frac{I_{\psi\lambda\lambda}}{I_{\lambda\lambda}},$$

and expressions for a_{21}, a_{22} and $A_{22} = Ea_{22}$ are given in (A3) and (A4). Combining these results gives

$$\begin{aligned}
El'_M(\psi) = \frac{1}{n}\Bigg\{ &\frac{1}{8I_{\lambda\lambda}^2}(3I_{\psi\lambda\lambda\lambda\lambda} + 8I_{\lambda,\psi\lambda\lambda\lambda} + 8I_{\psi\lambda,\lambda\lambda\lambda} + 12I_{\lambda\lambda,\psi\lambda\lambda} + 8I_{\psi,\psi\lambda,\lambda\lambda} + 4I_{\lambda,\lambda,\psi\lambda\lambda}) \\
&+ \frac{1}{12I_{\lambda\lambda}^3}[I_{\lambda\lambda\lambda}(4I_{\psi\lambda\lambda\lambda} + 15I_{\psi\lambda,\lambda\lambda} - 9I_{\lambda,\lambda,\psi\lambda}) \\
&\qquad\qquad\qquad - 24I_{\lambda,\lambda\lambda}(I_{\lambda,\psi\lambda\lambda} + I_{\lambda,\lambda,\psi\lambda}) \\
&\quad + I_{\psi\lambda\lambda}(-I_{\lambda\lambda\lambda\lambda} + 2I_{\lambda,\lambda\lambda\lambda} - 12I_{\lambda,\lambda,\lambda\lambda} - 4I_{\lambda,\lambda,\lambda,\lambda})] \\
&+ \frac{I_{\psi\lambda\lambda}}{8I_{\lambda\lambda}^4}(7I_{\lambda\lambda\lambda}^2 + 36I_{\lambda\lambda\lambda}I_{\lambda,\lambda\lambda} + 24I_{\lambda,\lambda\lambda}^2)\Bigg\} + O(n^{-2}),
\end{aligned} \tag{21.8}$$

showing that the score statistic from the modified profile likelihood is unbiased to $O(n^{-1})$. One interpretation of this is that $l_M(\psi)$ has more nearly the properties of a genuine log-likelihood. Another interpretation is that the score equation is more nearly an unbiased estimating equation for ψ. The unbiasedness result was obtained by Liang (1987): his derivation uses an indirect argument adapted from Cox and Reid (1987) based on conditioning on the observed score statistic (essentially Z_ψ above). McCullagh and Tibshirani (1990) developed modifications to $l_P(\psi)$ that guarantee, at least to first order, $El_P'(\psi) = 0$, although they evaluated the expectation at $(\psi, \hat{\lambda}_\psi)$, whereas we are working at the notional true value (ψ, λ). As outlined at the end of this section, the unbiasedness holds only if ψ and λ are orthogonal, so one effect of their modification is in effect to find approximate orthogonal parameters. This unbiasedness result is also discussed in the special context of full exponential families by Levin and Kong (1990).

It is of course also of relevance to examine the variance properties of the score statistic. Of particular interest is whether or not the variance of U_M is equal to the expected value of its derivative, as would be true if l_M were a proper likelihood function. McCullagh and Tibshirani (1990) proposed a scale adjustment to l_P to ensure that this property holds, at least to first order.

A derivation similar to that outlined above shows that

$$-El_P''(\psi) - \mathrm{Var}\, l_P'(\psi) = \frac{1}{2I_{\lambda\lambda}} (I_{\psi\psi\lambda\lambda} - 2I_{\psi\lambda,\psi\lambda} - 2I_{\psi,\lambda,\psi\lambda})$$

$$- \frac{I_{\psi\psi\lambda}}{2I_{\lambda\lambda}^2}(I_{\lambda,\lambda\lambda} + I_{\lambda,\lambda,\lambda}) + \frac{I_{\psi\lambda\lambda}}{2I_{\lambda\lambda}^2}(2I_{\psi,\lambda\lambda} + I_{\psi\lambda\lambda}) + O(n^{-1})$$

$$(21.9)$$

and

$$-El_M''(\psi) - \mathrm{Var}\, l_M'(\psi) = \frac{1}{I_{\lambda\lambda}} (I_{\psi\psi\lambda\lambda} + I_{\lambda,\psi\psi\lambda}) + \frac{I_{\psi\psi\lambda}}{I_{\lambda\lambda}^2}(I_{\lambda\lambda\lambda} + I_{\lambda,\lambda\lambda}) + O(n^{-1}),$$

$$(21.10)$$

so both the profile and modified profile likelihood fail to have the property that the derivative of the score has mean value equal to the variance of the score statistic, to second order. Although the right-hand side of (21.10) has a simpler structure than that of (21.9), it is difficult to make any general conclusions. We show in Example 2.3 that if ψ is the canonical parameter of a full exponential family, then (21.10) is zero, but if ψ is the expectation parameter, it is not zero except possibly in special cases. Because the right-hand side of (21.10) is not zero in general, it follows that the adjusted likelihood of McCullagh and Tibshirani (1990) will not equal the modified profile likelihood l_M, except in special cases.

Liang (1987) considered examining the mean-square error of the modified score function U_M, again in a conditional sense, and showed that U_M was preferred to U_P on this basis. Another aspect of the second-order properties of U_M that may be relevant for the theory of estimating equations is the quantity

$$\frac{E\{U_M^2(\psi)\}}{[E\{-U_M'(\psi)\}]^2},$$

as Godambe (1976) showed that this quantity is minimized by the optimal estimating equation. The expression for this in terms of expected likelihood derivatives is

$$\frac{1}{nI_{\psi\psi}}\left[1 + \frac{1}{n}\left\{\frac{I_{\psi\lambda,\psi\lambda}}{I_{\lambda\lambda}} - \frac{1}{2}\left(\frac{I_{\psi\lambda\lambda}^2}{I_{\lambda\lambda}^2}\right)\right\} + O(n^{-2})\right]. \tag{21.11}$$

Example 21.1   *The inverse Gaussian distribution*   We write the density in the form

$$f(y; \mu, \sigma) = \frac{1}{\sqrt{(2\pi\sigma)}} y^{-3/2} \exp -\frac{1}{2\sigma\mu^2 y}(y - \mu)^2, \qquad y, \mu, \sigma > 0.$$

We take μ as the parameter of interest, and σ the nuisance parameter. In the exponential family form of the density, the canonical parameters are $(-1/(2\sigma\mu^2), -1/(2\sigma))$, with sufficient statistics $(y, 1/y)$, showing that μ and σ are orthogonal, since $Ey = \mu$, and canonical parameters are orthogonal to expectation parameters in exponential families. The log-likelihood function based on a sample of size n is

$$l(\mu, \sigma) = -\frac{n}{2}\log \sigma - \frac{1}{2\sigma\mu^2}\sum \frac{(y_i - \mu)^2}{y_i}$$

with $\hat{\mu} = \bar{y}$ and $\hat{\sigma}_\mu = \sum (y_i - \mu)^2/y_i n\mu^2$. The profile log-likelihood function (21.1) is

$$l_P(\mu) = -(n/2)\log \hat{\sigma}_\mu$$

and the modified profile log-likelihood function (21.2) is

$$l_M(\mu) = -((n-2)/2)\log \hat{\sigma}_\mu$$

showing that l_P and l_M give the same point estimate $\hat{\mu}$, from the estimating equations

$$U_P(\mu) = -\frac{n(\bar{y} - \mu)}{\hat{\sigma}_\mu \mu^3} = 0$$

and

$$U_M(\mu) = \left(\frac{n-2}{n}\right) U_P(\mu) = 0.$$

Although the mean and variance of the two estimating equations could be computed directly, we used formulas (21.9) and (21.10) to show that

$$-EU'_P - \text{Var } U_P = -2/(\sigma\mu^3) + O(n^{-1})$$

and

$$-EU'_M - \text{Var } U_M = O(n^{-1}).$$

It can also be verified that the $O(n^{-1})$ term in the bias, given in (21.8), is zero for this example. The same results obtain for the normal (μ, τ) distribution with μ the parameter of interest and τ the nuisance parameter (Ferguson 1989).

Example 21.2 *Exponential family* We write

$$f(y; \psi, \phi) = \exp\{s\psi + t\phi - k(\psi, \phi)\} f(y)$$

and assume that the components of Y are independent and identically distributed. The orthogonal version of ϕ is the expectation parameter $\lambda = ET = k_{01}(\psi, \phi)$, where $k_{ij} = \partial^{i+j} k(\psi, \phi)/\partial\psi^i \partial\phi^j$. It follows on differentiation of the definition of λ with respect to λ and ψ that

$$1 = k_{02}(\partial\phi/\partial\lambda)$$

$$0 = k_{11} + k_{02}(\partial\phi/\partial\psi),$$

etc. Thus all the terms in the right-hand side of (21.10) are functions of the derivatives of $k(\psi, \phi(\psi, \lambda))$. It turns out that $I_{\psi\psi\lambda} = 0$ and $I_{\psi\psi\lambda\lambda} = -I_{\lambda, \psi\psi\lambda}$, showing that $-El''_M(\psi) = \text{Var } l'_M(\psi)$. It is not that case that the right-hand side of (21.9) equals 0, however. It follows that problems close to this exponential family form have (21.10) nearer to 0 than (21.9), and in this sense show an advantage of l_M over l_P.

If instead the parameter of interest is a component of the expectation parameter, and the nuisance parameter is the canonical parameter, as in Example 21.1 above, expression (21.10) is calculated with the roles of ψ and λ reversed. Rather than redefine the exponential family, we temporarily use the notation λ for the parameter of interest and ψ for the nuisance parameter. The $O(1)$ term in the right-hand side of (21.10) becomes

$$\frac{1}{I_{\psi\psi}}(I_{\lambda\lambda\psi\psi} + I_{\psi, \lambda\lambda\psi}) + \frac{I_{\lambda\lambda\psi}}{I_{\psi\psi}^2}(I_{\psi\psi\psi} + I_{\psi, \psi\psi}), \tag{21.12}$$

where again the Is are all functions of the derivatives of $k(\psi, \phi(\psi, \lambda))$. In this case $I_{\psi, \lambda\lambda\psi} = 0$ and $I_{\psi, \psi\psi} = 0$, but (21.12) is not zero in general, and is a rather complicated function of the derivatives of k, given in the Appendix as equation (A5).

Example 21.3 *Components of variance* We now consider a simplified version of a components-of-variance problem, with Y_i independently

$N(\mu, \tau + v_i)$, for $i = 1, \ldots, n$, and v_i known. This corresponds to an unbalanced one-way classification in which the within-group variance component is estimated with a large number of degrees of freedom. Taking τ as the parameter of interest, we have

$$l_P(\tau) = -\frac{1}{2} \sum_{i=1}^n \log(\tau + v_i) - \frac{1}{2} \sum_{i=1}^n \frac{y_i^2}{\tau + v_i} + \frac{1}{2} \frac{\{\sum (\tau + v_i)^{-1}\}^2}{\sum (\tau + v_i)^{-1}}$$

and

$$l_M(\tau) = l_P(\tau) - \tfrac{1}{2} \log \left\{ \sum_{i=1}^n (\tau + v_i)^{-1} \right\}.$$

Although the estimating equations for $\hat{\psi}$ and $\hat{\psi}_M$ are complicated, it is apparent from the form of the correction term that they will not give the same point estimate. It is possible to verify that the right-hand side of (21.10) is zero, using definitions of I appropriate to the non-identically distributed case. We have

$$I_{\mu\mu} = n^{-1} \sum (\tau + v_i)^{-1},$$

$$I_{\tau\mu\mu} = n^{-1} \sum (\tau + v_i)^{-2},$$

$$I_{\mu,\tau\tau\mu} = -I_{\tau\tau\mu\mu} = 2n^{-1} \sum (\tau + v_i)^{-3},$$

and

$$I_{\tau\mu,\tau\mu} = -I_{\tau,\mu,\tau\mu} = n^{-1} \sum (\tau + v_i)^{-3},$$

with the other Is appearing in (21.9) or (21.10) equal to zero.

Switching the roles of the parameters, so that μ becomes the parameter of interest, leads to quite complicated estimating equations from U_P and U_M, because the expression for $\hat{\tau}_\mu$ does not have a convenient closed-form expression. However, it is relatively easy to check that the right-hand side of (21.10) is 0 but that the right-hand side of (21.9) is

$$-2 \frac{\sum (\tau + v_i)^{-3}}{\sum (\tau + v_i)^{-2}}.$$

The derivation in the case of $p \times 1$ vector λ follows the above with minor notational modifications. Vectors are distinguished by boldface type, matrices by the presence of two dots in the subscript, and three- and four-way arrays by the presence of three and four dots, respectively, in the subscript. Thus, for example,

$$I_{\lambda\lambda..} = (I_{\lambda\lambda..}(j, k))$$

$$= \frac{1}{n} \left(E \left\{ -\frac{\partial^2}{\partial \lambda_j \, \partial \lambda_k} l(\psi, \lambda) \right\} \right), \qquad j, k = 1, \ldots, p, \qquad (21.13)$$

and is the matrix counterpart of $I_{\lambda\lambda}$. Similar correspondences hold for other

*I*s used. The random *Z*s are modified in a similar fashion:

$$\mathbf{Z}_{\psi\lambda} = (Z_{\psi\lambda_j}) = \left(\frac{1}{\sqrt{n}} \frac{\partial^2}{\partial\psi \, \partial\lambda_j} l(\psi, \lambda)\right), \qquad j = 1, \ldots, p$$

corresponds to the earlier $Z_{\psi\lambda}$ and so on.

Following the method used for scalar λ, expansion of $(\partial/\partial\psi)l(\psi, \hat{\lambda}_\psi)$ about (ψ, λ) yields

$$l_P'(\psi) = \sqrt{n}\, Z_\psi + \mathbf{Z}_\lambda' I_{\lambda\lambda..}^{-1} \mathbf{Z}_{\psi\lambda}$$

$$+ \tfrac{1}{2} \mathbf{Z}_\lambda' I_{\lambda\lambda..}^{-1} I_{\psi\lambda\lambda..} I_{\lambda\lambda..}^{-1} \mathbf{Z}_\lambda + \frac{1}{\sqrt{n}} a_{13}^* + \frac{1}{n} a_{14}^* + O_p(n^{-3/2}), \quad (21.14)$$

which has the expected value

$$El_P'(\psi) = -\tfrac{1}{2}\, tr(I_{\lambda\lambda..}^{-1} I_{\psi\lambda\lambda..}) + O(n^{-1}).$$

As in the scalar case, the bias in the profile score is again $O(1)$.

Expansion of the remaining element in (21.2) proceeds as in the scalar case to give

$$\frac{\partial}{\partial\psi} \log\{\det nj_{\lambda\lambda}(\psi, \hat{\lambda}_\psi)\} = -tr(I_{\lambda\lambda..}^{-1} I_{\psi\lambda\lambda..}) + \frac{1}{\sqrt{n}} a_{21}^* + \frac{1}{n} a_{22}^* + O_p(n^{-3/2}),$$

$$(21.15)$$

which again has a non-random leading term. Combining (21.14) and (21.15) yields

$$El_M'(\psi) = O(n^{-1}) \qquad (21.16)$$

again showing that the modified profile score is unbiased to one higher order than the profile score.

Similar procedures verify that

$$-El_P''(\psi) - \mathrm{Var}\, l_P'(\psi) = O(1) \qquad (21.17)$$

and

$$-El_M''(\psi) - \mathrm{Var}\, l_M'(\psi) = O(1), \qquad (21.18)$$

paralleling the results of the scalar case. The vector version of (21.10) is

$$-El_M''(\psi) - \mathrm{Var}\, l_M'(\psi) = tr\{I_{\lambda\lambda..}^{-1}(I_{\psi\psi\lambda\lambda..} + I_{\lambda,\psi\psi\lambda..})\}$$

$$+ tr\{I_{\lambda\lambda..}^{-1}[I_{\lambda,\lambda\lambda...} \circ (I_{\lambda\lambda..}^{-1} \mathbf{I}_{\psi\psi\lambda})]$$

$$+ I_{\lambda\lambda..}^{-1}[I_{\lambda\lambda\lambda...} \circ I_{\lambda\lambda..}^{-1} \mathbf{I}_{\psi\psi\lambda})]\} + O(n^{-1}) \qquad (21.19)$$

where

$$A_{...} \circ y = \sum_k a_{ijk} y_k.$$

The derivation of equation (21.19) makes extensive use of several multivariate Bartlett relations, which allows substantial simplification under parameter orthogonality. The $1/\sqrt{n}$ and $1/n$ terms in the stochastic expansions of l_M' and l_P' are given in the Appendix.

It is essential for the vanishing of the $O(n^{-1})$ term in the bias of $U_M(\psi)$ that λ be orthogonal to ψ. If ψ and λ are not orthogonal parameters, $I_{\psi\lambda} \neq 0$ so that (21.4) becomes

$$Z_{\psi\lambda} = \frac{1}{\sqrt{n}} \frac{\partial^2}{\partial\psi\,\partial\lambda} l(\psi, \lambda) + \sqrt{n}\, I_{\psi\lambda}.$$

In this case,

$$l_P'(\psi) = \sqrt{n}\left(Z_\psi - \frac{Z_\lambda I_{\psi\lambda}}{I_{\lambda\lambda}}\right)$$

$$+ \left(\frac{Z_\lambda Z_{\psi\lambda}}{I_{\lambda\lambda}} + \frac{Z_\lambda^2 I_{\psi\lambda\lambda}}{2I_{\lambda\lambda}^2} - \frac{Z_\lambda Z_{\lambda\lambda} I_{\psi\lambda}}{I_{\lambda\lambda}^2} - \frac{Z_\lambda^2 I_{\psi\lambda} I_{\lambda\lambda\lambda}}{2I_{\lambda\lambda}^3}\right) + O_p(n^{-1/2}), \quad (21.20)$$

having expected value

$$El_P'(\psi) = \frac{I_{\lambda,\psi\lambda}}{I_{\lambda\lambda}} - \frac{I_{\psi\lambda\lambda}}{2I_{\lambda\lambda}} + \frac{I_{\psi\lambda}}{2I_{\lambda\lambda}^2}(I_{\lambda,\lambda\lambda} + I_{\lambda,\lambda,\lambda}) + O(n^{-1}). \quad (21.21)$$

Using the methods of this Section, it can be shown that

$$\frac{\partial}{\partial\psi} \log|nj_{\lambda\lambda}(\psi, \hat{\lambda}_\psi)| = -\frac{I_{\psi\lambda\lambda}}{I_{\lambda\lambda}} + \frac{I_{\psi\lambda} I_{\lambda\lambda\lambda}}{I_{\lambda\lambda}^2} + O_p(n^{-1/2}). \quad (21.22)$$

Combining (21.20) and (21.22) gives

$$El_M'(\psi) = \frac{1}{I_{\lambda\lambda}}(I_{\lambda,\psi\lambda} + I_{\psi\lambda\lambda}) + \frac{I_{\psi\lambda}}{2I_{\lambda\lambda}^2}(I_{\lambda,\lambda\lambda} + I_{\lambda,\lambda,\lambda} - I_{\lambda\lambda\lambda}) + O(n^{-1}). \quad (21.23)$$

The lack of orthogonality thus increases the bias in the modified profile score from $O(n^{-1})$ to $O(1)$.

21.3   Discussion

The modified profile likelihood function is not a genuine likelihood function, meaning it does not represent the density or probability for an observed random vector, so it cannot have all the properties of a likelihood function, although it is disappointing that the additional terms on the right-hand side of (21.10) are not zero in general.

The simplest way to verify that l_M is not a log-likelihood is to note that it is not invariant under reparametrizations of the nuisance parameter λ. One question is whether or not the right-hand side of (21.10) can be set equal to zero by suitable choice of orthogonal parameter λ^*, say. It can be verified that the right-hand side of (21.10) in the new parametrization is

$$\frac{1}{I_{\lambda\lambda}^*}(I_{\psi\psi\lambda\lambda}^* + I_{\lambda,\psi\psi\lambda}^*) + \frac{I_{\psi\psi\lambda}^*}{I_{\lambda\lambda}^{*2}}(I_{\lambda\lambda\lambda}^* + I_{\lambda,\lambda\lambda}^*) - \frac{I_{\psi\psi\lambda}^*}{I_{\lambda\lambda}^*}\frac{\lambda^{*''}}{(\lambda^{*'})^2}.$$

The detailed formulae of Section 21.2 are likely to be central to the development of a more refined asymptotic distribution for associated test statistics or estimating functions constructed using adjusted profile likelihoods. Although the arguments involved are local, the ultimate objective would be to cast the results in an equivalent globally formulated version.

References

Barndorff-Nielsen, O. E. (1983). On a formula for the distribution of the maximum likelihood estimator. *Biometrika*, **70**, 343–65.

Barndorff-Nielsen, O. E. (1986). Inference on full or partial parameters, based on the standardized log likelihood ratio. *Biometrika*, **73**, 307–33.

Barndorff-Nielsen, O. E. (1987). Contribution to the discussion of 'Parameter orthogonality and approximate conditional inference' by D. R. Cox and N. Reid. *J. Roy. Stat. Soc.* B, **49**, 18–20.

Barndorff-Nielsen, O. E. (1988). *Parametric statistical models and likelihood.* Lecture Notes in Statistics No. 50, Springer-Verlag, New York.

Bartlett, M. S. (1936). The information available in small samples. *Proc. Camb. Phil. Soc.*, **34**, 33–40.

Cox, D. R. and Reid, N. (1987). Parameter orthogonality and approximate conditional inference (with discussion). *J. Roy. Stat. Soc. B.*, **49**, 1–39.

Cox, D. R. and Reid, N. (1989). Stability of maximum likelihood estimators of orthogonal parameters. *Canad. J. Stat.*, **17**, 229–34.

Cox, D. R. and Reid, N. (1990). Approximate computation of orthogonal parameters. Preprint.

Ferguson, H. (1989). Asymptotic properties of a conditional maximum likelihood estimator. Ph.D. Thesis, Dept. of Statistics, University of Toronto.

Firth, D. (1987). Contribution to the discussion of 'Parameter orthogonality and approximate conditional inference' by D. R. Cox and N. Reid. *J. Roy. Stat. Soc.* B., **49**, 22–23.

Fraser, D. A. S. and Reid, N. (1989). Adjustments to profile likelihood. *Biometrika*, **76**, 477–88.

Godambe, V. P. (1976). Conditional likelihood and unconditional optimum estimating equations. *Biometrika*, **63**, 277–84.

Kalbfleisch, J. D. and Sprott, D. A. (1970). Application of likelihood methods to models involving large numbers of parameters (with discussion). *J. Roy. Stat. Soc.* B., **32**, 175–208.

Levin, B. and Kong, F. (1990). Bartlett's bias correction to the profile score function is a saddlepoint correction. *Biometrika*, **77**, 219–21.

Liang, K. Y. (1987). Estimating functions and approximate conditional likelihood. *Biometrika*, **74**, 695–702.

McCullagh, P. and Tibshirani, R. (1990). A simple method for the adjustment of profile likelihoods. *J. Roy. Stat. Soc.* B., **52**, 325–44.

Neyman, J. and Scott, E. L. (1948). Consistent estimates based on partially consistent observations. *Econometrica*, **16**, 1–32.

Appendix

We record here some of the more lengthy formulae used to derive the results in Section 21.2. In the expansion of $l'_p(\psi)$ in the case of a scalar nuisance parameter, the $1/\sqrt{n}$ and $1/n$ terms are

$$
a_{13} = \frac{Z_\lambda^2 Z_{\psi\lambda\lambda}}{2I_{\lambda\lambda}^2} + \frac{Z_\lambda Z_{\lambda\lambda} Z_{\psi\lambda}}{I_{\lambda\lambda}^2} + \frac{Z_\lambda^3 I_{\psi\lambda\lambda\lambda}}{6I_{\lambda\lambda}^3}
$$

$$
+ \frac{Z_\lambda^2 Z_{\psi\lambda} I_{\lambda\lambda\lambda}}{2I_{\lambda\lambda}^3} + \frac{Z_\lambda^2 Z_{\lambda\lambda} I_{\psi\lambda\lambda}}{I_{\lambda\lambda}^3} + \frac{Z_\lambda^3 I_{\psi\lambda\lambda} I_{\lambda\lambda\lambda}}{2I_{\lambda\lambda}^4}
$$

$$
a_{14} = \frac{Z_\lambda^3 Z_{\psi\lambda\lambda\lambda}}{6I_{\lambda\lambda}^3} + \frac{Z_\lambda^2 Z_{\psi\lambda} Z_{\lambda\lambda\lambda}}{2I_{\lambda\lambda}^3} + \frac{Z_\lambda^2 Z_{\lambda\lambda} Z_{\psi\lambda\lambda}}{I_{\lambda\lambda}^3}
$$

$$
+ \frac{Z_\lambda Z_{\psi\lambda} Z_{\psi\lambda} Z_{\lambda\lambda}^2}{I_{\lambda\lambda}^3} + \frac{Z_\lambda^4 I_{\psi\lambda\lambda\lambda\lambda}}{24I_{\lambda\lambda}^4} + \frac{Z_\lambda^3 Z_{\psi\lambda} I_{\lambda\lambda\lambda\lambda}}{6I_{\lambda\lambda}^4}
$$

$$
+ \frac{Z_\lambda^3 Z_{\lambda\lambda\lambda} I_{\psi\lambda\lambda\lambda}}{2I_{\lambda\lambda}^4} + \frac{Z_\lambda^3 Z_{\psi\lambda\lambda} I_{\lambda\lambda\lambda}}{2I_{\lambda\lambda}^4} + \frac{Z_\lambda^3 Z_{\lambda\lambda\lambda} I_{\psi\lambda\lambda}}{2I_{\lambda\lambda}^4}
$$

$$
+ \frac{3Z_\lambda^2 Z_{\psi\lambda} Z_{\lambda\lambda} I_{\lambda\lambda\lambda}}{2I_{\lambda\lambda}^4} + \frac{3Z_\lambda^2 Z_{\lambda\lambda}^2 I_{\psi\lambda\lambda}}{2I_{\lambda\lambda}^4}
$$

$$
+ \frac{Z_\lambda^4 I_{\psi\lambda\lambda} I_{\lambda\lambda\lambda\lambda}}{6I_{\lambda\lambda}^5} + \frac{Z_\lambda^4 I_{\lambda\lambda\lambda} I_{\psi\lambda\lambda\lambda}}{4I_{\lambda\lambda}^5} + \frac{Z_\lambda^3 Z_{\psi\lambda} I_{\lambda\lambda\lambda}^2}{2I_{\lambda\lambda}^5}
$$

$$
+ \frac{2Z_\lambda^3 Z_{\lambda\lambda} I_{\psi\lambda\lambda} I_{\lambda\lambda\lambda}}{I_{\lambda\lambda}^5} + \frac{5Z_\lambda^4 I_{\psi\lambda\lambda} I_{\lambda\lambda\lambda}^2}{8I_{\lambda\lambda}^6}
$$

$$\text{(A1)}$$

and $Ea_{13} = 0$,

$$
A_{14} = Ea_{14}
$$

$$
= \frac{1}{8I_{\lambda\lambda}^2}(I_{\psi\lambda\lambda\lambda\lambda} + 4I_{\lambda,\psi\lambda\lambda\lambda} + 4I_{\psi\lambda,\lambda\lambda\lambda} + 8I_{\lambda\lambda,\psi\lambda\lambda} + 8I_{\lambda,\psi\lambda,\lambda\lambda} + 4I_{\lambda,\lambda,\psi\lambda\lambda})
$$

$$
+ \frac{1}{12I_{\lambda\lambda}^3}[I_{\lambda\lambda\lambda}(4I_{\lambda,\psi\lambda\lambda} + 11I_{\psi\lambda,\lambda\lambda} - I_{\lambda,\lambda,\psi\lambda})
$$

$$
+ 12I_{\lambda,\lambda\lambda}(I_{\psi\lambda,\lambda\lambda} - I_{\lambda,\lambda,\psi\lambda}) - 2I_{\psi\lambda\lambda}(I_{\lambda\lambda\lambda\lambda} + I_{\lambda,\lambda\lambda\lambda} + I_{\lambda,\lambda,\lambda,\lambda})]
$$

$$
+ \frac{I_{\psi\lambda\lambda}}{8I_{\lambda\lambda}^4}(24I_{\lambda,\lambda\lambda}^2 + 12I_{\lambda\lambda\lambda} I_{\lambda,\lambda\lambda} - I_{\lambda\lambda\lambda}^2).
$$

$$\text{(A2)}$$

The $1/\sqrt{n}$ and $1/n$ terms in the expansion of $l'_P(\psi)$ in the case of a vector nuisance parameter are

$$
\begin{aligned}
a^*_{13} = &\tfrac{1}{2}\mathbf{Z}'_\lambda I^{-1}_{\lambda\lambda..} \mathbf{Z}_{\psi\lambda..} I^{-1}_{\lambda\lambda..}\mathbf{Z}_\lambda + \mathbf{Z}'_\lambda I^{-1}_{\lambda\lambda..} \mathbf{Z}_{\lambda\lambda..} I^{-1}_{\lambda\lambda..}\mathbf{Z}_{\psi\lambda} + \tfrac{1}{6}I_{\psi\lambda\lambda\lambda...} \circ (I^{-1}_{\lambda\lambda..}\mathbf{Z}_\lambda)^3 \\
&+ \tfrac{1}{2}I_{\lambda\lambda\lambda...} \circ (I^{-1}_{\lambda\lambda..}\mathbf{Z}_\lambda)^2 \circ (I^{-1}_{\lambda\lambda..}\mathbf{Z}_{\psi\lambda}) + \mathbf{Z}'_\lambda I^{-1}_{\lambda\lambda..}I_{\psi\lambda\lambda..} I^{-1}_{\lambda\lambda..}\mathbf{Z}_{\lambda\lambda..} I^{-1}_{\lambda\lambda..}\mathbf{Z}_\lambda \\
&+ \tfrac{1}{2}I_{\psi\lambda\lambda..}I^{-1}_{\lambda\lambda..}\mathbf{Z}_\lambda \circ I^{-1}_{\lambda\lambda..}[I_{\lambda\lambda\lambda...} \circ (I^{-1}_{\lambda\lambda..}\mathbf{Z}_\lambda)^2]
\end{aligned}
$$

and

$$
\begin{aligned}
a^*_{14} = &\tfrac{1}{6}\mathbf{Z}_{\psi\lambda\lambda\lambda...} \circ (I^{-1}_{\lambda\lambda..}\mathbf{Z}_\lambda)^3 + \tfrac{1}{2}\mathbf{Z}_{\lambda\lambda\lambda...} \circ (I^{-1}_{\lambda\lambda..}\mathbf{Z}_\lambda)^2 \circ (I^{-1}_{\lambda\lambda..}\mathbf{Z}_{\psi\lambda}) \\
&+ \mathbf{Z}'_\lambda I^{-1}_{\lambda\lambda..}\mathbf{Z}_{\psi\lambda\lambda..} I^{-1}_{\lambda\lambda..}\mathbf{Z}_{\lambda\lambda..} I^{-1}_{\lambda\lambda..}\mathbf{Z}_\lambda + \mathbf{Z}'_\lambda I^{-1}_{\lambda\lambda..}\mathbf{Z}_{\lambda\lambda..} I^{-1}_{\lambda\lambda..}\mathbf{Z}_{\lambda\lambda..} I^{-1}_{\lambda\lambda..}\mathbf{Z}_{\psi\lambda} \\
&+ \tfrac{1}{24}I_{\psi\lambda\lambda\lambda\lambda....} \circ (I^{-1}_{\lambda\lambda..}\mathbf{Z}_\lambda)^4 + I_{\lambda\lambda\lambda\lambda....} \circ (I^{-1}_{\lambda\lambda..}\mathbf{Z}_\lambda)^3 \circ (I^{-1}_{\lambda\lambda..}\mathbf{Z}_{\psi\lambda}) \\
&+ \tfrac{1}{2}I_{\psi\lambda\lambda\lambda...} \circ (I^{-1}_{\lambda\lambda..}\mathbf{Z}_{\lambda\lambda..} I^{-1}_{\lambda\lambda..}\mathbf{Z}_\lambda) \circ (I^{-1}_{\lambda\lambda..}\mathbf{Z}_\lambda)^2 \\
&+ \tfrac{1}{2}\mathbf{Z}_{\psi\lambda\lambda..} \circ (I^{-1}_{\lambda\lambda..}\mathbf{Z}_\lambda) \circ I^{-1}_{\lambda\lambda..}[I_{\lambda\lambda\lambda...} \circ (I^{-1}_{\lambda\lambda..}\mathbf{Z}_\lambda)^2] \\
&+ \tfrac{1}{2}I_{\psi\lambda\lambda..} \circ (I^{-1}_{\lambda\lambda..}\mathbf{Z}_\lambda) \circ I^{-1}_{\lambda\lambda..}[\mathbf{Z}_{\lambda\lambda\lambda...} \circ (I^{-1}_{\lambda\lambda..}\mathbf{Z}_\lambda)^2] \\
&+ \tfrac{3}{2}I_{\lambda\lambda\lambda...} \circ (I^{-1}_{\lambda\lambda..}\mathbf{Z}_\lambda)^2 \circ (I^{-1}_{\lambda\lambda..}\mathbf{Z}_{\lambda\lambda..} I^{-1}_{\lambda\lambda..}\mathbf{Z}_{\psi\lambda}) \\
&+ \tfrac{3}{2}\mathbf{Z}'_\lambda I^{-1}_{\lambda\lambda..}I_{\psi\lambda\lambda..} I^{-1}_{\lambda\lambda..}\mathbf{Z}_{\lambda\lambda..} I^{-1}_{\lambda\lambda..}\mathbf{Z}_{\lambda\lambda..} I^{-1}_{\lambda\lambda..}\mathbf{Z}_\lambda \\
&+ \tfrac{1}{6}I_{\psi\lambda\lambda..}I^{-1}_{\lambda\lambda..}\mathbf{Z}_\lambda \circ I^{-1}_{\lambda\lambda..}[I_{\lambda\lambda\lambda\lambda....} \circ (I^{-1}_{\lambda\lambda..}\mathbf{Z}_\lambda)^3] \\
&+ \tfrac{1}{4}I_{\psi\lambda\lambda\lambda...} \circ (I^{-1}_{\lambda\lambda..}\mathbf{Z}_\lambda)^2 \circ I^{-1}_{\lambda\lambda..}[I_{\lambda\lambda\lambda...} \circ (I^{-1}_{\lambda\lambda..}\mathbf{Z}_\lambda)^2] \\
&+ \tfrac{1}{2}I_{\lambda\lambda\lambda...} \circ (I^{-1}_{\lambda\lambda..}\mathbf{Z}_\lambda) \circ I^{-1}_{\lambda\lambda..}[I_{\lambda\lambda\lambda...} \circ (I^{-1}_{\lambda\lambda..}\mathbf{Z}_\lambda)^2] \circ (I^{-1}_{\lambda\lambda..}\mathbf{Z}_{\psi\lambda}) \\
&+ 2I_{\psi\lambda\lambda..}I^{-1}_{\lambda\lambda..}\mathbf{Z}_{\lambda\lambda..} I^{-1}_{\lambda\lambda..}\mathbf{Z}_\lambda \circ I^{-1}_{\lambda\lambda..}[I_{\lambda\lambda\lambda...} \circ (I^{-1}_{\lambda\lambda..}\mathbf{Z}_\lambda)^2] \\
&+ \tfrac{5}{8}I_{\psi\lambda\lambda..}(I^{-1}_{\lambda\lambda..}[I_{\lambda\lambda\lambda...} \circ (I^{-1}_{\lambda\lambda..}\mathbf{Z}_\lambda)^2])^2.
\end{aligned}
$$

Similarly for the expansion of $l'_M(\psi)$ we need

$$
-a_{21} = \frac{Z_{\psi\lambda\lambda}}{I_{\lambda\lambda}} + \frac{Z_\lambda I_{\psi\lambda\lambda\lambda}}{I^2_{\lambda\lambda}} + \frac{Z_{\psi\lambda} I_{\lambda\lambda\lambda}}{I^2_{\lambda\lambda}}
$$

$$
+ \frac{Z_{\lambda\lambda} I_{\psi\lambda\lambda}}{I^2_{\lambda\lambda}} + \frac{2 Z_\lambda I_{\psi\lambda\lambda} I_{\lambda\lambda\lambda}}{I^3_{\lambda\lambda}} \tag{A3}
$$

$$
-a_{22} = \frac{Z_\lambda Z_{\psi\lambda\lambda\lambda}}{I^2_{\lambda\lambda}} + \frac{Z_{\psi\lambda} Z_{\lambda\lambda\lambda}}{I^2_{\lambda\lambda}} + \frac{Z_{\lambda\lambda} Z_{\psi\lambda\lambda}}{I^2_{\lambda\lambda}}
$$

$$
+ \frac{Z^2_\lambda I_{\psi\lambda\lambda\lambda\lambda}}{2 I^3_{\lambda\lambda}} + \frac{Z_\lambda Z_{\psi\lambda} I_{\lambda\lambda\lambda\lambda}}{I^3_{\lambda\lambda}} + \frac{2 Z_\lambda Z_{\lambda\lambda} I_{\psi\lambda\lambda\lambda}}{I^3_{\lambda\lambda}}
$$

$$+ \frac{2Z_\lambda Z_{\psi\lambda\lambda} I_{\lambda\lambda\lambda}}{I_{\lambda\lambda}^3} + \frac{2Z_\lambda Z_{\lambda\lambda\lambda} I_{\psi\lambda\lambda}}{I_{\lambda\lambda}^3} + \frac{2Z_{\psi\lambda} Z_{\lambda\lambda} I_{\lambda\lambda\lambda}}{I_{\lambda\lambda}^3}$$

$$+ \frac{Z_{\lambda\lambda}^2 I_{\psi\lambda\lambda}}{I_{\lambda\lambda}^3} + \frac{3Z_\lambda^2 I_{\psi\lambda\lambda} I_{\lambda\lambda\lambda\lambda}}{2I_{\lambda\lambda}^4} + \frac{2Z_\lambda^2 I_{\lambda\lambda\lambda} I_{\psi\lambda\lambda\lambda}}{I_{\lambda\lambda}^4}$$

$$+ \frac{2Z_\lambda Z_{\psi\lambda} I_{\lambda\lambda\lambda}^2}{I_{\lambda\lambda}^4} + \frac{6Z_\lambda Z_{\lambda\lambda} I_{\psi\lambda\lambda} I_{\lambda\lambda\lambda}}{I_{\lambda\lambda}^4} + \frac{4Z_\lambda^2 I_{\psi\lambda\lambda} I_{\lambda\lambda\lambda}^2}{I_{\lambda\lambda}^5}$$

and

$$A_{22} = Ea_{22} = -\frac{1}{2I_{\lambda\lambda}^2} (I_{\psi\lambda\lambda\lambda\lambda} + 2I_{\lambda,\psi\lambda\lambda\lambda} + 2I_{\psi\lambda,\lambda\lambda\lambda} + 2I_{\lambda\lambda,\psi\lambda\lambda})$$

$$- \frac{1}{2I_{\lambda\lambda}^3} [4I_{\lambda\lambda\lambda}(I_{\psi\lambda\lambda\lambda} + I_{\lambda,\psi\lambda\lambda} + I_{\psi\lambda,\lambda\lambda})$$

$$+ 4I_{\lambda,\lambda\lambda} I_{\psi\lambda\lambda\lambda} + I_{\psi\lambda\lambda}(I_{\lambda\lambda\lambda\lambda} + 4I_{\lambda,\lambda\lambda\lambda} + 2I_{\lambda\lambda,\lambda\lambda})]$$

$$- \frac{2I_{\psi\lambda\lambda}}{I_{\lambda\lambda}^4} (I_{\lambda\lambda\lambda}^2 + 3I_{\lambda\lambda\lambda} I_{\lambda,\lambda\lambda}). \tag{A4}$$

In the vector case, the expressions corresponding to (A3) are

$$-a_{21}^* = \text{tr}\{I_{\lambda\lambda..}^{-1} Z_{\psi\lambda\lambda..}\} + \text{tr}\{I_{\lambda\lambda..}^{-1} I_{\psi\lambda\lambda\lambda...} \circ (I_{\lambda\lambda..}^{-1}\mathbf{Z}_\lambda)\}$$

$$+ \text{tr}\{I_{\lambda\lambda..}^{-1} [I_{\lambda\lambda\lambda...} \circ (I_{\lambda\lambda..}^{-1}\mathbf{Z}_{\psi\lambda})]\} + \text{tr}\{Z_{\lambda\lambda..} I_{\lambda\lambda..}^{-1} I_{\psi\lambda\lambda..} I_{\lambda\lambda..}^{-1}\}$$

$$+ \text{tr}\{I_{\lambda\lambda..}^{-1} I_{\psi\lambda\lambda..} I_{\lambda\lambda..}^{-1} [I_{\lambda\lambda\lambda...} \circ (I_{\lambda\lambda..}^{-1}\mathbf{Z}_\lambda)]\}$$

$$+ \text{tr}\{I_{\lambda\lambda..}^{-1} [I_{\lambda\lambda\lambda...} \circ I_{\lambda\lambda..}^{-1} I_{\psi\lambda\lambda..} I_{\lambda\lambda..}^{-1}\mathbf{Z}_\lambda]\}$$

and

$$-a_{22}^* = \text{tr}\{I_{\lambda\lambda..}^{-1} [Z_{\psi\lambda\lambda\lambda...} \circ (I_{\lambda\lambda..}^{-1}\mathbf{Z}_\lambda)]\} + \text{tr}[I_{\lambda\lambda..}^{-1} [Z_{\lambda\lambda\lambda...} \circ (I_{\lambda\lambda..}^{-1}\mathbf{Z}_{\psi\lambda})]\}$$

$$+ \text{tr}\{Z_{\lambda\lambda..} I_{\lambda\lambda..}^{-1} Z_{\psi\lambda\lambda..} I_{\lambda\lambda..}^{-1}\} + \tfrac{1}{2}\text{tr}\{I_{\lambda\lambda..}^{-1} [I_{\psi\lambda\lambda\lambda\lambda....} \circ (I_{\lambda\lambda..}^{-1}\mathbf{Z}_\lambda)^2]\}$$

$$+ \text{tr}\{I_{\lambda\lambda..}^{-1} [I_{\lambda\lambda\lambda\lambda....} \circ (I_{\lambda\lambda..}^{-1}\mathbf{Z}_{\psi\lambda}) \circ (I_{\lambda\lambda..}^{-1}\mathbf{Z}_\lambda)]\}$$

$$+ \text{tr}\{I_{\lambda\lambda..}^{-1} [I_{\psi\lambda\lambda\lambda...} \circ (I_{\lambda\lambda..}^{-1} Z_{\lambda\lambda..} I_{\lambda\lambda..}^{-1}\mathbf{Z}_\lambda)]\}$$

$$+ \text{tr}\{I_{\lambda\lambda..}^{-1} Z_{\lambda\lambda..} I_{\lambda\lambda..}^{-1} [I_{\psi\lambda\lambda\lambda...} \circ (I_{\lambda\lambda..}^{-1}\mathbf{Z}_\lambda)]\}$$

$$+ \text{tr}\{I_{\lambda\lambda..}^{-1} Z_{\psi\lambda\lambda..} I_{\lambda\lambda..}^{-1} [I_{\lambda\lambda\lambda...} \circ (I_{\lambda\lambda..}^{-1}\mathbf{Z}_\lambda)]\}$$

$$+ \text{tr}\{I_{\lambda\lambda..}^{-1} [I_{\lambda\lambda\lambda...} \circ I_{\lambda\lambda..}^{-1} Z_{\psi\lambda\lambda..} I_{\lambda\lambda..}^{-1}\mathbf{Z}_\lambda]\}$$

$$+ \text{tr}\{I_{\lambda\lambda..}^{-1} I_{\psi\lambda\lambda..} I_{\lambda\lambda..}^{-1} [Z_{\lambda\lambda\lambda...} \circ I_{\lambda\lambda..}^{-1}\mathbf{Z}_\lambda)]\}$$

$$+ \text{tr}\{I_{\lambda\lambda..}^{-1} [Z_{\lambda\lambda\lambda...} \circ I_{\lambda\lambda..}^{-1} I_{\psi\lambda\lambda..} I_{\lambda\lambda..}^{-1}\mathbf{Z}_\lambda]\}$$

$$+ \text{tr}\{I_{\lambda\lambda..}^{-1} [I_{\lambda\lambda\lambda...} \circ (I_{\lambda\lambda..}^{-1} Z_{\lambda\lambda..} I_{\lambda\lambda..}^{-1}\mathbf{Z}_{\psi\lambda})]\}$$

$$+ \operatorname{tr}\{I_{\lambda\lambda..}^{-1} Z_{\lambda\lambda..} I_{\lambda\lambda..}^{-1}[I_{\lambda\lambda\lambda...} \circ (I_{\lambda\lambda..}^{-1} \mathbf{Z}_{\psi\lambda})]\}$$

$$+ \operatorname{tr}\{I_{\psi\lambda\lambda..} I_{\lambda\lambda..}^{-1} Z_{\lambda\lambda..} I_{\lambda\lambda..}^{-1} Z_{\lambda\lambda..} I_{\lambda\lambda..}^{-1}\}$$

$$+ \tfrac{1}{2} \operatorname{tr}\{I_{\lambda\lambda..}^{-1} I_{\psi\lambda\lambda..} I_{\lambda\lambda..}^{-1}[I_{\lambda\lambda\lambda....} \circ (I_{\lambda\lambda..}^{-1} \mathbf{Z}_{\lambda})^2]\}$$

$$+ \operatorname{tr}\{I_{\lambda\lambda..}^{-1}[I_{\lambda\lambda\lambda....} \circ (I_{\lambda\lambda..}^{-1} I_{\psi\lambda\lambda..} I_{\lambda\lambda..}^{-1} \mathbf{Z}_{\lambda}) \circ (I_{\lambda\lambda..}^{-1} \mathbf{Z}_{\lambda})]\}$$

$$+ \tfrac{1}{2} \operatorname{tr}\{I_{\lambda\lambda..}^{-1}[I_{\psi\lambda\lambda\lambda...} \circ I_{\lambda\lambda..}^{-1}(I_{\lambda\lambda\lambda...} \circ (I_{\lambda\lambda..}^{-1} \mathbf{Z}_{\lambda})^2)]\}$$

$$+ \tfrac{1}{2} \operatorname{tr}\{I_{\lambda\lambda..}^{-1}[I_{\lambda\lambda\lambda...} \circ I_{\lambda\lambda..}^{-1}(I_{\psi\lambda\lambda\lambda...} \circ (I_{\lambda\lambda..}^{-1} \mathbf{Z}_{\lambda})^2)]\}$$

$$+ \operatorname{tr}\{I_{\lambda\lambda..}^{-1}[I_{\psi\lambda\lambda\lambda...} \circ (I_{\lambda\lambda..}^{-1} \mathbf{Z}_{\lambda}) \circ I_{\lambda\lambda..}^{-1}(I_{\lambda\lambda\lambda...} \circ (I_{\lambda\lambda..}^{-1} \mathbf{Z}_{\lambda}))]\}$$

$$+ \operatorname{tr}\{I_{\lambda\lambda..}^{-1}[I_{\lambda\lambda\lambda...} \circ I_{\lambda\lambda..}^{-1}(I_{\lambda\lambda\lambda...} \circ (I_{\lambda\lambda..}^{-1} \mathbf{Z}_{\lambda}) \circ (I_{\lambda\lambda..}^{-1} \mathbf{Z}_{\psi\lambda}))]\}$$

$$+ \operatorname{tr}\{I_{\lambda\lambda..}^{-1}[I_{\lambda\lambda\lambda...} \circ (I_{\lambda\lambda..}^{-1} \mathbf{Z}_{\psi\lambda}) \circ I_{\lambda\lambda..}^{-1}(I_{\lambda\lambda\lambda...} \circ (I_{\lambda\lambda..}^{-1} \mathbf{Z}_{\lambda}))]\}$$

$$+ \operatorname{tr}\{I_{\lambda\lambda..}^{-1} I_{\psi\lambda\lambda..} I_{\lambda\lambda..}^{-1}[I_{\lambda\lambda\lambda...} \circ (I_{\lambda\lambda..}^{-1} Z_{\lambda\lambda..} I_{\lambda\lambda..}^{-1} \mathbf{Z}_{\lambda})]\}$$

$$+ 2 \operatorname{tr}\{I_{\lambda\lambda..}^{-1} I_{\psi\lambda\lambda..} I_{\lambda\lambda..}^{-1} Z_{\lambda\lambda..} I_{\lambda\lambda..}^{-1}[I_{\lambda\lambda\lambda...} \circ (I_{\lambda\lambda..}^{-1} \mathbf{Z}_{\lambda})]\}$$

$$+ \operatorname{tr}\{I_{\lambda\lambda..}^{-1}[I_{\lambda\lambda\lambda...} \circ I_{\lambda\lambda..}^{-1} I_{\psi\lambda\lambda..} I_{\lambda\lambda..}^{-1} Z_{\lambda\lambda..} I_{\lambda\lambda..}^{-1} \mathbf{Z}_{\lambda}]\}$$

$$+ \operatorname{tr}\{I_{\lambda\lambda..}^{-1}[I_{\lambda\lambda\lambda...} \circ I_{\lambda\lambda..}^{-1} Z_{\lambda\lambda..} I_{\lambda\lambda..}^{-1} I_{\psi\lambda\lambda..} I_{\lambda\lambda..}^{-1} \mathbf{Z}_{\lambda}]\}$$

$$+ \operatorname{tr}\{I_{\lambda\lambda..}^{-1} Z_{\lambda\lambda..} I_{\lambda\lambda..}^{-1}[I_{\lambda\lambda\lambda...} \circ I_{\lambda\lambda..}^{-1} I_{\psi\lambda\lambda..} I_{\lambda\lambda..}^{-1} \mathbf{Z}_{\lambda}]\}$$

$$+ \tfrac{1}{2} \operatorname{tr}\{I_{\lambda\lambda..}^{-1} I_{\psi\lambda\lambda..} I_{\lambda\lambda..}^{-1}[I_{\lambda\lambda\lambda...} \circ I_{\lambda\lambda..}^{-1}(I_{\lambda\lambda\lambda...} \circ (I_{\lambda\lambda..}^{-1} \mathbf{Z}_{\lambda})^2)]\}$$

$$+ \tfrac{1}{2} \operatorname{tr}\{I_{\lambda\lambda..}^{-1}[I_{\lambda\lambda\lambda...} \circ I_{\lambda\lambda..}^{-1} I_{\psi\lambda\lambda..} I_{\lambda\lambda..}^{-1}(I_{\lambda\lambda\lambda...} \circ (I_{\lambda\lambda..}^{-1} \mathbf{Z}_{\lambda})^2)]\}$$

$$+ \operatorname{tr}\{I_{\lambda\lambda..}^{-1}[I_{\lambda\lambda\lambda...} \circ I_{\lambda\lambda..}^{-1}[I_{\lambda\lambda\lambda...} \circ I_{\lambda\lambda..}^{-1} \mathbf{Z}_{\lambda} \circ I_{\lambda\lambda..}^{-1} I_{\psi\lambda\lambda..} I_{\lambda\lambda..}^{-1} \mathbf{Z}_{\lambda})]\}$$

$$+ \operatorname{tr}\{I_{\lambda\lambda..}^{-1} I_{\psi\lambda\lambda..}(I_{\lambda\lambda..}^{-1}[I_{\lambda\lambda\lambda...} \circ (I_{\lambda\lambda..}^{-1} \mathbf{Z}_{\lambda})])^2\}$$

$$+ \operatorname{tr}\{I_{\lambda\lambda..}^{-1}[I_{\lambda\lambda\lambda...} \circ I_{\lambda\lambda..}^{-1} I_{\psi\lambda\lambda..} I_{\lambda\lambda..}^{-1} \mathbf{Z}_{\lambda} \circ I_{\lambda\lambda..}^{-1}(I_{\lambda\lambda\lambda...} \circ (I_{\lambda\lambda..}^{-1} \mathbf{Z}_{\lambda}))]\}.$$

In the (2, 2) exponential family, discussed in Example 21.2, we found that the right-hand side of (21.10) is a function of the mixed derivatives of the function $k = k(\psi, \phi(\psi, \lambda))$. Two expression of it are

$$\frac{1}{I_{\psi\psi}}\left\{-\frac{\partial^3 \phi}{\partial \psi^2 \, \partial \lambda}\right\} + \frac{1}{I_{\psi\psi}^2}\left\{-\frac{\partial^2 \phi}{\partial \psi \, \partial \lambda}\right\}\left\{-k_{30} + \frac{3k_{21}k_{11}}{k_{02}} - \frac{3k_{12}k_{11}^2}{k_{02}^2} + \frac{k_{03}k_{11}^3}{k_{02}^3}\right\}$$

$$\text{(A5a)}$$

and

$$[k_{02}^4(k_{12}k_{30} - k_{20}k_{22})$$

$$+ k_{02}^3\{k_{11}^2 k_{22} + k_{11}(2k_{13}k_{20} - 3k_{12}k_{21}) + 2k_{12}^2 k_{20} - k_{03}(k_{11}k_{30} - k_{20}k_{21})\}$$

$$+ k_{02}^2 k_{11}\{2k_{03}(k_{11}k_{21} - 3k_{12}k_{20}) - k_{11}(k_{04}k_{20} + 2k_{11}k_{13} - k_{12}^2)\}$$

$$+ k_{02}k_{11}^2\{3k_{03}^2 k_{20} + 2k_{03}k_{11}k_{12} + k_{04}k_{11}^2\} - 2k_{03}^2 k_{11}^4]/\{k_{02}^4(k_{02}k_{20} - k_{11}^2)^2\}.$$

$$\text{(A5b)}$$

22
Resampling using estimating equations
S. Lele

ABSTRACT

This paper surveys resampling methods for a sequence of non-independent, non-identically distributed random variables. The jackknife method is extended to such a sequence through the use of linear estimating equations. In this paper, we extend Wu's bootstrap to dependent data through the use of linear estimating equations. The main idea is to perturb each component estimation equation by another easy-to-generate sequence of estimating equations with proper mean, variance, and correlation structure. Some validity results are provided.

22.1 Introduction

Resampling methods such as bootstrap and jackknife have proved to be extremely useful in various practical situations. These methods were developed mainly for independent and identically distributed random variables. Recently there have been several attempts at generalizing these methods for situations other than the i.i.d. random variables. For example, Wu (1986) studied the regression problem where the random variables are independent but not identically distributed. Liu (1988) proved the validity and higher-order properties of the bootstrap technique for independent but not identically distributed random variables. She also studied the properties of Wu's weighted bootstrap (see Section 7 of Wu's paper) and showed how one can improve its higher-order properties.

Another generalization of resampling techniques is for the situations where the random variables are dependent. Since there are various ways in which dependence can occur, there are various ways in which the resampling techniques can be generalized.

Bose (1988), Basawa *et al.* (1989) among others, work with autoregressive type models where the errors are i.i.d. random variables. They exploit this structure to generalize the bootstrap. On the other hand, Rajarshi (1990) assumes that the underlying process is stationary and Markovian of the known order. He then uses the kernel density estimator to estimate the transition density and uses it to generate the bootstrap samples. Lele (1989) generalizes this to Markov random fields. Kunsch (1989) assumes stationarity.

He estimates the m_n-dimensional marginal distributions using empirical distribution function and uses it to generate blocks of size m_n, joined together as if they were independent. He proves the validity of his bootstrap for $m_n \to \infty$ and $n \to \infty$, such that $(m_n/n) \to 0$. See also Hall (1985).

Lele (1991) for generalizing the jackknife method to dependent data, assumes that the parameter of interest satisfies a linear estimating equation $G(\mathbf{X}, \theta) = 0$ where $G(\mathbf{X}, \theta)$ can be written as $\sum_{i=1}^{n} g_i(\mathbf{X}, \theta) = 0$. Note that Godambe (1985) developed the foundations of statistical inference for stochastic processes using such estimating equations. The key property of these estimating equations for time-dependent sequence is that they are uncorrelated with each other. For the space-dependent sequence, such as the maximum pseudo-likelihood estimating equations (Besag 1974), these equations are correlated with each other within a neighbourhood, outside which they are uncorrelated. This property is exploited to generalize the jackknife technique to dependent data which are not necessarily stationary.

The main purpose of this paper is to suggest an approach to generalize the bootstrap technique for dependent data and semiparametric situations, using estimating equations. Due to the technical nature of the proofs, only heuristic arguments are presented. In the next section, I will briefly review the jackknife technique as applied to the estimating equations. In Section 22.3, I will motivate the need for developing the bootstrap technique for estimating equations. Section 22.4 describes the main idea behind the generalization. Section 22.5 describes how the technique can be used for the class of estimating equations discussed in Godambe (1985). In the discussion, extension to other situations will be indicated.

22.2 Jackknifing linear estimating equations

The following is a brief description of the technique described in Lele (1991). Let $X_1, X_2, \ldots, X_n$ be a sequence of possibly dependent random variables. Let θ be the parameter of interest. Following Godambe (1985), let θ be estimated using a linear estimating equation of the form

$$\sum_{i=1}^{n} g_i(\mathbf{X}, \theta) = 0,$$

where functions $g_i(\cdot)$ are such that:

(i) $E_\theta(g_i(\mathbf{X}, \theta)) = 0$ for all i.

(ii) $E_\theta(g_i(\mathbf{X}, \theta)g_j(\mathbf{X}, \theta)) = 0$ if $|i - j| > d$
    for some known d. Usually for a time-dependent sequence $d > 0$. This is the orthogonality property.

(iii) $\left| E_\theta\left(\dfrac{\partial}{\partial \theta} g_i(\mathbf{X}, \theta) \right) \right| > 0$ for all i.

Then under proper regularity conditions, it can be shown that

$$\sqrt{n}\,(\hat{\theta} - \theta) \to N(O, V_\theta),$$

where

$$V_\theta = \lim_{n \to \infty} \frac{nE_\theta \left(\sum_{i=1}^{n} g_i(\mathbf{X}, \theta) \right)^2}{\left| E_\theta \left(\sum_{i=1}^{n} \frac{\partial}{\partial \theta} g_i(\mathbf{X},\theta) \right) \right|^2}.$$

The jackknife estimator of V_θ is obtained in the following fashion:

Step 1: Delete the jth summand $g_j(\mathbf{X}, \theta)$ and solve $\sum_{i \neq j} g_i(\mathbf{X}, \theta) = 0$ to obtain $\hat{\theta}_{n, -j}$.

Step 2: $\hat{V}_\theta = (n - 1) \sum_{i=1}^{n} \sum_{j=i-d}^{i+d} (\hat{\theta}_{n, -j} - \hat{\theta}_{(\cdot)})(\theta_{n, -1} - \hat{\theta}_{(\cdot)})$

In Lele (1991) it is shown that, under certain regularity conditions:

(i) $\hat{V}_\theta \to V_\theta$ in probability;

(ii) $\dfrac{\sqrt{n}\,(\hat{\theta} - \theta)}{\sqrt{\hat{V}_\theta}} \to N(0, 1)$ in distribution.

One can use these results to obtain the confidence interval for θ. Generalization to the vector-parameter case is straightforward.

22.3   Motivation for bootstrap

From the description of the jackknife procedure, it is clear that for the purpose of studying the behaviour of $\hat{\theta}$ obtained by solving the linear estimating equation $\sum_{i=1}^{n} g_i(\mathbf{X}, \theta) = 0$, one can consider the estimating equations sequence $g_1(\mathbf{X}, \theta), g_2(\mathbf{X}, \theta), \ldots, g_n(\mathbf{X}, \theta)$ instead of the original sequence $X_1, X_2, \ldots, X_n$. The jackknife procedure for the i.i.d. random variables can then be used in a straightforward fashion for the sequence $g_1, g_2, \ldots, g_n$. A natural next step to try would be to bootstrap the estimating equations sequence. If the g_i are independent, following Liu (1988), the asymptotic validity and higher-order properties for such a bootstrap can be proved easily. In fact the uncorrelatedness of the g_i is sufficient to prove the validity of such a bootstrap. However, in this case the higher-order properties do not follow. It turns out that one can generalize Wu's bootstrap for the $\{g_i\}$ sequence and obtain the higher-order corrections at least partially. This idea is described in the next section. But before that discussion it seems necessary, in the light of the availability of various techniques, to motivate the necessity of another technique based on estimating equations.

First of all, note that almost all the bootstrap techniques described in the introduction rely on the assumption that the underlying process is stationary. Since Wu's bootstrap was developed particularly for non-identically distributed random variables, its extension also does not require stationarity.

Secondly consider the situations where one is using some semi-parametric structure, for example Cox's proportional hazards model or generalized linear model for time-series regression. It would be obviously advantageous to exploit such a structure to do resampling instead of adopting a completely non-parametric method.

This brings me to the third aspect of utilizing the estimating equations structure. In the discussion of Wu's paper, it was clear that the usual bootstrap procedure for regression gives unconditionally correct confidence intervals, whereas Wu's jackknife and bootstrap procedures give conditionally correct confidence intervals, conditional on the observed covariates. Similar issue can be raised in the above semiparametric models. Since the method suggested here is an extension of Wu's method, it also gives conditionally correct confidence intervals.

The fourth reason is the higher-order corrections related to the third moment. It can be argued that using Wu's method one can obtain these corrections at least partially for a dependent sequence which is not stationary.

The next two reasons are mainly computational. Note that if the $\{g_i\}$ sequence is such that g_i are correlated, then the variance estimator in Section 22.2 is not necessarily positive for a particular sample. One can do some sort of a smoothing to get a non-negative estimator (Lele 1991). It turns out that the procedure suggested here accomplishes such a smoothing in a natural fashion.

Suppose the underlying process is a Markov random field. Then one can, at least in principle, resort to parametric bootstrap method to get the confidence intervals. However, these methods are extremely computer intensive. Using the estimating equations structure and unilateral construction of some Markov random fields (Pickard 1977) this burden can be reduced substantially.

22.4 Bootstrapping estimating equations

Let $X_1, X_2, \ldots, X_n$ be a sequence of (possibly dependent) random variables. Let the parameter θ be such that it satisfies the linear estimating equation

$$E_\theta[g_i(\mathbf{X}, \theta)] = 0 \qquad \text{for all } i \text{ and } \theta \in \theta.$$

Let $\hat{\theta}$ be a solution to the equation

$$\sum_{i=1}^{n} g_i(\mathbf{X}, \theta) = 0.$$

We are interest in calculating a confidence interval for θ.

For the sake of simplicity, we will assume that θ is scalar, but extension to vector-valued θ is straightforward.

A standard method to calculate the confidence intervals is the following:

Step 1: Use the approximation

$$\left(-\frac{1}{n} \sum_{i=1}^{n} \frac{\partial}{\partial\theta} g_i(\mathbf{X}, \theta) \right) \sqrt{n} \, (\hat{\theta} - \theta) = \frac{1}{\sqrt{n}} \sum_{i=1}^{n} g_i(\mathbf{X}, \theta).$$

Step 2: Apply the central limit theorem on the right-hand side to get the result

$$\sqrt{n} \, (\hat{\theta} - \theta) \approx N \left(0, \frac{n \, \mathrm{Var}\left(\sum_{i=1}^{n} g_i(\mathbf{X}, 0) \right)}{E \left(\sum_{i=1}^{n} \frac{\partial}{\partial\theta} g_i(\mathbf{X}, \theta) \right)^2} \right).$$

Step 3: Estimate the variance of this normal distribution and use

$$\hat{\theta} \pm Z_{\alpha}^{/2} \sqrt{\left(\frac{\hat{V}_\theta}{n} \right)}$$

as the $100(1 - \alpha)$ per cent confidence interval for θ. For example, see Royall (1986). Clearly if one can improve upon the normal approximation in Step, 2, one can improve the confidence interval in Step 3. This is where bootstrapping can come in handy.

Note that the usual Taylor expansion argument for estimating equations uses the approximation

$$0 = \frac{1}{n} \sum_{1}^{n} g_i(\mathbf{X}, \theta) + \sqrt{n} \, (\hat{\theta} - \theta) \frac{1}{n} \sum_{i=1}^{n} \frac{\partial}{\partial\theta} g_i(\mathbf{X}, \theta)$$

$$+ (\sqrt{n} \, (\hat{\theta} - \theta))^2 \frac{1}{n^{3/2}} \left(\sum_{i=1}^{n} \frac{\partial^2}{\partial\theta^2} g_i(\mathbf{X}, \theta) \right).$$

In this paper we concentrate on improving the normal approximation to the first summand and do not consider the third term.

As noted earlier, we will consider the sequence $\{g_i(\mathbf{X}, \hat{\theta})\}_{i=1, n}$ instead of the sequence $\{X_i\}_{i=1, n}$. Since we are going to modify Wu's bootstrap, we will briefly describe his method first and then show how it can be modified. We follow the description in Liu (1988).

22.4.1 WU'S WEIGHTED BOOTSTRAP

Let $X_1, X_2, \ldots, X_n$ be independent but possibly not identical random variables. Let X_i have the same mean μ but possibly different variance and other moments. The problem is to calculate the confidence interval for μ. Let us define

$$Y_i = \bar{X}_n + (X_i - \bar{X}_n)t_i$$

where the t_i are i.i.d. random variables with mean 0 and variance 1. The random variables $Y_1, Y_2, \ldots, Y_n$ form a bootstrap sample. Then Liu proves that

Theorem 22.1   If $E|t_1|^3 < \infty$, $E|X_i|^{2+\delta} \leqslant k < \infty$ for some $\delta > 0$ and $i = 1, 2, \ldots, n$, $\lim_{n \to \infty} (1/n) \sum_1^n \sigma_i^2 > 0$, then

$$\lim_{n \to \infty} \| P^*(\sqrt{n} \, (\bar{Y}_n - \bar{X}_n) \leqslant x) - P(\sqrt{n} \, (\bar{X}_n - \mu) \leqslant x) \|_\infty = 0$$

almost surely, thus proving the validity of Wu's bootstrap. Moreover, if one takes t_i such that $E(t_i^3) = 1$, then the first three cumulants of the studentized statistics $\sqrt{n} \, (\bar{X}_n - \mu)/V_n$ and $\sqrt{n} \, (\bar{Y}_n - \bar{X}_n)/V_n^*$ (conditional on X_i) where

$$V_n = \left[\frac{1}{n} \sum_1^n (X_i - X)^2 \right]^{1/2} \quad \text{and} \quad V_n^* = \left[\frac{1}{n} \sum_1^n (Y_i - \bar{Y}_n)^2 \right]^{1/2},$$

match up to $o(n^{-1/2})$. Consequently there is a total $n^{-1/2}$ term correction by the bootstrap.

Estimating equations sequence. Now consider the estimating equation sequence $\{g_i(\hat{\theta})\}$ and a new sequence $\{Y_i\}_{i=1,n}$ defined by

$$Y_i = g_i(\hat{\theta})t_i. \tag{*}$$

(Note that $g_i(\hat{\theta})$ replaces $(X_i - \bar{X}_n)$ and $\sum_{i=1}^n g_i(\hat{\theta}) = 0$ replaces $\bar{X}_n$.) Suppose $\{g_i(\hat{\theta})\}_{i=1,n}$ is such that g_i are independent of each other, then it is clear from Liu's results that the following theorem holds.

Theorem 22.2   If $E|t_i|^3 < \infty$, $E|g_i(\hat{\theta})|^{2+\delta} \leqslant k < \infty$ for some $\delta > 0$ and $i = 1, 2, \ldots, n$, $\lim(1/n) \sum_i^n \mathrm{Var}(g_i(\theta)) > 0$, then

$$\lim_{n \to \infty} \left\| P^*(\sqrt{n} \, (\bar{Y}_n) \leqslant x) - P\left(\sqrt{n} \left(\frac{1}{n} \sum_1^n g_i(\theta) \right) \leqslant x \right) \right\|_\infty = 0 \quad \text{a.s.}$$

Moreover if the t_i are such that $E(t_i^3) = 1$, then there is a total $n^{-1/2}$ term correction to the distribution of $(1/\sqrt{n}) \sum_1^n g_i(X, \theta)$. This thus will improve the performance of the confidence intervals.

Suppose the sequence $\{g_i(\theta)\}$ is only uncorrelated and not independent. The result described in the theorem still holds. The higher-order terms, however, are not corrected completely. This turns out to be the case because $E(g_i^2 g_j)$ and $E(g_i g_j g_k)$ for $i \neq j \neq k$ may not be zero in general.

An obvious solution to this would be to choose the sequence $\{t_i\}_{i=1,n}$ in such a fashion that it reflects this higher-order dependence. In the next section we will consider an important general situation where such a choice can be made easily. Given such a sequence, one can construct the confidence interval for θ in the following fashion:

Step 1: Generate $Y_1^j, Y_2^j, \ldots, Y_n^j$; $j = 1, 2, \ldots, B$, using (*).
Step 2: Calculate the empirical distribution function $H(\cdot)$ of $\bar{Y}_n^j, j = 1, 2, \ldots, B$.
Let the lower and upper cut-off points be denoted by

$$C_l = H^{-1}(\alpha/2) \quad \text{and} \quad C_u = H^{-1}(1 - \alpha/2).$$

Step 3: The $100(1 - \alpha)$ per cent confidence interval for θ is then given by

$$\left(\hat{\theta} + \frac{C_l}{I(\theta)}, \hat{\theta} + \frac{C_u}{I(\theta)} \right), \quad \text{where } I(\hat{\theta}) = \left| \frac{1}{n} \sum_{i=1}^{n} \frac{\partial}{\partial \theta} g_i(X, \hat{\theta}) \right|.$$

Since one is replacing the normal approximation in Step 2 by its bootstrap estimate (which is better according to the results in Lie (1988)), one should get better confidence intervals.

22.5 Selection of the sequence $\{t_i\}_{i=1,n}$

We now consider an important general situation and describe the construction of the sequence $\{t_i\}$.

Godambe (1985) considered the use of estimating equations for inference in stochastic processes. The class he restricts his attention to is such that

$$E_\theta[g_i | X_1, X_2, \ldots, X_{i-1}] = 0 \quad \text{for all } i.$$

There are several situations, such as conditional least squares or Cox's partial likelihood for proportional hazards model, where the property is satisfied. In fact, something more holds true, viz.

(a) $E_\theta(g_i g_j) = 0$ for all $i \neq j$

(b) $E_\theta(g_i g_j g_k) = 0$ for all $i \neq j \neq k$

(c) $E_\theta(g_i^2 g_j) = 0$ for $j > i$.

The above properties follow easily by conditioning on the past. The consequence of this is

$$E_\theta \left(\sum_{i=1}^{n} g_i(X, \theta) \right)^3 = E_\theta \left(\sum_{i=1}^{n} g_i \right)^3 = \sum_{i=1}^{n} E[g_i^3] + \sum_{i > j} E_\theta[g_i^2 g_j].$$

It is clear from Liu's discussion of Wu's weighted bootstrap that if the sequence $\{t_i\}_{i=1,n}$ can be selected such that

(a) $E(t_i) = 0$

(b) $E(t_i^2) = 1$ for all i

(c) $E(t_i t_j) = 0$ for all $i \neq j$

(d) $E(t_i^2 t_j) = 0$ for $i < j$

(e) $E(t_i t_j t_k) = 0$ for $i \neq j \neq k$

(f) $E(t_i^2 t_j) = 1$ for $i \geq j$,

then one gets a higher-order correction for the distribution of

$$\frac{1}{\sqrt{n}} \sum_{i=1}^{n} g_i(X, \theta)$$

using the bootstrap procedure. Given such a sequence, the argument for the validity and higher-order correction follows identically on the lines of Theorem 5 and its corollary in Liu (1988), provided there exists a valid Edgeworth expansion for $(1/\sqrt{n}) \sum_{i=1}^{n} g_i(X, \theta)$.

In the following we will construct a sequence $\{t_i\}_{i=1,n}$ such that properties (a)–(e) are satisfied but (f) is not. Using such a sequence, one can thus get a partial correction. Let $\{Z_i\}$ be a process such that

$$Z_{i+1} = \zeta Z_i + \varepsilon_{i+1}, \qquad |\zeta| < 1,$$

where ε_i are i.i.d. random variables with mean 0, variance 1 and third moment 1. Consider

$$t_i = (Z_i - \zeta Z_{i-1})(Z_{i-1})(1 - \zeta^2)^{1/2}.$$

Then it is easy to check that

(a) $E(t_i) = 0$

(b) $E(t_i t_j) = 0$ for all $i \neq j$

(c) $E(t_i^2) = 1$

(d) $E(t_i t_j t_k) = 0$ for $i \neq j \neq k$

(e) $E(t_i^2 t_j) = 0$ for $i < j$

(f) $E(t_i^2 t_j) = f(\zeta, |i - j|)$ for $i \geq j$,

where $f(\zeta, |i - j|)$ is a decreasing function in $|i - j|$.

Note that since conditions (a)–(c) are satisfied, the validity of such a bootstrap follows. The function $f(\zeta, |i - j|)$ dictates the correction for the third comulants. Closer this function is to 1, the better is correction.

Note, for example, in the above situation,

$$E(t_i^3) = [E(\varepsilon_i^3)]^2 (1 - \zeta^2)^{3/2} / (1 - \zeta^3)$$

$$= \frac{(1 - \zeta^2)^{3/2}}{(1 - \zeta^3)}$$

$$E(t_{i+1}^2 t_i) = 2\zeta(1 - \zeta^2)$$

$$E(t_{i+2}^2 t_i) = 2\zeta^3, \quad \text{etc.}$$

By considering the higher-order autoregressive processes one might be able to have these moments closer to 1.

Note that the sequence $\{t_i\}_{i=1,n}$ in general should be such that:

(a) it mimics the first three moments of the $\{g_i\}_{i=1,n}$ sequence as closely as possible;
(b) it is easy to generate on the computer.

Any such sequence would serve the purpose. A simple way to generate such a sequence is to construct an estimating equation sequences for an easy-to-generate stochastic process, such as the autoregressive process. Unilateral spatial process described in Pickard (1977) can be used to generate $\{t_i\}_{i=1,n}$ sequence for bootstrapping the maximum pseudo-likelihood estimating equations for Markov random fields.

22.6 Discussion

This paper suggests an extension of Wu's bootstrap to estimating equations sequence. The main difficulty in this extension is the selection of the sequence $\{t_i\}$. This sequence $\{t_i\}$ has to mimic the moments of the estimating-equation sequence. A simple way to achieve this is to construct this sequence as an estimating-equation sequence from a simple-to-generate stochastic process. For the time-dependent case, autoregressive processes are natural candidates. For the spatial processes, one can use the process defined in Pickard (1977). This has a unilateral construction thus not needing the computer-intensive technique of the Gibbs sampler. In this paper, I perturb only a single g_i at a time. One can improve the performance of this bootstrap by perturbing a block of g_i at a time and letting the block size go to infinity at a proper rate. This would be a hybrid of Kunsch's blockwise bootstrap and Wu's bootstrap. Wu also suggests the use of weights to achieve some unbiasedness properties. Selection of such weights in the situations described in this paper is worth further study.

The referee has suggested the following alternative scheme for bootstrapping estimating equations. Consider the sequences $\{t_i^j\}_{i=1}^n$, $j = 1, 2, \ldots, B$, as discussed earlier. Now solve the estimating equation

$$\sum_{i=1}^{n} g_i(\theta)t_i^j = 0$$

to get $\hat{\theta}^j$ for $j = 1, 2, \ldots, B$. Distribution of $\hat{\theta}^j$ presumably would approximate the distribution of $\hat{\theta}$. This scheme potentially seems easy to apply, intuitive, and needs further study.

22.7   Acknowledgements

I am grateful to Professor M. B. Rajarshi and the referee for their helpful comments. I also acknowledge the kind encouragement of Professor Godambe.

References

Basawa, I. V., Malik, A. K., McCormic, W. P., and Taylor, R. L. (1989). Bootstrapping explosive autoregressive processes. *Ann. Stat.*, **17**, 1479–86.

Besag, J. E. (1975). Spatial interaction and the statistical analysis of lattice systems (with discussion). *J. Roy. Stat. Soc.*, Ser. B, **34**, 192–236.

Bose, A. (1988). Edgeworth correction by bootstrap in autoregressions. *Ann. Stat.*, **16**, 1709–22.

Godambe, V. P. (1985). The foundations of finite sample estimation in stochastic processes. *Biometrika*, **72**, 419–28.

Hall, P. (1985). Couting methods for inference in binary mosaics. *Biometrics*, **41**, 1049–52.

Kunsch, H. (1989). The jackknife and the bootstrap for general stationary observations. *Ann. Stat.*, **17**, 1217–41.

Lele, S. R. (1989). Nonparametric bootstrap for spatial processes. Technical Report, The Johns Hopkins University.

Lele, S. R. (1991). Jackknifing linear estimating equations: asymptotic theory and applications in stochastic processes. *J. Roy. Stat. Soc.*, Ser. B., **53**, 253–67.

Liu, R. (1988). Bootstrap procedures under some non i.i.d. models. *Ann. Stat.*, **16**, 1696–1708.

Pickard, D. (1977). A curious binary lattice process. *J. Appl. Prob.*, **16**, 12–24.

Rajarshi, M. B. (1990). Bootstrap in Markov sequences based on estimate of transition density. *Ann. Inst. Math. Stat.*, **42**, 253–68.

Royall, R. (1986). Model robust confidence intervals using maximum likelihood estimators. *Int. Stat. Rev.*, **54**, 221–6.

Wu, C. F. J. (1986). Jackknife, bootstrap and other resampling methods in regression analysis. *Ann. Stat.*, **14**, 1261–95.

23
On using bivariate moment equations in mixed normal problems
B. G. Lindsay and P. Basak

ABSTRACT

It is shown how to construct a set of bivariate moment equations that can be used to identify a discrete distribution with a specified number of support points in the plane. From these equations, one can construct method of moment estimators in problems involving a mixture of bivariate normal distributions with common covariance matrix. These estimators have the advantage over previous moment estimators of being computationally simple for any prespecified number of component distributions in the mixing distribution and of having known uniqueness and identifiability properties. The computational speed and consistency of these methods makes them useful in diagnostics, simulation, and in the construction of initial values for maximum likelihood algorithms.

23.1 Introduction and summary

It is well known that there are important computational difficulties involved in the use of maximum likelihood methods in the normal mixture model. To cite a recent example, in Finch *et al.* (1989) it was found, in an investigation of starting values and the probability of convergence, that more than 8 per cent of the starting values resulted in a failure to converge in 750 quasi-Newton iterations, and 25 per cent of the starting points resulted in a solution that was not the global maximum.

The ramifications of this go beyond simply the ability to compute estimates. The modern statistician often relies on such computationally intensive tools as bootstrap and simulation to provide a more realistic picture of the actual variability in a problem. The aforementioned difficulties render these tools virtually unusable.

Thus it would appear there is good reason to attempt to find estimates that would provide good starting values for a maximum likelihood iterative algorithm. In Lindsay (1989) it was shown that in the univariate normal mixture problem one could construct method of moments equations that had the following desirable features:

(1) *Computational speed.* Although an iterative algorithm is needed for σ^2, a simple, fast, reliable univariate bisection algorithm can be used. In the all important two-support-point problem the remaining estimates are obtained through explicit expressions.

(2) *Mathematical identifiability.* Using the theory of moments, it is possible to show that the estimates are unique and that the estimating functions do indeed identify the unknown parameters.

Earlier investigations by Day (1969) indicated that the univariate moment equations yielded estimates with reasonable efficiency properties. This would suggest that they also provide reasonable starting values. Indeed, Furman (1989) has shown by simulation that the moment estimates provided starting values for the EM algorithm—used for finding a set of maximum likelihood estimates—that worked as well as the *true values*, in the sense that the likelihoods were larger at the start with moment estimates and the moment and true-value starting values proceeded almost universally to the same mode of the likelihood. It was also shown by Furman that one could have a 'fast' version of the likelihood ratio test (LRT) in which one replaced in the ratio the maximum likelihood estimates by the corresponding moment estimates. This has a distribution that can (unlike the LRT) be quickly simulated, and that seemed to be nearly equivalent in power to the LRT.

This investigation started with the goal of extending these tools to the multivariate normal mixture problem. There were some immediate initial obstacles.

In Section 23.2 it is shown that there are certain inherent difficulties in constructing from moments a method that simultaneously has desirable equivariance properties and gives unique solutions. In addition, Day (1969) derived a set of moment equations and found them to be quite unsuccessful at estimating the parameters in the two-point bivariate normal mixture

$$(X, Y) \sim \pi N(\mu_1, \Sigma) + (1 - \pi)N(\mu_2, \Sigma).$$

There was still some hope of success, however, for the following reason. If we consider moments up to order 4 in X and Y, we have the expectations of X, Y, X^2, XY, Y^2, X^3, X^2Y, XY^2, Y^3, X^4, X^3Y, X^2Y^2, XY^3, and Y^4. However, there are only eight unknown parameters in this system: π, two bivariate means μ_1, μ_2 and the variance–covariance parameters in Σ. Clearly there is an issue here of how to choose a set of eight equations to solve. The set we choose here are chosen, unlike those in Day (1969), to satisfy the twin goals of computational ease and identifiability. (We note that Fukunaga and Flick (1983) also give a method of moments for this problem that appears to offer neither advantage.)

This paper constructs a method of moments for the bivariate normal model only, as in this case there are certain natural moment systems for the complex plane that can be used as a theoretical foundation.

Section 23.2 briefly reviews the univariate moment problem and introduces the premier difficulty facing a satisfactory moment approach in higher dimensions. In Section 23.3, we describe a method to solve for the support points of a p-point distribution and see that our solution is equivariant under rotation and multiplication by a positive scalar. Section 23.4 deals with unbiased estimates for a mixing distribution in the multivariate normal mixture problem $N(Q, \Sigma)$ with Σ known and Section 23.5 deals with $N(Q, \sigma^2 V)$ with V known. This method-of-moments analysis is extended to the mixture problem $N(Q, \Sigma)$ with Σ completely unknown in Section 23.6.

23.2 The moment technology

Let Q_p be a discrete distribution on the parameter set $\{\theta\}$ with p points of support $\theta_1, \theta_2, \ldots, \theta_p$ and associated probabilities $\pi_1, \pi_2, \ldots, \pi_p$. The approach taken to the problem of estimating Q_p by Lindsay (1989) can be summarized as follows. Given a selected function of θ, say, $\mu(\theta)$, one can first consider the problem of estimating the $2p - 1$ moments of Q_p defined by

$$m_k(Q) = \int \mu(\theta)^k \, dQ_p(\theta).$$

Suppose we can construct consistent estimates $\hat{m}_1, \ldots, \hat{m}_{2p-1}$ for this sequence—in many exponential family problems, there is a natural choice of $\mu(\theta)$ such that the estimates $\hat{m}_k$ are unbiased linear combinations of sample moments. If one would take these estimated moments $\hat{m}_1, \ldots, \hat{m}_{2p-1}$ and solve for the distribution $\hat{Q}_p$ that has those moments, one has a solution to the estimation problem.

The classical theory of moments (e.g., Uspensky 1938; Widder 1947) provides a set of tools to solve this problem. Most importantly, one has the following result: given a sequence of numbers $m_1, \ldots, m_{2p-1}$, one can determine if it is a sequence of moments for a distribution Q with p or more points of support, and if it is, then there exists a unique p-point distribution Q_p that has those numbers $m_1, \ldots, m_{2p-1}$ as moments. In addition, there are straightforward methods to calculate the support and masses of Q_p from the moments. Many of the ideas and techniques of this paper are based on extending these classical results to a bivariate structure.

The classical moment theory is closely related to the study of the zeros of univariate real polynomials of degree p in x, $a_0 + a_1 x + \cdots + a_p x^p$. In this paper, the comparable study is of complex 'polynomials' in variable $Z = X + iY$, here called *radial polynomials*, of the form

$$G(Z) = \alpha_0 + \alpha_1 Z + \alpha_2 Z\bar{Z} + \cdots + \alpha_p b_p(Z),$$

where $\alpha_0, \alpha_1, \ldots, \alpha_p$ are complex coefficients, $\bar{Z}$ is the conjugate of Z, and $b_{2k} = (Z\bar{Z})^k$ and $b_{2k+1} = (Z\bar{Z})^k Z$.

We now summarize our basic identifiability results, postponing the proofs to the following section. Let $\beta_k(Q) = \int b_k(Z)\, dQ(Z)$. The first result indicates that we can uniquely solve for a p-point distribution Q_p on the complex plane from the *radial moment* sequence $\beta_1(Q_p), \beta_2(Q_p), \ldots, \beta_{2p-1}(Q_p)$, provided that the support points of Q_p are at distinct distances from the origin.

Theorem 23.1 (Identifiability)   *If $Q_p = \sum \pi_j \delta\zeta_j$, then one can solve for Q_p from $E[b_k(Z)]$, $k = 1, 2, \ldots, 2p - 1$, as follows. From any origin from which $|\zeta_1|, |\zeta_2|, \ldots, |\zeta_p|$ are distinct*

(a) *$\{\zeta_1, \zeta_2, \ldots, \zeta_p\}$ is the solution set to $d_p(t) = 0$, defined in (23.1).*
(b) *The masses $\pi_1, \pi_2, \ldots, \pi_p$ are the unique solutions to the full rank system*

$$\pi_1 b_k(\zeta_1) + \pi_2 b_k(\zeta_2) + \cdots + \pi_p b_k(\zeta_p) = E[b_k(Z)]$$

$$\text{for } k = 0, 1, \ldots, p - 1.$$

The preceding theorem tells us that we can solve for Q_p from its radial moments; the next theorem indicates that for any Q with p are more points we can solve for a p-point approximant Q_p which has the same radial moment sequence.

Theorem 23.2 (Moment approximant)   *Suppose the distribution Q is such that $d_p(t) = 0$ (defined in 23.1 in the following section) has exactly p distinct roots. Then there exists a unique p-point distribution Q_p with the same radial moment sequence: $\beta_k(Q) = \beta_k(Q_p)$, $k = 0, 1, \ldots, 2p - 1$.*

The proof is given in Section 23.3.

Remark   One central difficulty in deriving a moment strategy for the multivariate problem is the conflict between certain desired equivariance properties and the desire to have identifiable systems. The following example illustrates this difficulty. Let the pair (X, Y) have a uniform distribution on the unit circle $\{(x, y): x^2 + y^2 = 1\}$ and consider the problem of approximating this with a two-point distribution Q_2. Now the distribution of (X, Y) is invariant under rotations about zero, so if the approximating distribution is *unique* and is *equivariant under rotations* about zero, then the two points must actually lie at the origin and the approximation has only one point. Similarly, any genuinely p-point approximant must either lack rotational equivariance or lack uniqueness.

The solution we propose here will have the characteristic of being dependent on the choice of origin. The two-point approximant for the unit circle will depend on the location of the circle relative to this origin. If the origin is not at the centre of the circle, then the approximant place masses $\frac{1}{2}$ at each of two points on the circle—the point nearest the origin and the point farthest away. This solution is clearly scalar multiplication and rotation-about-origin equivariant but is not translation equivariant.

The equivariance properties of the radial system will be summarized in Proposition 23.3.

23.3 The radial moment system

In this section we derive a system of radial moments from which one can determine the support points $\zeta_1, \zeta_2, \ldots, \zeta_p$ and probabilities $\pi_1, \pi_2, \ldots, \pi_p$ of any p-point bivariate distribution meeting a certain asymmetry requirement. We note that $2p + p - 1 = 3p - 1$ free parameters are needed to describe a p-point distribution and the set of moments we will use has exactly that many free elements. We represent the bivariate variable (X, Y) as a complex variable $Z = X + iY$, where $i = \sqrt{(-1)}$. Let $\bar{Z} = X - iY$ be the complex conjugate of Z. For any integer k let $b_{2k}(Z) = (Z\bar{Z})^k$ and $b_{2k+1}(Z) = (Z\bar{Z})^k Z$. Let

$$V_p(Z) = (1, b_1(Z), b_2(Z), \ldots, b_p(Z))' = (1, Z, Z\bar{Z}, Z^2\bar{Z}, \ldots)'$$

and let the moment matrix $\mathbf{M}_p$ equal $E[V_p(Z)\overline{V_p(Z)'}]$. For example,

$$\mathbf{M}_2 = E\begin{pmatrix} 1 & \bar{Z} & Z\bar{Z} \\ Z & Z\bar{Z} & Z^2\bar{Z} \\ Z\bar{Z} & Z\bar{Z}^2 & Z^2\bar{Z}^2 \end{pmatrix} = \begin{pmatrix} 1 & \bar{\beta}_1 & \bar{\beta}_2 \\ \beta_1 & \beta_2 & \bar{\beta}_3 \\ \beta_2 & \bar{\beta}_3 & \beta_4 \end{pmatrix}$$

Let the matrix $\mathbf{M}_p(t)$, for complex variable t, be equal to $\mathbf{M}_p$ with the last column replaced by $V_p(t)$. We first show that $\mathbf{M}_p(t)$ can be used to identify support points.

Proposition 23.1 *If Q_p is a p-point distribution, then it has support points $\zeta_1, \zeta_2, \ldots, \zeta_p$ among the complex solutions to*

$$d_p(t) = \det[\mathbf{M}_p(t)] = 0. \tag{23.1}$$

Proof Represent the matrix $\mathbf{M}_p(t)$ as

$$E[V_p(Z_0), \bar{Z}_1 V_p(Z_1), \ldots, b_{p-1}(\bar{Z}_{p-1})V_p(Z_{p-1}), V_p(t)],$$

where $Z_0, Z_1, \ldots, Z_{p-1}$ are an i.i.d. sample from Q_p. The determinant is zero if any repeats occur in the sequence $Z_0, Z_1, \ldots, Z_{p-1}$. It is also zero if $Z_0, Z_1, \ldots, Z_{p-1}$ take on distinct values $\zeta_1, \zeta_2, \ldots, \zeta_p$ and t takes on one of these same values.

The function $d_p(t)$ is a polynomial with complex coefficients in the variables $1, t, t\bar{t}, \ldots, b_p(t)$. We note that the coefficient of $d_p(t)$, the highest-order term in $d_p(t)$, is $\det \mathbf{M}_{p-1}$; since $\mathbf{M}_p$ is Hermitian, the determinant is real-valued, but it could be zero. In the univariate case, the analogous determinant is positive if and only if Q has p or more points of support. In order to consider this issue here, we will need to consider the nature of the solutions to complex polynomial equations

$$G(Z) = \alpha_0 + \alpha_1 Z + \alpha_2 Z\bar{Z} + \cdots + \alpha_p b_p(Z) = 0.$$

If $\alpha_p \neq 0$, we will say $G(Z)$ is of degree p. We can write such a polynomial as

$$G(Z) = P_0(Z\bar{Z}) + Z P_1(Z\bar{Z}),$$

where the polynomials P_0 and P_1 have the form

$$P_0(t) = \alpha_0 + \alpha_2 t + \alpha_4 t^2 + \cdots \qquad \text{and}$$

$$P_1(t) = \alpha_1 + \alpha_3 t + \alpha_5 t^2 + \cdots. \qquad \text{with } \alpha_k = 0 \text{ for } k > p.$$

Lemma 23.1 $\mathbf{M}_p$ *is non-negative definite, and is positive definite if and only if there exists no radial polynomial $G_p(Z)$ of degree p or less that is zero with probability 1 under Q.*

Proof $\qquad \alpha' \mathbf{M}_p \bar{\alpha} = E[G_p(Z)\overline{G_p(Z)}] \geq 0, \qquad \alpha = (\alpha_0, \alpha_1, \ldots, \alpha_p)'.$

We now describe a method for solving the polynomial equations $G_p(Z) = 0$. It turns out to be as easy as finding the real roots of a real polynomial.

Proposition 23.2 *Let $G_p(Z) = P_0(Z\bar{Z}) + Z P_1(Z\bar{Z})$ be of degree p.*
(a) *The polynomial $H(r) = P_0(r)\overline{P_0(r)} - r P_1(r)\overline{P_1(r)}$ in real variable r has real-valued coefficients and is of degree p in r.*
(b) *Suppose $H(r)$ has non-negative real roots $r_1, r_2, \ldots, r_k$. Then there is a set of solutions to $G(Z) = 0$ for each value $r_j, j = 1, 2, \ldots, k$ such that: If $P_1(r^j) \neq 0$, the solution set is the single point $\zeta_j = -P_0(r^j)/P_1(r_j)$. If $P_1(r_j) = 0$, then the set consists of all points ζ satisfying $\zeta\bar{\zeta} = r_j$.*
(c) *If $G_p(Z)$ has p roots $\zeta_1, \zeta_2, \ldots, \zeta_p$ with distinct radii $|\zeta_j|$, then there are no other solutions to $G_p(Z) = 0$.*

Proof (a) That the highest-order term in $H(r)$ is $\alpha_p \bar{\alpha}_p r^p$ can be verified by checking the two cases p even and p odd.

(b) To solve the equation $P_0(Z\bar{Z}) + Z P_1(Z\bar{Z}) = 0$, we first find the solution set on the ball $\{Z: Z\bar{Z} = r\}$. If $P_1(r) \neq 0$, we get a unique potential solution

$$Z_r^* = -P_0(r)/P_1(r).$$

In order for this to be on the ball, we must also have $Z_r^* \bar{Z}_r^* = r$ and hence $H(r) = 0$. If r_j is a root of $H(r)$, and $P_1(r_j) \neq 0$, then this argument can be retraced to show that $Z_{r_j}^* = \zeta_j$ is a solution. On the other hand, if r_j satisfies $H(r_j) = 0$, and $P_1(r_j) = 0$, then $P_0(r_j) = 0$, and so by inspection

$$P_0(Z\bar{Z}) + ZP_1(Z\bar{Z}) = 0 \qquad \text{for all } Z \text{ satisfying } Z\bar{Z} = r_j.$$

(c) If $P_1(r_j) = 0$, and $P_0(r_j) = 0$, then r_j must be a root of multiplicity at least two of $H(r)$, in which case there cannot be p distinct $|\zeta_j|$ values.

Corollary 23.1 *Suppose that Q_p has support points $\zeta_1, \zeta_2, \ldots, \zeta_p$ with distinct radii. Then $\mathbf{M}_p$ is a rank-p matrix. Further, each of $\mathbf{M}_1, \mathbf{M}_2, \ldots, \mathbf{M}_{p-1}$ are non-singular, so $\det \mathbf{M}_1 > 0$, $\det \mathbf{M}_2 > 0, \ldots, \det \mathbf{M}_{p-1} > 0$ and let $\det \mathbf{M}_p = 0$.*

Proof The column space of $\mathbf{M}_p$ is spanned by the p-vectors $V_p(\zeta_1), V_p(\zeta_2), \ldots, V_p(\zeta_p)$, hence rank $\leq p$. On the other hand, from Lemma 23.1 and Proposition 23.2, we know the embedded matrices $\mathbf{M}_{p-k}$ are full rank, as there is no degree of $p - k$ radial polynomial having p radially distinct roots.

Corollary 23.2 *Suppose that the p-point distribution Q_p puts mass on $\zeta_1, \zeta_2, \ldots, \zeta_p$ with distinct radii $|\zeta_j|$. Then $d_p(t) = 0$ has exactly the p solutions $\zeta_1, \zeta_2, \ldots, \zeta_p$.*

Proof We know that $d_p(t) = 0$ at the values of ζ_j by Proposition 23.1. It has degree p by Corollary 23.1 and so by part (c) of Proposition 23.2, these are all the solutions.

We can now prove Theorems 23.1 and 23.2.

Proof of Theorem 23.2 (a) Follows from Proposition 23.1.
(b) The matrix

$$\mathbf{W}_p = \begin{pmatrix} 1 & 1 & \cdots & 1 \\ b_1(\zeta_1) & b_1(\zeta_2) & \cdots & b_1(\zeta_p) \\ b_2(\zeta_1) & b_2(\zeta_2) & \cdots & b_2(\zeta_p) \\ \vdots & \vdots & & \vdots \\ b_{p-1}(\zeta_1) & b_{p-1}(\zeta_2) & \cdots & b_{p-1}(\zeta_p) \end{pmatrix}$$

is full rank provided $\det \mathbf{M}_{p-1} > 0$, as one can write

$$\det \mathbf{M}_{p-1} = \pi_1 \pi_2 \ldots \pi_p \det \mathbf{W}_p \det \bar{\mathbf{W}}_p / p!,$$

using the methods in the appendix of Lindsay (1989). But $\det \mathbf{M}_{p-1} > 0$ by Corollary 23.1.

Proof of Theorem 23.2   The claim amounts to showing that the $2p \times (p + 1)$ matrix

$$[V_{2p-1}(\zeta_1), V_{2p-1}(\zeta_2), \ldots, V_{2p-1}(\zeta_p), E\{V_{2p-1}(Z)\}]$$

has less than full column rank, as we know that the first p columns are full rank. We start by considering the upper $(p + 1) \times (p + 1)$ submatrix. Since $d_p(t) = 0$ for $t = \zeta_1, \zeta_2, \ldots, \zeta_p$, the column vectors $V_p(\zeta_1), V_p(\zeta_2), \ldots, V_p(\zeta_p)$ are all in a p-dimensional subspace that also contains $E[V_p(Z)]$. Since the upper square matrix is less than full column rank, it is also less than full row rank, and hence the row $p + 1$ is in the space spanned by the first p rows. We can now shift down to the square matrix having rows 2 to $p + 2$. Again, it can be shown using $d_p(t) = 0$ that it less than full rank. So row $p + 2$ is in the space spanned by rows 2 to $p + 1$. Hence it is also spanned by rows 1 to p, as follows using the above property of row $p + 1$. One can continue in this fashion, adding a row at a time down to row $2p$.

Finally, the next result describes the equivariance properties of our system so that we can better understand its behaviour as a method of approximating arbitrary distributions by matching moments.

Proposition 23.3   (a) *The solutions to $d_p(t) = 0$ are equivariant under a rotation of Z and multiplication by a positive scalar c.*
(b) *The system of moments $E [b_k(Z + a)]$, $k = 1, 2, \ldots, 2p - 1$ cannot be obtained by transformation of $E [b_k(Z)]$, $k = 1, 2, \ldots, 2p - 1$, except for $p = 1$. Thus a translation of the data alters the set of moments being used and so translation equivariance does not hold.*

Proof   (a) This is easily verified, as rotation with scalar multiplication amounts to multiplication of Z by $c\, e^{i\theta}$.
(b) Can be checked by seeing that, for example, computing $E(Z + a)(\bar{Z} + \bar{a})^2$ requires known $aE[\bar{Z}^2]$.

23.4   Unbiased estimates

We now turn to the estimation problem. The first objective is to construct a sequence of estimates $\hat{\beta}_1, \hat{\beta}_2, \ldots, \hat{\beta}_{2p-1}$ for the radial moments of the unknown p-point mixing distribution Q on the two means (μ_1, μ_2) in the bivariate normal mixture $(X, Y) \sim MN(Q, \Sigma)$, treating Σ as known. The theory of Sections 23.2 and 23.3 then provides a method of solving for Q_p from these radial moments in the Σ known case.

What is required, then, are the estimates of the moments of Q of the form

$E[Z^k \bar{Z}^1]$, with $Z = \mu_1 + i\mu_2$. These in turn can be represented as linear combinations of moments of the form $E[\mu_1^k \mu_2^l]$. It is particularly easy to construct unbiased estimates of these moments. If (X, Y) have a mixed normal distribution, with mixing distribution Q on (μ_1, μ_2), then the moment generating function for (X, Y) can be written as

$$E[\exp(t_1 X + t_2 Y)] = E[\exp(t'\mu)] \exp(t'\Sigma t/2),$$

$$\text{where } t = (t_1, t_2)', \ \mu = (\mu_1, \mu_2)',$$

and so the moment generating function for Q can be written as

$$E[\exp(t'\mu)] = E[\exp(t_1 X + t_2 Y - it_1 Z_1 - it_2 Z_2)],$$

where (Z_1, Z_2) are jointly bivariate normal, mean vector 0 and covariance matrix Σ, independent of (X, Y). By differentiating both sides with respect to (t_1, t_2), one can solve for the moments of μ_1 and μ_2 in terms of those of X and Y. For example, using

$$\frac{\partial^2}{\partial t_1^2} \frac{\partial^2}{\partial t_2^2}$$

gives

$$E[\mu_1^2 \mu_2^2] = E[(X - iZ_1)^2 (Y - iZ_2)^2]$$

$$= E[X^2 - Z_1^2 - 2iXZ_1)(Y^2 - Z_2^2 - 2iYZ_2)]$$

$$= E[X^2 Y^2 + Z_1^2 Z_2^2 - X^2 Z_2^2 - Y^2 Z_1^2 - 4XYZ_1 Z_2$$

$$\quad - 2i\{X^2 YZ_2 - YZ_1^2 Z_2 + XY^2 Z_1 - XZ_1 Z_2^2\}]$$

$$= E[X^2 Y^2 + E(Z_1^2 Z_2^2) - X^2 W(Z_2^2) - Y^2 E(Z_1^2) - 4XYE(Z_1 Z_2)$$

$$\quad - 2i\{X^2 YE(Z_2) - YE(Z_1^2 Z_2) + XY^2 E(Z_1) - XE(Z_1 Z_2^2)\}]$$

Now substitute in

$$E[Z_1^2] = \sigma_1^2, \quad E[Z_2^2] = \sigma_2^2, \quad E[Z_1 Z_2] = \rho\sigma_1 \sigma_2,$$

$$E(Z_1) = E(Z_2) = E(Z_1^2 Z_2) = E(Z_1 Z_2^2) = 0, \quad \text{and} \quad E[Z_1^2 Z_2^2] = \sigma_1^2 \sigma_2^2 (1 + 3\rho^2)$$

to show that

$$\frac{1}{n}\Sigma X_i^2 Y_i^2 + \sigma_1^2 \sigma_2^2 (1 + 3\rho^2) - \sigma_1^2 \frac{1}{n}\Sigma Y_i^2 - \sigma_2^2 \frac{1}{n}\Sigma X_i^2 - 4\rho\sigma_1 \sigma_2 \frac{1}{n}\Sigma X_i Y_i$$

is an unbiased estimator of $E_Q[\mu_1^2 \mu_2^2]$. Note that the unbiased estimator of β_k depends on the elements of Σ. and so it will de denoted by $\hat{\beta}_k(\Sigma)$.

23.5   Estimation with specified shape V

Let us now suppose that we wish to estimate Q_p and σ^2 from a sample from the distribution $N(Q_p, \sigma^2 V)$, where the shape V is specified, and Q_p is a p-point distribution. We can do so in a way completely analogous to the solution to the univariate normal problem $N(Q_p, \sigma^2)$ (Lindsay 1989). That is, we can construct the estimates $\hat{\beta}_k(\sigma^2 V)$ of the moments for our polynomial basis $\beta_k = E[b_k(Z)]$ using the techniques of the last section for $k = 1, 2, \ldots, 2p$, with Σ set to $\sigma^2 V$. These estimated moments depend on σ^2. The estimator $\hat{\sigma}_p^2$ of σ^2 is then determined by forcing the estimated moment matrix $\hat{\mathbf{M}}_p(\sigma^2)$ to be non-negative definite with determinant zero, as this is necessary for the system of estimated moments $\hat{\beta}_1(\sigma_p^2), \hat{\beta}_2(\sigma_p^2), \ldots, \hat{\beta}_{2p-1}(\sigma_p^2)$ to be the moments of a p-point distribution. We now need to establish that such a method identifies the true value of σ^2. Let $\mathbf{M}_p(\sigma^2)$ be the matrix formally obtained by substituting the true moments of (X, Y) in for the sample moments in $\hat{\mathbf{M}}_p(\sigma^2)$.

Lemma 23.2   *Suppose that Q_p is a p-point distribution with support points at $\zeta_1, \zeta_2, \ldots, \zeta_p$ which have distinct radii relative to the chosen origin. If $(X, Y) \sim N(Q_p, \sigma^2 V)$, then σ^2 can be found as the first non-negative root of $\det \mathbf{M}_p(\sigma^2) = 0$.*

Proof   See Lindsay (1989).

Provided the origin was chosen so that the support points have different radii, the resulting p-point method of moments estimators of $\sigma^2, \hat{\sigma}_p^2$ is consistent as per Lindsay (1989). Once $\hat{\sigma}_p^2$ is chosen, this can be used to create estimates of the radial moments of Q, viz., $\hat{\beta}_k = \hat{\beta}_k(\hat{\sigma}_p^2)$, and these moments then generate, by the technique of Section 23.3, a set of estimated locations $\hat{\zeta}_1, \hat{\zeta}_2, \ldots, \hat{\zeta}_p$ and weights $\hat{\pi}_1, \hat{\pi}_2, \ldots, \hat{\pi}_p$, which are consistent.

23.6   The general case

When the Σ is completely unknown, there are too many parameters to be estimated using just the radial moment system. We note that the two second-order moments $E[Z^2]$ and $E[\bar{Z}]^2$ have not yet been used, and they provide the additional two equations needed to estimate the full covariance matrix. These are the natural moments to add, as by using them the methods described here will match the sample covariance to the covariance matrix of the model.

We need to slightly expand out radial moment theory in order to incorporate these new moments. Let $V_p^*(Z) = (1, Z, \bar{Z}, Z\bar{Z}, \ldots, b_p(Z))'$ and let $\mathbf{M}_p^* = E[V_p^*(Z)V_p^*(\bar{Z})']$. For example,

$$\mathbf{M}_2^* = E\begin{pmatrix} 1 & \bar{Z} & Z & Z\bar{Z} \\ Z & Z\bar{Z} & Z^2 & Z^2\bar{Z} \\ \bar{Z} & \bar{Z}^2 & Z\bar{Z} & Z\bar{Z}^2 \\ Z\bar{Z} & Z\bar{Z}^2 & Z^2\bar{Z} & Z^2\bar{Z}^2 \end{pmatrix}.$$

Here it takes eight real-valued entries to fill in $\mathbf{M}_2^*$, and in general it takes $3p + 2$ entries to fill in $\mathbf{M}_p^*$. Since a p-point distribution depends on $3p - 1$ parameters, there must be three constraints on the entries of the matrix. These can be identified by the following.

Lemma 23.3 *If $\mathbf{M}_p^*$ contains the moments of a p-point distribution with radially distinct support points, then it is of rank p (rank deficient by 2).*

Proof The matrix $\mathbf{M}_p$ is a submatrix of $\mathbf{M}_p^*$, so the rank is at least p. Each column of the matrix $\mathbf{M}_p^*$ can be written as a linear combination of the p vectors $V_p^*(\zeta_j)$, $j = 1, 2, \ldots, p$.

Now for $\mathbf{M}_p^*$ to be of rank p, every $(p + 1) \times (p + 1)$ submatrix must have determinant zero. This gives a system of constraint equations satisfied by the elements of $\mathbf{M}_p^*$.

This also gives the method-of-moment stratagem: using estimated moments $\hat{\beta}_k(\Sigma)$, $k = 1, 2, \ldots, 2p$, choose $\hat{\Sigma}$ in such a way that $\hat{\mathbf{M}}_p^*(\hat{\Sigma})$ is of rank p. Next, use the estimated radial moment subsystem $\hat{\beta}_k(\hat{\Sigma})$ to solve for the support points and weights of the distribution Q_p.

We now consider the case $p = 2$ in greater detail. A natural first step is to consider the implication of forcing the 3×3 submatrix of $\hat{\mathbf{M}}_2(\Sigma)$ defined by

$$\hat{\mathbf{M}}_1^*(\Sigma) = \hat{E}\begin{pmatrix} 1 & \bar{Z} & Z \\ Z & Z\bar{Z} & Z^2 \\ \bar{Z} & Z^2 & Z\bar{Z} \end{pmatrix}.$$

to have rank 2. Here the symbol $\hat{E}$ indicates that we are using in the matrix the estimated moments of Q. It can be shown that $\det \hat{\mathbf{M}}_1^* = 0$ if and only if the correlation between μ_1 and μ_2 is one and so all the support points lie on a line. This in turn implies that the distribution Q has a covariance matrix of the form

$$\Sigma_Q = \lambda \begin{pmatrix} \cos^2 \theta & \cos \theta \sin \theta \\ \cos \theta \sin \theta & \sin^2 \theta \end{pmatrix} = \lambda \Gamma.$$

It follows if we set $\Sigma = \mathbf{S}^2 - \lambda \Gamma$, where $\mathbf{S}^2$ is the sample covariance matrix, then $\hat{\mathbf{M}}_1(\Sigma)$ has determinant zero as required. Thus this rank-2 constraint reduces the estimation problem from the three parameters of Σ to two unknown parameters, λ and θ.

Next, we suppose that we have chosen the sample mean point as origin in the radial moment system. This offers a great simplification at the next step, as well as demonstrating a potential hazard. Now consider the rank-2 constraint determined by the 3×3 submatrix of $\hat{M}_2(\Sigma)$ defined by

$$\hat{\mathbf{B}} = \hat{E} \begin{pmatrix} Z\bar{Z} & \bar{Z}^2 & Z\bar{Z}^2 \\ Z^2 & Z\bar{Z} & Z^2\bar{Z} \\ Z^2\bar{Z} & Z\bar{Z}^2 & Z^2\bar{Z}^2 \end{pmatrix}.$$

Here setting $\Sigma = \mathbf{S}^2 - \lambda \Gamma$ puts the upper 2×2 matrix of $\hat{\mathbf{B}}$ into the form

$$\begin{pmatrix} \lambda & \lambda(\cos 2\theta - i \sin 2\theta) \\ \lambda(\cos 2\theta + i \sin 2\theta) & \lambda \end{pmatrix}.$$

This has determinant zero. It follows that λ drops out of the ensuing calculation of the equation $\det \hat{\mathbf{B}} = 0$ leaving an equation for θ which can be solved as

$$\cos \hat{\theta} = \mathrm{Re}(\hat{E}Z\bar{Z}^2)/|\hat{E}(Z\bar{Z}^2)|.$$

Now we are left with a single unknown λ. One can now use the methods of Section 23.5 to solve for λ, as the covariance matrix $\hat{\Sigma}(\lambda) = S^2 - \lambda \hat{\Gamma}$ has a single remaining unknown parameter.

We note that the described estimating system has the following liability. The solution in θ depends critically on $|E(Z\bar{Z}^2)| \neq 0$. In particular, under the assumption that the mixing distribution has equal mass at two points ζ_1, ζ_2 and one uses the mean as origin, then the theoretical value of $|E(Z\bar{Z}^2)|$ is zero. This indicates that in this case the solution will be inconsistent and the estimated angles will simply represent random noise.

23.7 Discussion

In this paper the groundwork has been laid for extending the method of moments for normal mixtures into higher dimensions. It has been shown that there exist certain potential limitations on any method of moments because of conflicts that necessarily arise between the statistical ideals of uniqueness, rotational equivariance, and translation equivariance. The approach taken here involved sacrificing translation equivariance, the achievement of which would have involved using larger moment systems that would no longer be in a one-to-one numerical relationship with the parameters to be estimated. In return for this sacrifice, we get uniqueness of solutions and a simple method for estimating the support points of the unknown mixing distribution.

A number of important steps need to be taken before these methods can be considered an important practical tool. Although estimation in the important bivariate normal model $N(Q, \sigma^2 V)$, V known, is straightforward and fast, it would be desirable to develop as well a fast way to solve for Σ in the equations for the multivariate $N(Q, \Sigma)$ model that does not require using the sample mean as origin. Another major step requires finding the extensions of these results beyond the complex plane. The algebraic machinery of complex numbers was used extensively in the analysis. There is no simple extension of this tool, but it is believed that important aspects of our treatment will carry over via the appropriate identification of a polynomial system.

References

Day, N. E. (1969). Estimating the components of a mixture of normal distributions. *Biometrika*, **56**, 463–74.

Finch, S. J., Mendell, N. R., and Thode, H. C., Jr. (1989). Probabilistic measures of adequacy of a numerical search for global maximum. *J. Amer. Stat. Assoc.*, **84**, 1020–3.

Fukunaga, K. and Flick, T. E. (1983). Estimation of the parameters of a Gaussian mixture using the method of moments. *IEEE Trans. Pat. Anal. Mach. Intel.*, **PAMI-5**, 410–16.

Furman, D. (1989). Unpublished Dissertation. The Pennsylvania State University.

Huber, P. J. (1985). Projection pursuit. *Ann. Stat.*, **13**, 435–75.

Lindsay, B. G. (1989). Moment matrices: applications in mixtures. *Ann. Stat.*, **17**, 722–40.

McLachlan, G. J. and Basford, K. E. (1988). *Mixture models: inference and applications to clustering*. Mercel Dekker, New York.

Uspensky, J. V. (1937). *Introduction to mathematical probability*. McGraw-Hill, New York.

Widder, D. V. (1947). *The Laplace transform*. Princeton University Press, Princeton, N.J.

24
Estimating functions in semi-parametric models
Ya'acov Ritov

ABSTRACT

Suppose we have a random sample from a distribution which is only known to belong to some non-parametric family and we wish to estimate some parameter of this distribution. Under some regularity conditions this parameter can be estimated in a $\sqrt{n}$ rate. In many cases there is an estimator which is the root of an estimating equation. Typically, this estimating equation is not of the form $n^{-1} \sum \phi(x_i; \theta)$, but it converges to such a function as $n \to \infty$. We discuss how special features of the problem can be used in the construction and the analysis of an estimator. In our discussion we follow some of the ideas in Bickel *et al.* (1991). In particular, we consider the special roles of (i) convexity of the estimating equation as a function of the parameter θ; (ii) convexity of the family of possible distributions (i.e., the family is closed under mixtures); and (iii) generalized maximum likelihood. Most of the discussion is on a non-formal level.

24.1 Introduction

Suppose we have a random sample from a distribution P, $P \in \mathscr{P}$, and $\mathscr{P}$ is a non-parametric (or semi-parametric) family of distributions. We wish to estimate some parameter $\zeta \equiv \zeta(P)$ of the data distribution which is the unique solution of $W(\zeta, P) = 0$ for some function $W(\cdot, \cdot)$. A familiar example is the likelihood equation:

$$W(\zeta, P) \equiv \int l(x; \zeta) \, dP(x) = 0,$$

where $\mathscr{P} = \{P\zeta : \zeta \in \Theta\}$, $\Theta \subset \mathbb{R}^d$ and $l(x; \zeta) = \partial/\partial\zeta \log(\partial P\zeta/\partial\mu)$ is the derivative of the log-likelihood with respect to some dominating measure μ.

Since P, the second argument of W, is unknown, we may wish to estimate ζ using the estimating equation $W(\cdot, \mathbb{P}_n) = 0$, where $\mathbb{P}_n$ is the empirical distribution function. More generally, we may consider the solution of $W_n(\cdot, \mathbb{P}_n) = 0$, for some approximation W_n of W. Indeed, the maximum likelihood estimator is the solution of

$$W_n(\zeta, \mathbb{P}_n) \equiv W(\zeta, \mathbb{P}_n) = \int l(x; \zeta) \, d\mathbb{P}_n(x) = 0.$$

The maximum likelihood estimator was generalized in Huber (1964) to the general M-estimators. Here the parameter is the solution of $\int \psi(x; \zeta) \, dP(x) = 0$ and is estimated by the solution of

$$W_n(\zeta, \mathbb{P}_n) \equiv \int \psi(x; \zeta) \, d\mathbb{P}_n(x) = 0.$$

Yet we may wish to generalize further and include functions W which cannot be represented as an average of a given function. A simple example is the M-estimator with nuisance parameter. We begin with some ψ_1 and ψ_2 such that $E_P\{\psi_k(X; \zeta, \eta)\} \equiv 0$, $k = 1, 2$, and then define

$$W(\zeta, P) \equiv E_P\{\psi_1(X; \zeta, \eta(\zeta, P))\},$$

where $\eta(\zeta, P)$ satisfies $E_P\{\psi_2(X; \zeta, \eta(\zeta, P))\} = 0$ for all ζ. The parameter of interest is then estimated by the solution of

$$W_n(\zeta, \mathbb{P}_n) \equiv \int \psi_1(x; \zeta, \hat{\eta}(\zeta)) \, d\mathbb{P}_n(x) = 0,$$

where $\hat{\eta}(\zeta)$ is a solution of $\int \psi_2(X; \zeta, \eta) \, d\mathbb{P}_n(x) = 0$. Some less trivial cases are the solution of U-statistic-type estimating equations, e.g.

$$\iint \psi(x_1, x_2; \zeta) \, dP(x_1) \, dP(x_2) = 0, \quad \text{or} \quad \zeta - \int p^2(x) \, dx = 0,$$

where p is the Lebesgue density of P. An example for a parameter defined by a U-statistic equation of the first type is the parameter defined by the Hodges–Lehmann estimator of the centre of location (ζ is the median of $P * P$ so that $W(\zeta, P) = \iint |(\mathbb{I}(x_1 + x_2 > \zeta) - 0.5] \, dP(x_1) \, dP(x_2))$ where $\mathbb{I}$ is the indicator function), the variance of this estimator is related to the solution of $\zeta - \int p^2(x) \, dx = 0$.

Not all 'natural' estimating equations have an exact solution. For example, the natural estimating equation for the median is

$$W_n(\zeta, \mathbb{P}_n) = n^{-1} \sum_{i=1}^{n} \mathbb{I}(X_i > \zeta) - 0.5 = 0.$$

Of course, this equation does not have a solution if n is an odd number. To be able to include such cases, an estimator $\tilde{\zeta}$ is defined to be a solution of the estimating equation if $W_n(\tilde{\zeta}, \mathbb{P}_n) = o_p(n^{-1/2})$.

Much of our attention will be devoted to models of the form

$$\mathbb{P} = \{P_{\zeta, \eta} : \zeta \in \mathbb{R}^d, \eta \in H\}, \tag{24.1}$$

where H is some large non-Euclidean space and $\zeta(P_{\zeta,\eta}) = \zeta$. Following Begun *et al.* (1983), such a $\mathscr{P}$ will be called a semi-parametric model. With some abuse of notation we may consider estimating equations of the form of either $W_n(\zeta, \hat{\eta}(\zeta)) = 0$ or $W_n(\zeta; \mathbb{P}_n) = 0$.

In the next section we will quote from Bickel *et al.* (1991) (which we will refer to as **BKRW**) a general result concerning estimating equations. The conditions of this theorem are hard to verify, and the rest of the paper is devoted to non-rigorous consideration of special cases where features of the specific problem can be used to assist in the construction of the estimating equation, or to simplify the theoretical analysis of the asymptotic behaviour of the estimator. In particular we will discuss different types of convexity and the special role of the generalized maximum likelihood estimator of the nuisance parameter.

Notational remark In the following P_0 will denote the true underlying distribution, and ζ_0 is the true value of the parameter $\zeta_0 = \zeta(P_0)$. We will use P and ζ to generate a general distribution and parameter value.

24.2 A general asymptotic theory

There are two possible ways to investigate the asymptotic behaviour of general estimating equations. In the first route, we try to give conditions that are easily verified and prove theorems which are relatively hard to prove, but then, unfortunately, we end with conditions that are hard to meet. We follow BKRW in preferring the second route, in which the theorems are easy to prove, while it is difficult to verify that the conditions are satisfied for a given particular example. We do that, since we feel that it is more useful, in this stage of the research, to give a conceptional frame for the general theory, while the details of each particular problem should be filled by *ad hoc* technicalities. The theorem we are going to state now is taken from BKRW. It is a generalization of the results of Huber (1967) and Pollard (1985).

Theorem 24.1 (GKRW) *Let $W_n(\cdot, \cdot) = 0$ be an estimating equation for the parameter $\zeta(P)$ defined by $W(\zeta(P), P) = 0$. Suppose that*

(A1) *For any sequence $\varepsilon_n \to 0$:*

$$\sup\left\{\sqrt{n}\,\frac{|W_n(\zeta, \mathbb{P}_n) - W_n(\zeta_0, \mathbb{P}_n) - W(\zeta, P_0)|}{1 + \sqrt{n}|\zeta - \zeta_0|} : |\zeta - \zeta_0| \leqslant \varepsilon_n\right\} = o_p(1).$$

(A2) *There is some function ψ such that*

$$\int \psi(x, P)\, \mathrm{d}P(x) = 0, \qquad \int \|\psi(x, P)\|^2\, \mathrm{d}P(x) < \infty$$

and

$$W_n(\zeta_0, \mathbb{P}_n) = n^{-1} \sum_{i=1}^{n} \psi(X_i, P_0) + o_p(n^{-1/2}).$$

(A3) $W(\cdot, P)$ *is continuously differentiable with non-singular matrix of derivatives* $\dot{W}(P)$ *at* $\zeta(P)$.

(A4)   *There exists* $\sqrt{n}$ *consistent estimator* $\hat{\zeta}$ *such that* $W_n(\hat{\zeta}, \mathbb{P}_n) = o_p(n^{-1/2})$.

Then

$$\sqrt{n}\,(\hat{\zeta} - \zeta_0) = n^{-1/2} \sum_{i=1}^{n} \dot{W}(P_0)^{-1}\psi(X_i, P_0) + o_p(1).$$

In particular, $\hat{\zeta}$ *is asymptotically normal.*

24.3   Convex functions, monotonicity, and consistency

Proving that the solution is $\sqrt{n}$-consistent may be the hardest step in proving the asymptotic of the estimator. Given consistency, we can expand W_n and W around ζ_0 to verify the asymptotic normality, etc. General non-parametric conditions that guarantee the global behaviour of an estimator, and hence its consistency, are hard to find. An important exception is the case where W and W_n are monotone.

A function $f: \mathbb{R}^d \to \mathbb{R}^d$ is a monotone field if, for any x in the domain of f and any $y \in \mathbb{R}^d$, the real function of the real variable λ, $y^T f(x + \lambda y)$ is monotone. The major example is the gradient of a convex function. So, let us assume that W and W_n are the gradients of the convex functions D and D_n respectively. There are two convenient facts about convex functions. The first is that with convex functions, pointwise convergence implies uniform convergence over compact spaces, the second is that controlling the behaviour of the estimating function over a small neighbourhood of the true value parameter is enough to obtain a good global behaviour of the estimator. To be more exact:

Theorem 24.2 (Brown 1985 and Ritov 1987)   *Suppose that* $W_n(\zeta, \mathbb{P}_n)$ *and* $W(\zeta, P)$ *are monotone.*

(1)   *Assume that for a compact set* $C \subset \mathbb{R}^d$ *and for any* $t \in C$:

$$\sqrt{n}\,\{W_n(\zeta_0 + t/\sqrt{n}, \mathbb{P}_n) - W_n(\zeta_0, \mathbb{P}_n) - W(\zeta_0 + t/\sqrt{n}, P)\} \xrightarrow{p} 0. \qquad (24.2)$$

Then
$$\sup_{t \in C} \sqrt{n}\,\{W_n(\zeta_0 + t/\sqrt{n}, \mathbb{P}_n) - W_n(\zeta_0, \mathbb{P}_n) - W(\zeta_0 + t/\sqrt{n}, P)\} \xrightarrow{p} 0.$$

(2)   *Suppose that conditions (A2) and (A3) of Theorem 24.1 are satisfied and that for any* M, *(24.2) holds for* $C = \{|t| \leqslant M\}$. *Then there is* $\hat{\zeta}$ *such that* $W_n(\hat{\zeta}, \mathbb{P}_n) = o_p(n^{-1/2})$ *and any such* $\hat{\zeta}$ *is* $\sqrt{n}$ *consistent. In particular, the conclusion of Theorem 24.1 holds.*

Proof For the first part see Ritov (1987). For the second part let $t_0 = \zeta_0$ and

$$t_{i+1} = t_i - \dot{W}^{-1}(P)W_n(t_i, \mathbb{P}_n) \tag{24.3}$$

where $\dot{W}(P) = \bar{W}(P_0)$. Then

$$\|t_{i+1} - \zeta_0\| \leqslant \|t_i - \zeta_0 - \dot{W}^{-1}(P)W(t_i, P_0)\|$$
$$+ \|\dot{W}^{-1}(P)[W_n(t_i, \mathbb{P}_n) - W_n(\zeta_0, \mathbb{P}_n) - W(t_i, P)]\|$$
$$+ \|\dot{W}^{-1}(P)W_n(\zeta_0, \mathbb{P}_n)\|$$
$$= R_1 + R_2 + R_3, \quad \text{say.} \tag{24.4}$$

Fix any $\varepsilon > 0$ and let M be large enough such that R_3 is, with probability of at least $1 - \varepsilon$, less than $0.5Mn^{-1/2}$. Suppose now that $\|t_i - \zeta_0\| \leqslant Mn^{-1/2}$. Then, since W is continuously differentiable

$$R_1 = \|t_i - \zeta_0 - \dot{W}^{-1}(P)[W(t_i, P_0) - W(\zeta_0, P_0)]\|$$
$$= \|[I - \dot{W}^{-1}(P)\dot{W}(t_i^*, P_0)](t_i - \zeta_0)\| = o_p(n^{-1/2}),$$

where t_i^* is some intermediate point. Finally, $R_2 = o_p(n^{-1/2})$ by the first part of the theorem. We conclude that $n^{1/2}\|t_{i+1} - \zeta_0\|$ is bounded by M as well. Since $t_0 = \zeta_0$, we obtain by induction that, with probability of at least $1 - \varepsilon$, $\sup\{n^{1/2}\|t_i - \zeta_0\|\}$ is bounded by M. Therefore, (24.3) and the first part imply that

$$\sup\|t_{i+2} - t_{i+1}\| = \sup\|t_{i+1} - t_i - \dot{W}^{-1}(P)(W_n(t_{i+1}, \mathbb{P}_n) - W_n(t_i, \mathbb{P}_n))\|$$
$$\leqslant \sup\|t_{i+1} - t_i - \dot{W}^{-1}(P)(W(t_{i+1}, P) - W(t_i, P))\| + o_p(n^{-1/2})$$
$$= o_p(n^{-1/2}) \tag{24.5}$$

by the continuity of the derivative of W. Note that the $o_p(n^{-1/2})$ is uniformly small for all i. We conclude from (24.3) and (24.5) that

$$\lim_i \sup\|W_n(t_i, \mathbb{P}_n)\| = \lim_i \sup\|n^{-1/2}\dot{W}(P)(t_{i+1} - t_i)\| = o_p(n^{-1/2}).$$

Since $\{t_i\}$ is bounded, we can take $\hat{\zeta}$ to be any limit point of $t_1, t_2, \ldots$. But then,

$$\|W_n(\hat{\zeta}, \mathbb{P}_n)\| \leqslant \lim_i \sup\|W_n(t_i, \mathbb{P}_n)\| = o_p(n^{-1/2}).$$

The existence of the generalized solution of the estimating equations was proved.

To prove that any solution is $\sqrt{n}$ consistent, suppose that for some estimator $\tilde{\zeta} = \tilde{\zeta}_n$:

$$W_n(\tilde{\zeta}, \mathbb{P}_n) = o_p(n^{-1/2}) \tag{24.6}$$

and that $\lim_{M \to \infty} \limsup_n P\{\|\tilde{\zeta} - \zeta_0\| > Mn^{-1/2}\} > 0$. Fix M, and define λ^* by $\|\lambda^*(\tilde{\zeta} - \zeta_0)\| = Mn^{-1/2}$, $\lambda^* > 0$. Note that

$$\limsup P(0 < \lambda^* < 1) > 0. \qquad (24.7)$$

Let

$$f_n(\lambda) = \frac{(\tilde{\zeta} - \zeta)^\mathrm{T}}{\|\tilde{\zeta} - \zeta\|} W_n(\zeta_0 + \lambda(\tilde{\zeta} - \zeta_0), \mathbb{P}_n).$$

We have by part (1) and then by (A3),

$$f_n(\lambda^*) = \frac{(\tilde{\zeta} - \zeta)^\mathrm{T}}{\|\tilde{\zeta} - \zeta\|} (W_n(\zeta_0, \mathbb{P}_n) + W(\zeta_0 + \lambda^*(\tilde{\zeta} - \zeta_0), P)) + o_p(n^{-1/2})$$

$$= \frac{(\tilde{\zeta} - \zeta)^\mathrm{T}}{\|\tilde{\zeta} - \zeta\|} W_n(\zeta_0, \mathbb{P}_n) + \frac{M(\tilde{\zeta} - \zeta)^\mathrm{T}}{\sqrt{n} \|\tilde{\zeta} - \zeta\|} \dot{W}(P) \frac{(\tilde{\zeta} - \zeta)}{\|\tilde{\zeta} - \zeta\|} + o_p(n^{-1/2}).$$

$$(24.8)$$

But, by assumption (A2), the first term on the RHS is $O_p(n^{-1/2})$ and M can be arbitrarily large, hence $n^{-1/2} f_n(\lambda^*) \overset{P}{\to} 0$. Since $n^{-1/2} f_n(0) = O_p(1)$, it follows from (24.7) and the monotonicity of f_n that $n^{-1/2} f_n(1) \overset{P}{\not\to} 0$ as well, contradicting (24.6).

Example 24.1 Distribution with known marginal.   Suppose that $X = (Y, Z)$ and for all possible P

$$E_P\{\psi(Y, Z; \zeta(P))\} = 0.$$

We wish to construct an estimator of ζ_0 that will use the fact that the marginal distributions of Y and Z are supposed to be known, that is

$$\mathscr{P}_0 = \left\{ \begin{array}{l} \text{all joint distributions with} \\ P_Y = P_Y^0 \text{ and } P_Z = P_Z^0 \end{array} \right\},$$

for some P_Y^0 and P_Z^0. Any M-estimation of the centre of location applied either to the difference between Y and Z or to their ratio can serve as an example. This model, whose asymptotics was studied by Bickel *et al.* (1991), may be used as a methodological introduction to cases where we have a moderate size sample from X, and independent and much larger samples from Y and Z.

An immediate application of our set-up will be to consider

$$W_n(\zeta, \mathbb{P}_n) \equiv \int \psi(y, z; \zeta) \, d\mathbb{P}_n(y, z), \qquad (24.9)$$

which does not use the *a priori* knowledge of the marginals. A more efficient estimator may be achieved when we use $W_n(\zeta, \mathbb{P}_n) \equiv \int \psi(y, z; \zeta) \, d\hat{P}_n(y, z)$, where $\hat{P}_n$ is an estimator of the joint distribution that takes into account the known marginal distributions.

Let the sample space be $\mathbf{Y} \times \mathbf{Z}$. Let $A_1^n, \ldots, A_{m(n)}^n, B_1^n, \ldots, B_{m(n)}^n$ be measurable partitions of $\mathbf{Y}$ and $\mathbf{Z}$ respectively. Suppose that

$$\mathscr{P} = \mathscr{P}_0 \cap \left\{ P: \inf \frac{P(A_j^n \times B_k^n)}{P_Y^0(A_j^n)P_Z^0(B_k^n)} > 0 \right\}.$$

Assume that the partitions are constructed such that

$$\max_j \{P(A_j^n), P(B_j^n): A_j^n, B_j^n \text{ not } P_Y^0 \times P_Z^0 \text{ atoms}\} \to 0$$

and

$$\sqrt{(n/\log n)} \min_j \{P(A_j^n), P(B_j^n)\} \to \infty$$

as $n \to \infty$. (This construction is always possible.) Then one possible construction of $\hat{P}_n$ is as follows (see Bickel *et al.* for details). Consider the contingency table with entries $n_{ij} = n\mathbb{P}_n(A_i^n \times B_j^n)$. Let $\{\hat{p}_{ij}\}$ be the constrained MLE of $p_{ij}^0 = P_0(A_i^n \times B_j^n)$ given n_{ij} and under the constraints $p_{i\cdot} = P_Y^0(A_i^n)$ and $p_{\cdot j} = P_Z^0(B_j^n)$. Finally, let

$$\hat{P}_n(C) = \sum_{i=1}^{m(n)} \sum_{j=1}^{m(n)} \hat{p}_{ij} \frac{\mathbb{P}_n(C \cap A_i^n \times B_j^n)}{\mathbb{P}_n(A_i^n \times B_j^n)}$$

for any measurable C. The rationale of the estimator will be discussed in the next section.

Take $W_n(\zeta, \mathbb{P}_n) \equiv \int \psi(y, z; \zeta) \, d\hat{P}_n(y, z)$ such that $\psi(y, z; \zeta)$ is a monotone random field. Then W_n is monotone and the above theorem can be used to verify the existence and the essential uniqueness of the generalized solution. Use Bickel *et al.* (1989) to verify the pointwise convergence. A simple example of this type is given by $\psi(y, z; \zeta) = \psi_H((y/z) - \zeta)$ where ψ_H is the influence function of the Huber's M-estimator. For another example, suppose that we want to find the 'best' fit of an exponential family to the true distribution (which is not assumed to belong to this exponential family). To be more precise, consider an s-dimensional exponential family of distributions with densities $\exp\{\sum_1^s \zeta_j T_j(\cdot) - A(\zeta)\}: \zeta \in \mathscr{Z}$ with respect to some measure μ. We may take $W(\zeta, P)$ to be the gradient of

$$D(\zeta, P) = \int \sum_{j=1}^s \zeta_j T_j(x) \, dP(x) - A(\zeta),$$

where $\zeta = (\zeta_1, \ldots, \zeta_s)^T$ and

$$\int \exp\left\{ \sum_{j=1}^s \zeta_i T_i(x) - A(\zeta) \right\} d\mu(x) = 1.$$

Again, it is well known that

$$D_n(\zeta, P) = \int \sum_{j=1}^{s} \zeta_j T_j(x) \, d\mathbb{P}_n(x) - A(\zeta),$$

is convex (cf. Lemann 1986, p. 66).

24.4 The particular role of the non-parametric MLE of the nuisance parameter

It may not be an easy task to find the efficient estimating equations for semi-parametric models, that estimating equation which yields a regular estimator with the lowest possible asymptotic variance. In general, it involves finding the least favourable model in the class, i.e. a finite-dimensional sub-model of P such that the asymptotic variance of the efficient estimator of ζ with respect to this sub-model is the largest. This may involve the following steps:

(1) find the Frechet derivative $\dot{\zeta}(P)$ of the function $\zeta(P)$;
(2) approximate the local structure of the $\mathscr{P}$ by a Hilbert space (the 'tangent space');
(3) find the projection $\chi(\cdot; P)$ of $\dot{\zeta}(P)$ on the tangent space;
(4) find an estimator with influence function $\psi(\cdot; P)$.

See BKRW or Pfanzagl (1982) for details. This may prove to be a tedious task, if possible at all. Luckily, it may be not necessary. A case in hand is the following example.

Example 24.2 (Censored regression)   Suppose that $T = \zeta Z + \varepsilon$ but we observe $X = (Y, Z, \Delta)$ where ε is independent of Z and C, $Y = T \wedge C$ and $\Delta = \mathbb{I}\{T \leqslant C\}$. (We denote the minimum of two variables by $\wedge$ and the indicator of an event is denoted by $\mathbb{I}$.) Clearly, if T is observed directly, ζ can be estimated through the solution of

$$\tilde{W}_n(\zeta, \mathbb{P}_n) \equiv n^{-1} \sum_{i=1}^{n} Z_i \psi(T_i - \zeta Z_i) = 0$$

for some function ψ, see Huber (1981). Ritov (1990), following Buckely and James (1979), suggested to extend this to the censored case by replacing T_i, when it is not observed (i.e. when $\Delta = 0$), by its conditional expectation given the observation X and under the 'assumption' that the $\zeta(P_0) = \zeta$. To be formal, let $\mathbb{F}_{n,\theta}$ be the product limit (the Kaplan–Meier) estimator of the censored variable based on the sample $(Y_i - \zeta^T Z_i, \Delta_i)$: $i = 1, 2, \ldots, n$. Then we consider

$$W_n(\zeta, \mathbb{P}_n) = n^{-1} \sum_{i=1}^{n} Z_i \left\{ \Delta_i \psi(Y_i - \zeta Z - i) + (1 - \Delta_i) \frac{\int_{Y_i - \zeta Z_i} \psi(t) \, d\mathbb{F}_{n,\theta}(t)}{1 - \mathbb{F}_{n,\theta}(Y_i - \zeta Z_i)} \right\}.$$

Ritov and Wellner (1988) gave the influence function of the asymptotically efficient estimators. Tsiatis (1988) considered a class of estimating equations which are functions of the order statistics. This class includes an efficient estimator. The motivation of his work stems from some tests of fit for ζ. Ritov (1990) proves the asymptotic equivalence of Tsiatis' class of estimators and the class suggested above.

Both Tsiatis (1988) and Buckley and James (1979) used a heuristic argument based on the philosophy of the EM algorithm. In essence they replaced the unknown nuisance parameter with a generalized maximum likelihood estimator (GMLE).

We argue that using the product limit estimator is not an arbitrary decision and being the GMLE of the distribution is really what makes it work. Ritov (1990) actually conjectured that using any other non-equivalent estimator would result in a non-regular estimator of ζ.

Suppose $\mathscr{P} = \{P_{\zeta,\eta}\}$, where $\eta \in \mathscr{H}$ is an infinite-dimensional nuisance parameter. Suppose that $W_n(\zeta, \eta)$ produces an estimating equation which can be used when it is known that the nuisance parameter equation is η. Then, solve $W_n(\zeta, \hat\eta(\zeta)) = o_p(n^{-1/2})$, where $\hat\eta(\zeta)$ is the generalized MLE of η under the assumption that $\zeta(P_0) = \zeta$. We follow in this discussion some of the ideas of Severini and Wong (1989). A similar argument restricted to parametric families is given in Randeles (1982).

A parametric family of distribution is called regular if (i) it is dominated by some measure μ; (ii) if its densities with respect to μ are $\{f_\zeta(\cdot)\}$ and $s_\zeta \equiv f_\zeta^{1/2}$ then s_ζ has a continuous L_2 derivative $\dot s_\zeta$ with respect to the parameter; (iii) $I_\zeta \equiv 4 \int \dot s_\zeta(x) \dot s_\zeta^{\mathrm{T}} \, d\mu$ is invertible. We call any regular parametric family

$$\mathscr{P}^* = \{P_{\zeta,\eta_v} : \zeta \in \Theta \subset R^d, v \in (-1, 1)\} \subset \mathscr{P} \qquad (24.10)$$

a regular sub-model of $\mathscr{P}$. The asymptotic variance of an estimator which is based on the assumption that $P_0 \in \mathscr{P}^*$ defines a lower bound for what can be achieved while assuming only that $P \in \mathscr{P}$—one cannot estimate ζ in the model $\mathscr{P}$ better than in the smaller model $\mathscr{P}^*$. This point of view, restricted for efficient estimators (i.e. regular estimator with the lowest possible asymptotic variance), is advocated in Stein (1956), Koshevnik and Levitt (1976), Pfanzagl (1982), Begun et al. (1983) and BKRW among others. The price for efficiency may be in terms of robustness and construction difficulties. The same idea, however, is useful whether or not we want to use an efficient estimator.

Suppose now that $\hat{\zeta}$ is an estimator with influence function $\psi(\cdot\,; \zeta, \eta)$. Then ψ should satisfy the following conditions with respect to any regular sub-model (24.10) (Klaassen 1987):

$$\int \psi(x; \zeta_0, \eta_0)\, dP_{\zeta_0, \eta_0}(x) = o_p(|v|) \qquad \text{as } v \to 0 \qquad (24.11)$$

and

$$\int \psi(x; \zeta_0, \eta_0)\, dP_{\zeta, \eta_{v_0}}(x) = \zeta - \zeta_0 + o_p(|\zeta - \zeta_0|) \qquad \text{as } \zeta \to \zeta_0. \qquad (24.12)$$

Suppose that $W_{\zeta_n}(\zeta, \mathbb{P}_n)$, an estimating equation appropriate for the model $\{P_{\zeta, \eta_0}\}$ has an influence function $\psi_\zeta(\cdot\,; \zeta, \eta)$ that satisfies only (24.12) while (24.11) fails. Then for any regular sub-model $\mathscr{P}^*$ there is some c such that

$$\psi^*(\cdot\,; \zeta, \eta_0) = \psi_\zeta(\cdot\,; \zeta, \eta_0) + c\,\frac{\partial}{\partial v}\log p_{\zeta, \eta_v}(\cdot)|_{v=0} \qquad (24.13)$$

will satisfy both (24.11) and (24.12). Suppose

$$\left\{ \frac{\partial}{\partial v}\log p_{\zeta, \eta_v}\colon p_{\zeta, \eta_v} \text{ a regular sub-model} \right\}$$

is a linear space. Then there is some $\mathscr{P}_0^*$ and c such that ψ^*, defined in (24.13), satisfies (24.11) and (24.12) for *any* parametric sub-model $\mathscr{P}^*$. Denote this influence function by

$$\psi_p \equiv \psi_\zeta + \psi_\eta,$$

where ψ_η is the v-derivative of the log-likelihood with respect to some parametric sub-model $\mathscr{P}_0^*$. One may wish now to modify his estimating function $W_{\zeta_n}(\cdot\,, \cdot) = 0$ to an estimating function appropriate to the full model $\mathscr{P}$ by considering

$$W_n(\cdot\,; \zeta, \eta) \equiv W_{\zeta_n}(\cdot\,; \zeta, \eta) + n^{-1}\sum_{i=1}^n \psi_\eta(\cdot\,; \zeta, \eta) = 0. \qquad (24.14)$$

Now the advantage in using the GMLE $\hat{\eta}_\zeta$ is that $\hat{\eta}(\zeta)$ is, in particular, the MLE with respect to the sub-model $\mathscr{P}_0^*$ and hence it satisfies the likelihood equation with respect to v. The second term on the RHS of (24.14) with $\eta = \hat{\eta}_\zeta$ is, therefore, zero, or

$$W_n(\cdot\,; \zeta, \eta) = W_{\zeta_n}(\cdot\,; \zeta, \hat{\eta}(\zeta)).$$

Explicit calculation of ψ_η was avoided. Of course, one should still prove that $\psi_{\hat{\eta}(\zeta)}(\cdot\,; \zeta, \hat{\eta}(\zeta))$ approximates $\psi_{\eta_0}(\cdot\,; \zeta, \eta_0)$.

Example 24.1 (continued)   Suppose we want to estimate $\zeta = E\{h(Y, Z)\}$ for some function h. Then we may use the function

$$W_n(\zeta; \mathbb{P}_n) = \sum_{i=1}^{n} h(Y_i, Z_i) - \zeta = \int [h(y, z) - \zeta]\, d\mathbb{P}_n. \qquad (24.15)$$

This does not use the information given by the known marginal distributions. There is a better estimation equation which takes this knowledge into account and it is given by

$$W_n(\zeta, \mathbb{P}_n) = \sum_{i=1}^{n} [h(Y_i, Z_i) - a(Y_i; P_0) - b(Z_i; P_0)] - \zeta + o_p(n^{-1/2})$$

$$= \int [h(y, z) - a(y; P_0) - b(z; P_0) - \zeta]\, d\mathbb{P}_n + o_p(n^{-1/2}), \qquad (24.16)$$

where a and b are a solution of the ACE type equations:

$$a(Y; P) = E_P(h(Y, Z)|Y) - E_P(b(Z; P)|Y)$$
$$b(Z; P) = E_P(h(Y, Z)|Z) - E_P(a(Y; P)|Z) \qquad (24.17)$$

(Bickel *et al.* 1989). In general, a and b are not uniquely defined (but their sum is). The problem is, of course, that $a + b$ depends on the unknown joint distribution P_0. Replacing P_0 by the empirical distribution function is of no help since equations (24.17) depend on the conditional expectations, while the conditional expectations of $\mathbb{P}_n$ are consistent only in the case that X has discrete marginals. One may try to solve equations (24.17) using another, more appropriate estimator of the underlying distribution. Following Bickel *et al.* (1989) we suggest to use the MLE $\hat{P}_n$ with respect to some sieve (see Grenander 1981 and Geman and Hwang 1982). The rational is that replacing P_0, $a(\cdot; P_0)$ and $b(\cdot; P_0)$ in (24.16) by $\hat{P}_n$, $a(\cdot; \hat{P}_n)$ and $b(\cdot; \hat{P}_n)$ respectively we obtain

$$\int [h(y, z) - a(y; \hat{P}_n) - b(z; \hat{P}) - \zeta]\, d\mathbb{P}_{nn} = \int [h(y, z) - \zeta]\, d\hat{P}_n + o_p(n^{-1/2})$$

(Bickel *et al.* 1989).

Example 24.3 (Regression model with biased sampling)   The following model was analysed by Bickel and Ritov (1991). We begin with a simple regression model $Y' = \zeta^T Z' + \varepsilon'$ where ε and Z' are independent. We observe a random sample from $X = (Y, Z, S)$, where S is a strata variable, $P(S = i) = \lambda_i$, $i = 1, \ldots, k$, and given $S = i$, (Y, Z) is a biased sample from (Y', Z') with a known weight function $w_i(y, z)$, i.e. $p \in \mathscr{P}$ if it has density with respect to Lebesgue measure $\times \mu \times$ counting measure

$$p_{\zeta, \lambda, G, H}(i, y, z) = \lambda_i w_i(y, z) g(y - \zeta z) h(z) / W_i(G, H),$$

where

$$W_i(G, H) = \int\int w_i(y, z)g(y - \zeta z)h(z) \, dy \, d\mu(z).$$

We assume that μ is finite supported. Let $\hat{\lambda}_\zeta$, $\hat{G}_\zeta$ and $\hat{h}_\zeta$ be the MLE of the λ, G, and h respectively, when it is assumed that the ζ_0 is equal to ζ. The asymptotic properties of estimators in the presence of biased sampling are discussed in Gill *et al.* (1988), BKRW, Jewel (1985), Quesenberry and Jewel (1986), and Bickel and Ritov (1991).

We begin with an estimator with an influence function $\psi(Z', \varepsilon')$, where $E\{\psi(Z', Y' - \zeta^T Z')\} = 0$. When biased sampling is introduced we should modify the influence function since, in general, $E\{\psi(Z, Y - \zeta^T Z)\} \neq 0$. A natural way to do that would be to consider

$$\psi_B(X; P) = \psi(Z, \varepsilon) - a(Z; P) - b(\varepsilon; P) - c(S; P). \qquad (24.18)$$

Where a, b, and c form the solution of the ACE type equations:

$$E_P(\psi_B(X; P)|Z) = 0$$
$$E_P(\psi_B(X; P)|\varepsilon) = 0$$
$$E_P(\psi_B(X; P)|S) = 0$$

In particular, if $\psi(\cdot; \cdot)$ is the influence function of the efficient estimator without biased sampling, then (24.18) defines the influence of an efficient estimator in the presence of bias sampling. Again, we could estimate a, b, and c (note that this time, since $\varepsilon = Y - \zeta^T Z$, we should estimate them separately for any value of ζ). Bickel and Ritov (1991) elect to by-pass the explicit calculation of this function, and this was done using the GMLE of the distribution. Begin with $\psi(\cdot, \cdot)$ as above and then use the estimation function

$$\int \psi(z, y - \zeta z) \, d\mathbb{P}_n(y, z, s) - \int \psi(z, y - \zeta z) \, d\hat{P}_\zeta(y, z, s). \qquad (24.19)$$

where $\hat{P}_\zeta = P_{\zeta, \hat{\lambda}_\zeta, \hat{G}_\zeta, \hat{H}_\zeta}$ and the hats define the GMLE of the given quantities. Again, we do not need to estimate the projection directly since (24.19) is exactly equal to

$$\int [\psi(z, y - \zeta^T z) - a(z; \hat{P}_\zeta) - b(y - \zeta^T z; \hat{P}_\zeta) - c(s; \hat{P}_\zeta)] \, d\mathbb{P}_n.$$

That is, we have replaced the unknown distribution with its GMLE to obtain an estimating equation that defines a $\sqrt{n}$-consistent asymptotically normal estimator. Moreover, the efficient estimator of ζ belong to this class.

One remark is in place. Using the GMLE let us find a convenient estimator. It usually does not help in proving the actual asymptotics of the estimator. This would involve verification of assumptions (A2) of Theorem 24.1. To do that we should know the influence function of the estimator and hence the theoretical projection of the initial influence on the 'tangent space'.

24.5 Convex families and the existence of simple estimating equations

Let $\mathscr{P}$ be a family of distribution and let the parameter ζ be defined explicitly by $W(\zeta(P), P) = 0$. The family $\mathscr{P}$ is ζ-convex if, for all possible values of ζ, the mixing of any two members of the section $\mathscr{P}_\zeta = \{P \in \mathscr{P}, W(\zeta, P) = 0\}$ is another member of this sub-family. The research of this class of models was probably motivated by the symmetric location model: the mixture of two distributions on the line which are symmetric about ζ is, again symmetric about ζ. See Bickel (1982).

Example 24.4 (Mixture of exponential families) Suppose that

$$\mathscr{P} = \left\{ P \colon P(x; \zeta, \eta) = \int Q(x; \zeta, \xi)\, d\eta(\xi), \quad \eta \text{ absolutely continuous} \right\}.$$

where $\mathscr{Q} = \{Q(\cdot; \zeta, \xi) \colon \xi \in \Xi\}$ is, for any given ζ, an exponential family with a complete sufficient statistic $T(X; \zeta)$. Clearly, this model is ζ-convex—a mixture of mixtures is a mixture.

This model was studied extensively. It is a reasonable interpretation of the model where $X_i \sim Q((\cdot; \zeta, \zeta_i)$, the X_i are independent and the ξ_i are unknown parameters. See, for example, Godambe (1976, 1980), Lindsey (1983a–c). Amari and Kumon (1988). A formal semi-parametric analysis of this class is given in BKRW and Pfanzagl (1982). A particular applied example that falls under this class is the classical errors in variables with normal errors.

This class of models is interesting because if the model belongs to a ζ-convex family, it is relatively easy to construct an estimating function W_n even if ζ was not defined initially through an estimating function $W(\zeta(P), P) = 0$. This is so, since, if $W_n(\zeta, \mathbb{P}_n) = 0$ can be used as an estimating function when it is known that the true distribution is a contiguous alternative to some P_1, then, essentially, it is useful for any true P.

Consider the semi-parametric model of (24.1). Let $\tilde{\zeta}$ be an estimator which is appropriate if it is known that the nuisance parameter is 'approximately' equal to some $\eta_0 \in \mathscr{H}$. By that we mean that the estimator is regular with respect to any parametric sub-model that passes through P_0. Denote the influence function of this estimator by $\psi_0(x; \zeta)$. Then ψ_0 satisfies (24.11) and (24.12). Suppose now that for any fixed $P \in \mathscr{P}$

$$\mathscr{P}^* = \{(1 - v)P_0 + vP \colon v \in (-1, 1)\} \subset \mathscr{P}$$

is a regular parametric sub-model. Then we obtain from (24.11) that for all $P \in \mathscr{P}_\zeta$

$$\int \psi_0(x; \zeta_0) \, dP(x) = 0.$$

Hence if W_n satisfies

$$W_n(\zeta, \mathbb{P}_n) = n^{-1} \sum_{i=1}^{n} \psi_0(X_i; \zeta) + o_p(n^{-1/2}),$$

then $W_n(\zeta, \mathbb{P}_n) = 0$ can be used whether or not η_0 approximates the true distribution. This is especially easy when adapative estimators are possible (see Bickel 1982), that is, when an efficient estimator has the same asymptotic variance whether or not we know the value of the nuisance parameter η. More generally, we may identify a simple parametric sub-model to which we can adapt. That is, write $\eta = (\gamma, \eta') \in \Gamma \times \mathscr{H}'$, where Γ is an open set of a Euclidean space, and suppose that we can adapt to η' unknown—i.e. the influence function of the MLE of ζ when γ is unknown and η' is assumed known satisfies (24.11) and (24.12). Then, we may use the MLE as our estimator for the model $\mathscr{P}$ pretending that the true value of η' is any η'_0 of convenience.

Example 24.5 (Group families)   Suppose that $X = (Y, Z)$ with density

$$p(y, z; \zeta, \eta) = \int \xi^2 \zeta \, e^{-\xi(y + \zeta z)} \, d\eta(\xi).$$

This model was investigated by Lindsey (1983a–c) in the context of exponential families of Example 24.4. (Here, clearly, $T(X; \zeta) = -(Y + \zeta Z)$.) This model can be extended, without losing any information to a large model. Namely, we don't need to assume anything about the distribution of X except that $(Y, \zeta Z)$ and $(\zeta Z, Y)$ have the same distribution.

The general class is the following. Let $X = (Y, Z)$, and for some parametric group $\mathscr{G} = \{g_\zeta\}$:

$$\mathscr{P} = \{P_{\zeta, \eta}: \mathscr{L}(g_\gamma Y, g_\zeta g_\gamma Z) = \eta \in \mathscr{H}, \quad \text{for some } g_\gamma \in \mathscr{G}\},$$

where $\mathscr{L}(X)$ denotes the law of X and

$$\mathscr{H} = \{\eta: \eta(Y, Z) = \eta(Z, Y)\}.$$

Some discussion of this model can be found in Pfangzagl (1982) and BKRW.

Let η_0 be any member of $\mathscr{H}$, and let $W(\zeta, \eta) = 0$ be the likelihood equation for ζ with respect to the sub-model:

$$\mathscr{P}_{\eta_0} = \{P_{\zeta, \eta_0}: \mathscr{L}(g_\gamma Y, g_\gamma g_\zeta Z) = \eta_0, \quad g_\gamma, g_\zeta \in \mathscr{G}\}. \tag{24.20}$$

We claim that this estimating equation may be used whether or not the true distribution $\mathbf{Z} \in \mathscr{P}$ belongs to $\mathscr{P}_{\eta_0}$. An asymptotically efficient estimator for the model $\mathscr{P}$ can be constructed by using $W_n(\zeta, \mathbb{P}_n) = W(\zeta, \hat{\eta})$ for some estimator $\hat{\eta}$.

One could hope that the MLE with respect to the sub-model

$$\mathscr{P}_{\eta_0} = \{P_{\zeta, \eta_0} : \mathscr{L}(Y, g_\zeta Z) = \eta_0\}$$

could be used. Unfortunately, this is not true.

We specialize now to the following particular example (see Pfanzagl 1981). Let $X = (Y, Z) \in R^2$, and

$$\mathscr{P} = \{P : P(Y > 0 \quad \text{and} \quad Z > 0) = 1, \mathscr{L}(Y, {}^\zeta Z) = \mathscr{L}({}^\zeta Z, Y)\}.$$

Let

$$\mathscr{P}_0 = \left\{P : P(Y > 0 \text{ and } Z > 0) = 1, \frac{\partial P}{\partial \mu} = p(y, z; \zeta, \gamma) = \gamma \zeta \, e^{-\gamma y - \gamma \zeta z}, \gamma > 0\right\},$$

where μ is the Lebesgue measure. The likelihood equations for (ζ, γ) under the model $\mathscr{P}_0$ can be easily found to be

$$0 = \frac{2}{\gamma} - \sum_{i=1}^{n} Y_i - \zeta \sum_{i=1}^{n} Z_i$$

$$0 = \frac{1}{\zeta} - \sum_{i=1}^{n} Y_i - \gamma \sum_{i=1}^{n} Z_i. \tag{24.21}$$

These equations yield the estimator $\hat{\zeta} = \sum_{i=1}^{n} Y_i / \sum_{i=1}^{n} Z_i$, which makes sense for the model $\mathscr{P}$. Note that if we take the further sub-model $\mathscr{P}_0' \subset \mathscr{P}_0$ with the restriction $\gamma = 1$, we get only (24.21), which does not define a consistent estimator with respect to $\mathscr{P}$. More generally we may use any family $\mathscr{P}_0 = \{P : \partial P/d\mu = q(\gamma y, \gamma \zeta z; \xi)\}$, where $q(y, z; \xi) \equiv q(y, z; \xi)$ such that the distribution of X_1 can be reasonably approximated by a member of $\mathscr{P}_0$ and find $\hat{\zeta}$ which is the MLE for ζ relative to the model $\mathscr{P}_0$.

We want to emphasize that it is not true that in the case of convex families we may use any estimator that is useful if the nuisance parameter is known. Only an estimator that is appropriate when the nuisance parameter is *approximately* known can be used. In other words, the estimator should be robust to the specification of the nuisance parameter.

The estimating equations obtained in this way are linear in the observations (i.e. $W_n(\zeta, \mathbb{P}_n) = n^{-1} \sum_{i=1}^{n} \psi(X_i; \zeta)$. However, estimating functions which are linear in the observations are not a complete class. In general, we can get a lower asymptotic variance if we try to adapt to the true underlying distribution.

Example 24.6 (Errors in variables)   This is a subclass of the models considered in Example 24.4. Here we assume that $X = (Y, Z)$, where

$$Y = \zeta\Xi + \varepsilon_Y$$
$$Z = \Xi + \varepsilon_Z,$$

$(\varepsilon_Y, \varepsilon_Z)^\mathrm{T}$ and Ξ are independent, $(\varepsilon_Y, \varepsilon_Z)^\mathrm{T}$ bivariate normal, and Ξ cannot be expressed as a convolution of a normal random variable with some other random variable. We cannot estimate ζ using the MLE estimator appropriate when the distribution of Ξ is known. Yet, the MLE of ζ, pretending that the distribution of Ξ belongs to some location-shift family, can be used, as long as the members of this family have a continuous component.

A better estimator (in terms of asymptotic variance) can be achieved if we try to estimate the influence function of the efficient estimator. This route was considered in Bickel and Ritov (1987) where an asymptotically efficient estimator was suggested. A small Monte Carlo experiment suggests, however, that in 'practical' situations the efficient estimator is hardly better than more intuitive estimators.

We conclude with the following example.

Example 24.7 (Signal with a location jitter)   Ritov (1989) considered the following problem. Suppose we observe

$$dX_i(t) = s(t - \tau_i; \zeta)\, dt + dB_i(t), \qquad t \in [0, T], \quad i = 1, \ldots, n,$$

where $X_1, \ldots, X_n$ are independent, $B_1(\cdot)$ is a standard Brownian motion, and $\tau_1, \ldots, \tau_n$ are unknown *a priori* time shifts. To make the analysis well defined we assume that the time shifts are i.i.d. random variables. The model becomes, under this assumption, a mixture model that does not belong to the exponential family structure of Example 21.4. We do not know any explicitly defined efficient estimator of ζ for this example. However, a simple regular estimator can be found under some assumptions. In particular assume that the support of $s(\cdot; \zeta)$ and the support of τ_1 are such that we can assume that the observations are on a circle. (That is, $P_0\{s(t - \tau_1; \zeta) = 0\} = 1$ for all $t \notin [0, T]$.) The model, with ζ fixed and τ varying, has an invariant structure. Any estimating equation which is equivariant for this model could be used when the distribution of the τs is unknown. This is achieved with the estimator which corresponds to the assumption that τ_1 is uniformly distributed over the circle is appropriate whether or not the random shift actually has this distribution. However, assuming any other distribution for τ_1 will not yield a consistent estimator, even if the assumed distribution is very close to the true distribution of the jitter.

24.6 Acknowledgements

All I know about this subject I got from the joint work with P. J. Bickel and J. A. Wellner. I want to thank a referee for a careful reading of the manuscript.

This research was done while the author was visiting the Department of Pennsylvania, Philadelphia.

References

Amari, S.-I. and Kumon, M. (1988). Estimation in the presence of infinitely many nuisance parameters—geometry of estimating equations, *Ann. Stat.*, **16**, 1044–88.

Begun, J., Hall, W., Huang, W., and Wellner, J. A. (1983). Information and asymptotic efficiency in parametric–semiparametric models. *Ann. Stat.*, **11**, 432–52.

Bickel, P. J. (1982). On adaptive estimation. *Ann. Stat.*, **10**, 431–44.

Bickel, P. J. and Ritov, Y. (1987). Efficient estimation in the error in variables model. *Ann. Stat.*, **15**, 513–40.

Bickel, P. J. and Ritov, Y. (1991). Large sample theory of estimation in biased sampling regression models I. *Ann. Stat.* (In press.)

Bickel, P. J., Ritov, Y., and Wellner, J. A. (1991). Efficient estimation of a probability measure *P* with known marginals. *Ann. Stat.* (In press.)

Bickel, P. J., Klaassen, C. A. J., Ritov, Y., and Wellner, J. A. (1991). *Efficient and adaptive estimation in semiparametric models.* Forthcoming monograph. John Hopkins University Press, Baltimore.

Brown, B. M. (1985). Multiparameter linearization theorems. *J. Roy. Stat. Soc.*, B **47**, 325–31.

Buckely, J. and James, I. R. (1979). Linear regression with censored data. *Biometrika*, **66**, 429–36.

Geman, S. and Hwang, C. R. (1982). Nonparametric maximum likelihood estimation by the method of sieves. *Ann. Stat.*, **10**, 410–14.

Gill, R. G., Vardi, Y., and Wellner, J. A. (1988). Large sample theory of empirical distributions in biased sampling models. *Ann. Stat.*, **16**, 1069–1112.

Godambe, V. P. (1976). Conditional likelihood and unconditional optimum estimating equations. *Biometrika*, **63**, 277–84.

Godambe, V. P. (1980). On the sufficiency and ancilarity in the presence of nuisance parameter. *Biometrika*, **67**, 419–28.

Grenander, U. (1981). *Abstract inference.* Wiley, New York.

Huber, P. J. (1964). Robust estimation of a location parameter. *Ann. Math. Stat.*, **35**, 73–101.

Huber, P. J. (1967). The behavior of maximum-likelihood estimates under nonstandard conditions. *Proc. Vth Berkeley symp.*, **1**, 221–33.

Jewel, N. P. (1985). Least squares regression with data arising from stratified samples of the dependent variable. *Biometrika*, **72**, 11–21.

Klaassen, C. A. J. (1987). Consistent estimation of the influence function of locally asymptotically linear estimators. *Ann. Stat.*, **15**, 1548–62.

Koshevnik and Levitt, B. Y. (1976). On nonparametric analogue of the information matrix. *Th. Prob. Appl.*, **21**, 738–53.

Lehmann, E. L. (1986). *Testing statistical hypotheses*. Wiley, New York.

Levitt, B. Y. (1978). Infinite dimensional information bounds. *Th. Prob. Appl.*, **23**, 723–40.

Lindsey, B. G. (1983a). Efficiency of the conditional score in a mixture setting. *Ann. Stat.*, **11**, 486–97.

Lindsey, B. G. (1983b). The geometry of the mixing likelihood: a general theory. *Ann. Stat.*, **11**, 86–94.

Lindsey, B. G. (1983c). The geometry of the mixing likelihood, part II: the exponential family. *Ann. Stat.*, **11**, 783–92.

Pfanzagl, J. (with Welfelmeyer, W.) (1982). *Contribution to a general asymptotic statistical theory*. Springer-Verlag, New York.

Pollard, D. (1985). New ways to prove central limit theorems. *Econometric Theory*, **1**, 295–314.

Quesenberry, C. P. and Jewel, N. P. (1986). Regression analysis based on stratified samples. *Biometrika*, **73**, 605–14.

Randeles, R. H. (1982). On the asymptotic normality of statistics with estimated parameters. *Ann. Stat.*, **10**, 462–74.

Ritov, Y. (1987). Tightness of monotone random fields. *J. Roy. Stat. Soc.* B **49**, 331–3.

Ritov, Y. (1989). Estimating a signal with nuisance parameters. *Biometrika*, **76**, 31–8.

Ritov, Y. (1990). Estimation in a linear regression model with censored data. *Ann. Stat.*, **18**, 303–28.

Ritov, Y. and Wellner, J. A. (1988). Censoring, martingales and the Cox model. In N. V. Prabhu (ed.), *Contemporary Mathematics (AMS)*, *Volume on Statistical Inference for Stochastic Processes* **80**, 191–219.

Severini, T. A. and Wong, W. H. (1989). On maximum likelihood estimation in infinite dimensional parameter spaces. Manuscript.

Stein, C. (1956). Efficient nonparametric testing and estimation. *Proc. 3rd Berkeley Symp.*, **1**.

Tsiatis, P. A. (1988). Estimating regression parameters using linear rank tests for censored data. Manuscript.

Author index

Subject index

accuracy 259, 260
ancillarity 16, 241–5
  and efficiency 268, 270–1, 273–5
  Fisher information 250–3
  nuisance parameters 246–50
  vectors 245–6
asymptotics 148
  confidence zones 167
  and efficiency 152, 153–7
  maximum likelihood estimators 173–8
  normality 164
  optimality 152–3
  and semi-parametric models 321–2, 330
augmented regression estimator 219–20
autoregression 261–2

beta-binomial distribution 75, 93–4
bias-corrected estimator 58–61
biased sampling 329–30
bias reduction 86
binomial regression 79
bivariate moment equations 305–17
bootstrap methods 295–301
branching processes 178–86
Brewer Hájek estimator 206

capture–recapture studies 65–87
coeffcient of variation 90
conditional inference 97–100
conditional least squares 140
  and maximum likelihood 171, 172, 178,
    180–6
conditional likelihood 69–70, 92
conditional maximum likelihood estimator
    98, 242
  Darroch's estimator 70, 71, 72
  unbiasedness 91, 92
conditional score functions 247, 248–9, 252–3
confidence intervals 161–8
  for quantiles 211–4
  and super-population 224, 233–5
  and variance estimation 220–1
consistency 322
continuous-time processes 187

convex functions 322–6, 331–4
convolution theorem 155–6
counting process models, partially
    specified 147–58
covariance misspecification 109–10
Cox's regression model 23–4, 100
Cramer–Rao inequality 12, 14, 16

Darroch's conditional maximum likelihood
    estimator 70–2

efficiency
  and fidelity 256–7
  geometrical aspects 267–75
  and martingales 87
  and measurement errors 53–4
  and super-population 230–1
errors
  in measurement 47–63
  in variables 334
estimating functions
  definition 4
  multiparametric case 13–14
  role 4–6
  solutions 149–51
  unbiasedness 91
exponentiality 285, 331
exponential models 103–16

factorization 91–2, 98, 99
fidelity 256–64
Fisher information 250–3
Fisher–Newton–Raphson method 13

gamma distribution 45, 94, 99
Gauss–Markov (GM) theorem 3–8, 17
  optimality 136, 137, 138
generalized normal model (GNM) 106,
    111–15, 116
generalized quadratic model (GQM) 103–16
generalized regression estimator 218–21
group families 332–3